D1418740

Introduct

PROBABILITY

and

RANDOM PROCESSES

LIBRARY

Communications and Signal Processing

Senior Consulting Editor

Stephen W. Director, University of Michigan, Ann Arbor

Auñón/Chandrasekar: *Introduction to Probability and Random Processes*
Antoniou: *Digital Filters: Analysis and Design*
Candy: *Signal Processing: The Model-Based Approach*
Candy: *Signal Processing: The Modern Approach*
Carlson: *Communications Systems: An Introduction to Signals and Noise in Electrical Communication*
Cherin: *An Introduction to Optical Fibers*
Collin: *Antennas and Radiowave Propagation*
Collin: *Foundations for Microwave Engineering*
Cooper and McGillem: *Modern Communications and Spread Spectrum*
Davenport: *Probability and Random Processes: An Introduction for Applied Scientists and Engineers*
Drake: *Fundamentals of Applied Probability Theory*
Huelsman and Allen: *Introduction to the Theory and Design of Active Filters*
Jong: *Method of Discrete Signal and System Analysis*
Keiser: *Local Area Networks*
Keiser: *Optical Fiber Communications*
Kershenbaum: *Telecommunications Network Design Algorithms*
Kraus: *Antennas*
Kuc: *Introduction to Digital Signal Processing*
Mitra: *Digital Signal Processing: A Computer-Based Approach*
Papoulis: *Probability, Random Variables, and Stochastic Processes*
Papoulis: *Signal Analysis*
Papoulis: *The Fourier Integral and Its Applications*
Peebles: *Probability, Random Variables, and Random Signal Principles*
Proakis: *Digital Communications*
Schwartz: *Information Transmission, Modulation, and Noise*
Schwartz and Shaw: *Signal Processing*
Siebert: *Circuits, Signals, and Systems*
Smith: *Modern Communication Circuits*
Taub and Schilling: *Principles of Communication Systems*
Taylor: *Principles of Signals and Systems*

Introduction to

PROBABILITY

and

RANDOM PROCESSES

Jorge Auñón
Texas Tech University

V. Chandrasekar
Colorado State University

The McGraw-Hill Companies, Inc.
New York St. Louis San Francisco Auckland Bogotá Caracas Lisbon
London Madrid Mexico City Milan Montreal New Delhi San Juan
Singapore Sydney Tokyo Toronto

McGraw-Hill

A Division of The McGraw-Hill Companies

INTRODUCTION TO PROBABILITY AND RANDOM PROCESSES
International Editions 1998

1 2 3 4 5 6 7 8 9 0 SLP PMP 9 8 7

This book was set in Palatino.
The editors were Lynn Cox and Terri Wicks.
The production supervisor was Louis Swaim.
The cover was designed by Nicole Leong.

Library of Congress Cataloging-in-Publication Data

Aunon, Jorge.
    Introduction to probability and random processes / Jorge Aunon,
V. Chandrasekar.
        p.   cm.
    Includes index.
    ISBN 0-07-001563-5
    1. Probabilities.   2. Stochastic processes.   I. Chandrasekar, V.
II. Title.
QA273.A86   1996
519.2–dc21

                                                    96-44042
                                                      CIP

http://www.mhcollege.com

When ordering this title, use ISBN 0-07-115649-6

Printed in Singapore

To my mother and father
Haydée Prieto and Fernando Auñón

To my beloved wife
Margaret

To my daughters
Christine, Melissa, Serena, and Maria

J.A.

I dedicate this book
to my "GURUS"
(all my teachers).

V.C.

Contents

Preface   xiii

Chapter 1   Introduction to Probability   1
    1.0   Basic Concepts of Probability   1
    1.1   Definitions   4
    1.2   Elementary Set Theory   6
    1.3   Axiomatic Approach to the Theory of Probability   10
    1.4   Discrete and Continuous Sample Spaces   11
    1.5   Conditional Probability and Bayes' Theorem   12
    1.6   Independence of Events   18
    1.7   Bernoulli Trials   21
    1.8   Chapter Summary   27
    1.9   Problems   28
  1.10   Computer Examples   33

Chapter 2   Random Variables   40
    2.0   Introduction   40
    2.1   Concept of a Random Variable   40
    2.2   Cumulative Distribution Function   41
    2.3   Probability Density Function   47

ix

2.4   Basic Statistical Properties of a Random Variable and   50
        Expected Values
2.5   Moments   52
2.6   Examples of Different Distributions   56
2.7   Moment-Generating Functions and Characteristic Functions 68
2.8   Transformation of a Random Variable   71
2.9   Conditional Probability Density   81
2.10   Chapter Summary   85
2.11   Problems   87
2.12   Computer Examples   93

Chapter 3   Multiple Random Variables   105

3.0   Introduction   105
3.1   Joint Cumulative Distribution Function of two Random   106
        Variables
3.2   Joint Probability Density Function of two Random   108
        Variables
3.3   Statistical Properties of Jointly Distributed Random   112
        Variables: Joint Moments
3.4   Jointly Distributed Gaussian Random Variables   118
3.5   Functions of Gaussian Jointly Distributed Random   119
        variables
3.6   Conditional Probability Density   121
3.7   Probability Density Function of Sum of two Random   124
        Variables
3.8   Expected Value of Sums of Random Variables   136
3.9   Variance of Sum of Random Variables   136
3.10   Brief Introduction to Central Limit Theorem   138
3.11   Estimate of Population Mean, Expected Value,   141
        and Variance of Estimate
3.12   Estimate of Population Variance   145
3.13   Computer Generation of Uniform Random Variables   147
3.14   Computer Generation of Gaussian Distributed Random   148
        Variables using the Central Limit Theorem
3.15   Testing the Equivalence of a Probability Density Function 148
        of Experimental Data to a Theoretical Density Function
3.16   Chapter Summary   157
3.17   Problems   159
3.18   Computer Examples   163

Chapter 4   Multiple Random Variables   203

4.0   Random Processes: Basic Definitions   203
4.1   Basic Properties of Random Processes   207
4.2   Autocorrelation Function   212
4.3   The Cross-Correlation Function   225

4.4   Estimate of Autocorrelation Function of Sample Records   228
        of Limited Duration
4.5   Chapter Summary   232
4.6   Problems   235
4.7   Computer Examples   241

Chapter 5   Spectral Density   268
5.0   Introduction   268
5.1   Basic Definition of the Spectral Density of a   269
        Random Process
5.2   Properties   272
5.3   White Noise   274
5.4   Time Domain and Frequency Domain Correspondence   277
5.5   Estimate of the Autocorrelation Function of a Discrete   277
        Function of Time using Frequency Domain Techniques
5.6   Estimation of the Spectral Density of Records of   282
        Limited Duration
5.7   Cross-Spectral Density   307
5.8   Estimation of Cross-Spectral Density of Records of   311
        Limited Duration
5.9   Coherence   315
5.10   Chapter Summary   321
5.11   Problems   324
5.12   Computer Examples   327

Chapter 6   Linear Systems   439
6.0   Introduction   439
6.1   Properties   441
6.2   Random Inputs   441
6.3   Estimate of the Response of Linear Systems using   448
        Frequency Domain Techniques
6.4   Matched Filter   453
6.5   Chapter Summary   462
6.6   Problems   463
6.7   Computer Examples   469

Chapter 7   Applications   492
7.0   Introduction   492
7.1   Radar Systems   493
7.2   Communication Systems   497
7.3   Solid-State Electronics   503
7.4   Computer Engineering   504
7.5   Biomedical Systems   507

Appendix   Introduction of the Fast Fourier Transform   513
   A.1   Introduction   513
   A.2   Relationship Between the Continuous Fourier   514
            Transform (CFT) and the Discrete Fourier Transform (DFT)
   A.3   The Inverse Discrete Fourier Transform (IDFT)   518
   A.4   The FFT Algorithm, Computational Speed, and Effects 518
            of Zero Padding
   A.5   Practical Aspects of the Use of the FFT Algorithm   525

            Bibliography   529
            Index   531

Preface

This book presents an introduction to the topics of statistics, random variables, and random processes. It is intended for junior and senior engineering students, and offers a unique, practical approach to the subject. With a carefully selected blend of both theoretical and real-data examples, it connects the classroom to real-world problem solving, and uses the computer to explore the subtleties of probability theory and its applications. The book includes a large number of computer examples using both Mathcad[1] and MATLAB.[2] It also contains traditional end-of-chapter problems. Rather than

---

[1]Mathcad is a registered trademark of MathSoft, Inc., 101 Main Street, Massachusetts, 02142. WWW:http://www.mathsoft.com. Printed by permission of MathSoft, Inc.

[2]MATLAB is a registered trademark of the MathWorks, Inc., 24 Prime Park Way, Natick, MA 01760-1500, Phone: 508-647-7000,WWW: http://www.mathworks.com.

emphasizing programming sophistication, the book shows how the computer can be used as a live interface to visualize problem solving. The use of Mathcad and MATLAB demonstrate a natural progression through the solution of a problem. Students are able to change parameters in equations, or change entire equations and have results immediately available for their use. The computer examples available with each chapter are *not* designed as Mathcad or MATLAB tutorials. They are written in a fairly simplistic manner in order that students will be able to modify them to solve classroom type problems.

Chapters 1 and 2 present a general introduction to probability theory and to the basic statistical properties of a random variable. Chapter 3 departs from traditional textbooks by introducing students to the computer generation of gaussian-distributed random variables. The run test and the chi-square test are explained to test the equivalence of a probability density function of experimental data to a theoretical density function. Both Mathcad and MATLAB computer examples are presented at the end of the chapter.

Chapter 4 introduces the concept of random processes and the autocorrelation function. The coverage includes estimating the autocorrelation function of records of limited duration. Following the book's central pedagogical strategy, this section begins with an explanation of theory and a working, theoretical example. Then, in order to make the transition between a theoretical result and its practical use, real time-limited data is presented and analyzed.

Chapter 5 introduces students to the concept of the spectral density of a random process, and explores a number of techniques for the estimation of spectral density. First, the spectral density is estimated as the Fourier transform of the autocorrelation function. The autocorrelation function of a time-limited record is estimated using the FFT, and then the Fourier transform of the autocorrelation is found. The chapter also explores a second technique, the periodogram. In this technique, the data is divided into overlapping segments; the computation of the Fourier transform of each segment; and the average of all segments obtained. The model-based approach is the third technique explored. It exposes students to the theory that a random process may be described as an autoregressive process. The order of the model and coefficients of the model are found, and the estimate of the spectral density is calculated. A large collection of Mathcad and MATLAB computer examples are presented at the end of the chapter.

Chapter 6 introduces the analysis of linear systems when their inputs are random processes. It also considers the special case of matched filters. Chapter 7 presents a wide variety of useful applications, ranging from biomedical systems to radar systems. Finally, the Appendix provides a concise yet complete presentation of the fast Fourier transform (FFT).

In closing, we would like to thank the reviewers who aided in the development of the book: Dr. Ronald A. Iltis, University of California at Santa Barbara; Dr. V. Krishnan, University of Massachusetts at Lowell; Dr. Steven A. Tretter, University of Maryland; and, Jitendra K. Tugnait, Auburn University. We would also like to thank our editor, Lynn B. Cox, and editing supervisor, Terri Wicks.

Jorge I. Auñón
V. Chandrasekar

About the Authors

Jorge I. Auñón is Dean of Engineering and a Professor of Electrical Engineering at Texas Tech University. He received his D.Sc. from The George Washington University in 1972. From 1973 until 1988, he taught electrical engineering at Purdue University. At Colorado State University, he served as Head of Electrical Engineering, and in 1995 he moved to Texas Tech University. Dr. Auñón is a Fellow of the IEEE and a Fulbright Scholar.

V. Chandrasekar received his Ph.D from Colorado State University, where he is currently an Associate Professor of Electrical Engineering. His main research interests are signal processing, radar systems, remote sensing polarimetry and neural networks. He has published over 120 papers on these topics. Dr. Chandrasekar has received several awards for his outstanding commitment to teaching and research. In 1996, he was a visiting professor at the National Research Council (CNR) of Italy.

Introduction to

PROBABILITY

and

RANDOM PROCESSES

Introduction to Probability

1

1.0 BASIC CONCEPTS OF PROBABILITY

The term *probability* is intimately involved with the term *uncertainty*. We say that it is improbable that there is life on the moon, that there is a 25 percent chance of rain tomorrow afternoon, etc. Practically, we are often confronted with situations where the exact truth is not known. We take chances and assume that a particular situation is likely to occur. Weather is a good example; the accurate prediction of the path of a hurricane has enabled many cities to adequately prepare for inclement weather. A tornado, however, is much more unpredictable, often having disastrous consequences. We are faced in everyday life with statements that have associated with them the word *probable*. For example, the probability of rolling a 1 using a fair die is 1/6; the probability of drawing a king of hearts out of a complete deck of cards is 1/52; etc. The first statement, addressing the probability of a 1's occurring when we toss a die, makes use of a number of assumptions. For example, it is assumed that the die is "fair"; we are also not counting other probable occurrences such as the loss of the die when it is tossed. It is usually assumed that when we toss a die, each side is equally likely to appear. One way of obtaining the number 1/6 is to toss the die many times and simply count the number of times in which a 1 appears. This concept leads to the well-known ratio

$$\frac{\text{Number of favorable occurrences}}{\text{Number of times the die was tossed}}$$

By evaluating this number we are actually evaluating the probability of a 1's occurring. This basic (and historic) definition of probability assumes that all possible outcomes, such as the occurrence of a specific face in the tossing of a die, are equally probable. Using this ratio, we conclude that every probability P is a number between 0 and 1:

$$0 \le P \le 1$$

Note that for P to be equal to 1, the numerator and denominator of the above ratio must equal each other. In such a case, when the probability of the event equals 1, we have a *certain* event. The relative frequency approach is essentially a counting approach. Basically, we are going to perform the experiment *many* times and then count the number of times that the particular event we are interested in occurs. For example, consider again the experiment of tossing a die. There are a number of possible events associated with this experiment. Let us mention just a few:

A 6 comes up.
An even number comes up.
A number less than 4 comes up.
A number between 1 and 6 (inclusive) comes up.

The fourth event is an interesting one, since there is nothing random associated with its outcome. Whenever we toss a die, the number *will be* between 1 and 6 (inclusive). This is what we call a certain event. Note that a certain event is not a random event, since we know ahead of time that it will occur. Therefore, the probability associated with this event is 1.

For any of the other events, the discrete probability associated with each may or may not be easy to calculate. The relative frequency approach provides us with a way to arrive at the probability associated with specific events. As mentioned earlier, this approach is essentially a counting approach. Suppose that we are interested in the probability that a 6 comes up when we toss a die. We perform the following experiment: We toss the die 1000 times and count each time that a 6 comes up. Then we approximate the probability associated with the event "a 6 comes up" as follows:

$$P(6) \approx \frac{\text{number of times a 6 appears}}{\text{total number of tosses}} = \frac{N_6}{N_{\text{trials}}} \qquad (1.1)$$

where

$P(6)$ = probability that a 6 occurs

N_6 = number of times a 6 occurs

N_{trials} = number of trials

and where $\approx$ indicates that ratio of two sides tends to unity as number of trials approaches infinity (symbolized by ∞)

With the present example, we say that the ratio given by Equation (1.1) approaches the probability that a 6 occurs. In the example, we tossed the die 1000 times, therefore, N_{trials} = 1000; and N_6 is the total number of times we observe that a 6 occurs. Intuitively, we can see that as the number of trials increases, we come closer and closer to the correct number. Therefore,

$$\frac{N_6}{N_{trials}} \text{ approaches the true value of } P(6) \text{ as } N_{trials} \to \infty \qquad (1.2)$$

Intuitively, we know that $P(6) = 1/6$. Although this approach is an easy one to understand, it may not be the most feasible one to use. Often the necessary experiment becomes cumbersome, and one has to perform it thousands of times before arriving at something that approaches the true answer. Later on, the axiomatic approach to the theory of probability will be explored. Meanwhile, the relative frequency approach provides us with some important clues concerning the nature of the numbers associated with probabilities. Note that both N_6 and N_{trials} are positive numbers and therefore, $P(6)$ will also have to be a positive number, but always equal to or less than 1.

The study of the theory of probability is extremely important in engineering problems since the measurement of most physical quantities involves some type of uncertainty. This uncertainty may be caused by variations in the quantity being measured, by the measuring instrument, or by both. Frequently we are only able to measure a quantity for a limited time. For example, suppose that we would like to measure the speech of a singer in order to ascertain some properties of the speech. The speech is recorded through a microphone into an audio tape recorder, and then measurements are performed on the recorded speech. We must, however, ask some important questions. How long a speech sample should we record? How do we know that this particular speech sample is representative of the speech patterns of the particular singer? Would it be possible to obtain the same measurements by recording for only half as long, or do we need to record for a long time before we have some assurance that the speech sample is stable? These are but representative questions facing an engineer required to digitally process speech samples.

1.1 DEFINITIONS

We define the *space of elementary events S* as the collection of all possible outcomes of an experiment. For example, the space S of elementary events that corresponds to the experiment of tossing a die will consist of 6 points. In this particular case, the space is finite. Likewise, the space S corresponding to the tossing of a coin will consist of 2 points. A subset of the space S we call an *event*. Note that the *certain event* mentioned earlier corresponds to the space S. The probability of the certain event S equals 1. This corresponds to the experiment of tossing a single die and computing the probability that a number between 1 and 6 inclusive will come up. An example of an infinite space S is the space containing all positive integers greater than 0. Note that in this case, although S is infinite, it is *discrete*. An example of an infinite continuous space S is the space containing all possible numbers between 5 and 6. Of equal importance to the *certain event S* is the *impossible event* $\emptyset$. This event cannot occur. This corresponds to the experiment of tossing a six-sided die and computing the probability that a 7 comes up.

Consider now a subset of space S. Let us use the familiar example of the toss of a die in order to illustrate a number of fundamental concepts. By definition,

$$S = \{1, 2, 3, 4, 5, 6\}$$

We also know that $P(S) = 1$, that is, the probability of the certain event equals 1.

Consider now the following subsets of S:

Let A be the event "an even number comes up."
Let B be the event "a number less than 4 comes up."

We therefore see that

$$A = \{2, 4, 6\}$$

and likewise

$$B = \{1, 2, 3\}$$

We now toss a die and ask whether the outcome lies in A or in B or possibly in both. This event is characterized by the subset consisting of all points that lie in A or in B or in both. This subset is given by the *union*, or *logical sum*, of events A and B. The symbol $\cup$ represents the operation of a logical sum.

$$A \cup B = \{2, 4, 6\} \cup \{1, 2, 3\} = \{1, 2, 3, 4, 6\} \tag{1.3}$$

This operation characterizes the event "A or B or both."

In a similar manner, one could ask whether a measurement lies in both A and B. This event is characterized by the subset consisting of all points that lie in A as well as in B. This subset is given by the *intersection*, or *logical product*, of events A and B. The symbol $\cap$ represents the operation of a logical product.

$$A \cap B = \{2,4,6\} \cap \{1,2,3\} = \{2\} \tag{1.4}$$

This operation characterizes the event "A and B."

Let us consider now the case when A and B have nothing in common. For example,

Let A be the event "an even number comes up."
Let B be the event "an odd number comes up."

Clearly,

$$A = \{2,4,6\}$$

and likewise

$$B = \{1,3,5\}$$

We can see that no measurement can possibly lie in subsets A and B at the same time. Therefore,

$$A \cap B = \{2,4,6\} \cap \{1,3,5\} = \varnothing \tag{1.5}$$

In this case the intersection of events A and B is the impossible event or empty set. When the intersection of two events yields the empty set, then we say that events A and B are *mutually exclusive*. Events which are mutually exclusive cannot occur simultaneously. For example, in the toss of a die, the outcome cannot belong to A and B simultaneously.

The *complement* of an event A is designated by the symbol A^C and is defined as all outcomes of the experiment which are not contained in A. For example,

$$A = \{2,4,6\} \quad \text{then} \quad A^C = \{1,3,5\} \tag{1.6}$$

1.2 ELEMENTARY SET THEORY

Using concepts of set theory, we will be able to define operations on the events of an experiment. Consider now the experiment of tossing a coin. Associated with this experiment is a sample space S. We call the sample space S the collection of all possible outcomes for this experiment. Within this largest set we now define subsets as smaller collections of elements, all of which belong to space S. For example, space S for the experiment of tossing a coin is $\{H, T\}$. Possible subsets are the following: $\{H\}$, $\{T\}$, $\{H, T\}$, $\varnothing$. The sample space S for the experiment of tossing a die is $\{1, 2, 3, 4, 5, 6\}$, and some of the possible subsets are $\{1\}$, $\{1, 2\}$, $\{2, 4, 6\}$, etc.

One advantage of using the concepts of set theory, subsets, and elements within a space is that a number of operations may be performed on the subsets which will help us visualize the probability concepts associated with the different sets. As defined earlier, the union of two events A and B, represented as $A \cup B$, is the collection of outcomes which contains all the elements that are in event A or in event B or in both. Similarly, we defined the intersection of two events A and B as the collection of outcomes that are both in A and in B. This is represented by $A \cap B$. We can represent many of these concepts in a pictorial manner. Consider space S associated with the toss of a die, shown in Figure 1.1.

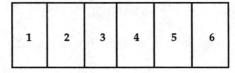

$$S$$

Figure 1.1 Sample space S.

Note that space S has been divided into six rectangles having the same area, the implication being that $P(S) = 1$, and the probability associated with each of the faces equals exactly $1/6$. In the previous section, we defined the events $A = \{2,4,6\}$, and $B = \{1,2,3\}$. We then defined the following operations:

$$A \cup B = \{2, 4, 6\} \cup \{1, 2, 3\} = \{1, 2, 3, 4, 6\}$$

We can represent these operations by using a pictorial representation similar to the one shown in Figure 1.1. These representations are called *Venn diagrams* (Figure 1.2).

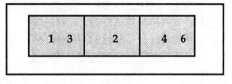

Figure 1.2 Venn diagram for the operation $A \cup B$.

The shaded area represents the event "A or B or both." Likewise, the Venn diagram representation for the intersection, or $A \cap B$, is shown in Figure 1.3.

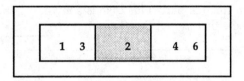

Figure 1.3 Venn diagram for the operation $A \cap B$.

The shaded area represents the event "A and B." Note that only when the outcome of the experiment is the number 2 does it satisfy the event "A and B."

Suppose now that events A and B are mutually exclusive; then the intersection of events A and B is called the *null set*:

Let A be the event that a number less than 3 appears, or $A = \{1, 2\}$.
Let B be the event that a number greater than 4 appears, or $A = \{5, 6\}$.

Note that $C = A \cap B = \varnothing$, i.e., the null set. The two mutually exclusive events $A = \{$a number less than 3$\}$ and $B = \{$a number greater than 4$\}$ are represented as two areas that do not intersect within space S. This is shown in Figure 1.4.

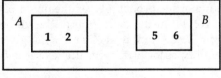

Figure 1.4 Two mutually exclusive events A and B.

Consider now the two events $A = \{$a number less than 4$\}$ and $B = \{$an even number less than 4$\}$. Note that since $B \subset A$, that is, B belongs to A, then $A \cap B = B$ (Figure 1.5).

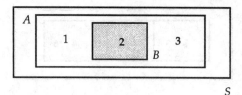

Figure 1.5 Event B is contained in event A: $B \subset A$.

In this particular case, event B is a subset of event A, and the two events are shown as one rectangle inside the other.

Another important case to consider is the event A and its complement A^C. The Venn diagram for this particular case is shown in Figure 1.6. Note that $A \cup A^C = S$. Also note that $A \cap A^C = \emptyset$ and that $P(A \cup A^C) = 1$.

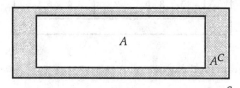

Figure 1.6 Venn diagram showing event A and its complement A^C.

When considering the above cases, we need to make the transition between the different events A and B, for example, and the *probability* associated with each of these events. The probability of any of the above events is directly related to the area it occupies within space S. It is obvious from the above discussions that $P(S) = 1$ and that for the cases represented by Figures 1.4 and 1.5, $P(A) \leq 1$ and $P(B) \leq 1$. For the case represented by Figure 1.5, $P(B) < P(A)$. For the case represented by Figure 1.4, one can observe that $P(C) = P(A \cap B) = 0$. This is true since, for this case, events A and B are mutually exclusive; therefore event C is the null set. For the case represented by Figure 1.3, the probability of the intersection of the two events is given by $P(C) = P(A \cap B) = P(2)$.

It remains to calculate the probability of the union of two events A and B illustrated by Figure 1.2. From the definition of the union of events A and B we can see that event C, the union of A and B, is given by $C = A \cup B = \{1,2,3,4,6\}$. The probability associated with this event is calculated as

$$P(C) = P(A \cup B) = P(A) + P(B) - P(A \cap B) \tag{1.7}$$

where $P(A \cap B)$ is the probability associated with the intersection of events A and B. Note that if events A and B had been mutually exclusive, then the probability of the intersection $P(A \cap B) = 0$. Also note that by definition

$$P(A \cap B) = P(B \cap A)$$

For the case of three arbitrary events, Equation (1.7) extends to the following:

$$\begin{aligned} P(D) &= P(A \cup B \cup C) \\ &= P(A) + P(B) + P(C) - P(A \cap B) - P(A \cap C) \\ &\quad - P(B \cap C) + P(A \cap B \cap C) \end{aligned} \tag{1.8}$$

Example 1.1 We are going to use a simple example to illustrate numerically the important concepts just mentioned.

If a single (fair) die is rolled, determine the probability of each of the following events:

(a) Obtaining the number 3.

Since the die we are using is fair, each number has an equal probability of occurring; therefore,

$$P(3) = 1/6$$

(b) Obtaining a number greater than 4.

In this case, the set of possible outcomes is given by $\{5, 6\}$. Note that these are all mutually exclusive events. Therefore,

$$P(5 \cup 6) = P(5) + P(6) = 1/6 + 1/6 = 2/6 = 1/3$$

(c) Obtaining a number less than 3 or obtaining a number greater than or equal to 2.

In this case, the two events we are considering are as follows:

$$A = \{a \text{ number less than } 3\} = \{1, 2\}$$

$$B = \{a \text{ number greater than or equal to } 2\} = \{2, 3, 4, 5, 6\}$$

Therefore, the event of interest is the following:

$$C = \{A \cup B\} = \{1, 2, 3, 4, 5, 6\}$$

$$P(C) = P(A \cup B) = P(A) + P(B) - P(A \cap B)$$

and the probability associated with event C is calculated as follows:

$$P(C) = P(1, \ 2) + P(2, 3, 4, 5, 6) - P(2)$$

$$P(C) = (1/6 + 1/6) + (1/6 + 1/6 + 1/6 + 1/6 + 1/6) - 1/6$$

Note that $P(2, 3, 4, 5, 6) = 1/6 + 1/6 + 1/6 + 1/6 + 1/6$ since all these events are mutually exclusive. Therefore the probability of event C is

$$P(C) = 2/6 + 5/6 - 1/6 = 1$$

1.3 AXIOMATIC APPROACH TO THE THEORY OF PROBABILITY

Consider now a number of events resulting from an experiment. We will assign a nonnegative number to the probability of each of the events. The probability associated with each of the events in space S is chosen so that it satisfies three axioms. Let A be an event associated with a certain experiment with sample space S. To this event A, we assign the probability value $P(A)$. This probability value $P(A)$ must satisfy the following *axioms*:

I. $0 \leq P(A) \leq 1$

This axiom states that the probability value lies between 0 and 1.

II. $P(S) = 1$

This axiom states that the probability of the certain event equals 1.

III. Given that events $A_1, A_2, \ldots, A_N, \ldots$ are mutually exclusive, then

$$P\left(\bigcup_{i=1}^{\infty} A_i\right) = \sum_{i=1}^{\infty} P(A_i)$$

This axiom is called the *addition rule for mutually exclusive events*. Note that this axiom applies to a sequence of mutually exclusive events. It follows from this axiom that for a finite collection of mutually exclusive events

$$P(A_1 \cup A_2 \cup \cdots \cup A_N) = P(A_1) + P(A_2) + \cdots + P(A_N)$$

These three axioms provide us with rules that any valid probability value associated with a set of events must satisfy. Many of the rules developed earlier in the chapter intuitively can be obtained from the above three axioms of probability. For example, consider event A and its complement A^C.

It was shown through set theory in the previous section that A and A^C are mutually exclusive and $A \cup A^C = S$. Subsequently, using axioms II and III, we obtain

$$1 = P(S) = P(A \cup A^C) = P(A) + P(A^C) \tag{1.9}$$

Similarly, the probability of the union of two events A and B deduced from set theory can be obtained from the probability axioms as follows: $A \cup B$ can be written as the union of two mutually exclusive events A and $A^C \cap B$. Therefore, using axiom III, we obtain

$$P(A \cup B) = P(A) + P(A^C \cap B) \tag{1.10}$$

In addition, event B can be written as

$$B = (A \cap B) \cup (A^C \cap B) \tag{1.11}$$

which implies

$$P(B) = P(A \cap B) + P(A^C \cap B) \tag{1.12}$$

Combining Equations (1.10) and (1.12), we get

$$P(A \cup B) = P(A) + P(B) - P(A \cap B) \tag{1.13}$$

which is the same result as that given by Equation (1.7).

Thus all the results developed intuitively in the past sections can be obtained from the three axioms of probability.

1.4 DISCRETE AND CONTINUOUS SAMPLE SPACES

Discrete sample spaces are perhaps easier to comprehend and analyze because they are usually countable. It gives us comfort to know that in a pinch we can always "count" the occurrences of a discrete event and actually calculate the required probabilities. Intuitively, one considers a sample space as having a "mass" equal to 1, the value of the probability of the certain event. Consequently, the probability of a specific event within S may be viewed as that portion of the mass (less than or equal to 1) corresponding to the probability of the event. Figure 1.1 exemplifies this concept. The probability of the certain event S equals 1. This is the probability that when a die is tossed, a number greater than or equal to 1 and less than or equal to 6 will show up. The "mass" of the space equals 1. Any one of the events shown, for example, "a 1 appears," has a probability of occurring equal to the proportion of its mass to the total mass. In this case, this value equals 1/6.

Discrete sample spaces may be infinite. For example, consider the experiment involving the selection of an integer number greater than 0.

Continuous sample spaces originate in experiments where the outcomes are real numbers. For example, consider the acquisition of a speech sample. Assume that we have an amplifier clipping the sounds at ±5 volts (V). At any one time, the value of the speech sample may be considered to be a random number between +5 and –5 V. Since there are an infinite number of possible values between +5 and –5 V, we can imagine that the probability of any value's being equal to, say, –1.2 V exactly would be 0, since there are an infinitely large number of possible outcomes between +5 and –5 V. Suppose that we now divide the range from +5 to –5 into 10 equal spaces. We could then ask the question, What is the probability that any one number falls within the interval of –1 to –2 V? Since the probability that the sample will fall within the range of +5 to –5 equals 1, we can imagine that the probability of the sample's being in the interval from –1 to –2 equals 1/10. This value reflects the assumption that every interval is equally likely to contain the sample. This is a very important concept that will be explored in greater detail in Chapter 2.

1.5 CONDITIONAL PROBABILITY AND BAYES' THEOREM

Random experiments and observations often involve cases where the probability of an event depends on the occurrence of some other event. For example, given a deck of 52 cards, suppose that we draw a heart as the first card. Without replacing the card in the deck, we may wish to calculate the probability that the next card drawn will also be a heart. There is now one less heart in the deck, so intuitively we think that the probability has decreased. This is a typical case where the concept of *conditional probability* must be used. The probability of the second card's being a heart is conditioned on the fact that the first card drawn was a heart. We now define the probability that event A occurs given that event B has occurred:

$$P(A/B) = \frac{P(A \cap B)}{P(B)} \quad P(B) > 0 \tag{1.14}$$

This equation represents the probability of event A, conditioned on the occurrence of event B. The probability of event A *given that* event B has occurred equals the probability of the intersection of events A and B divided by the probability of event B. Clearly, $P(B)$ must be greater than 0. Note that if events A and B are mutually exclusive, then the value of the conditional probability equals 0.

Example 1.2 Given a deck of cards, what is the probability of drawing a heart on the first draw and a heart on the second draw? This is a case where the events are conditioned by one another. In this case, a no-replacement strategy is assumed. We now define the events as follows:

$$A = \{\text{a heart is drawn second}\}$$
$$B = \{\text{a heart is drawn first}\}$$

intersection

$P(A \cap B) =$ probability that heart is drawn first *and* heart is drawn second

$P(A/B) =$ probability of drawing a heart on second draw *given that* a heart was drawn on first draw

where

$P(B) = 13/52$ (that is, 13 cards in a deck of 52 are hearts)

$P(A/B) = 12/51$ (now 51 cards left in deck, out of which 12 are hearts)

Consequently,

$$P(A \cap B) = P(A/B)P(B) = (12/51)(1/4) = 12/204$$

Example 1.3 Consider a box where we have five resistors. Three of the resistors are rated at 100 ohms (abbreviated Ω), and two of the resistors at 1000 Ω. We remove from the box two resistors in succession. What is the probability that the first resistor is 100 Ω and the second 1000 Ω? Let $P = \{\text{draw a 100-}\Omega \text{ resistor}\}$, and let $R = \{\text{draw a 1000-}\Omega \text{ resistor}\}$. Then P_1 means drawing a 100-Ω resistor first. We are interested in the event $\{\text{draw a 100-}\Omega$ resistor first and a 1000-Ω resistor second$\}$, or $\{P_1, R_2\}$. We now express the problem as follows:

$P(100 \ \Omega \text{ first and } 1000 \ \Omega \text{ second}) =$
$\qquad P(1000 \ \Omega \text{ second given that } 100 \ \Omega \text{ drawn first}) \cdot P(100 \ \Omega \ \text{ first})$

which equals

$$P(P_1 \cap R_2) = P(R_2/P_1)P(P_1)$$

We now compute these probabilities:

$P(P_1) = 3/5$ since three of the five resistors are rated at 100 Ω

$P(R_2/P_1) = 2/4$ since two of remaining four resistors are rated at 1000 Ω

Therefore,

$$P(P_1 \cap R_2) = P(R_2/P_1)P(P_1) = (3/5)\cdot(2/4) = 3/10$$

Example 1.4 The Birthday Problem This is an interesting problem since the answer to this example is not immediately obvious. The problem is essentially the following: Given N students in a class, how large is the probability that at least two students have their birthdays on the same day? Let us consider a specific example, let $N = 30$. One assumption we need to make is that all the 365 days in a year are equally likely as birthdays. Let us also ignore leap years. These assumptions must be made to simplify the problem. One way to solve this problem is to find out the probability that "no two people will have their birthdays on the same date." Let's call this event A. Then the probability of event A equals

$$P(A) = \frac{\text{no. of favorable cases}}{\text{no. of possible cases}} = \frac{365 \cdot 364 \cdots (365-N+1)}{365^N}$$

In our case $N = 30$; therefore,

$$P(A) = \frac{365 \cdot 364 \cdots 336}{365^{30}}$$

Note that we are really interested in $1 - P(A)$, that is, we are interested in the probability that two people will have their birthdays on the same date:

$$P = 1 - P(A) = 1 - \frac{365 \cdot 364 \cdots 336}{365^{30}} = .71$$

This is actually a fairly high number. This implies that if we make a bet that in a room of 30 people at least 2 will share the same birthday, we will have a 71 percent chance of winning the bet. Note that we have ignored leap years, and as pointed out earlier, some days have higher birth probability than other days.

Bayes' Theorem

Let us first consider the law of *total probability*. Consider the sample space S and a set of mutually exclusive events which is a partition of S. This may be expressed as follows:

Let $A_1, A_2, \ldots, A_N$ belong to S; that is, $A_1 \subset S$, and let $A_2 \subset S, \ldots, A_N \subset S$ be mutually exclusive events, that is, $P(A_i \cap A_j) = \varnothing$ for all i, j and $A_1 \cup A_2 \cdots A_N = S$. In addition, we know that $P(S) = A_1 \cup A_2 \cup \cdots \cup A_N = P(A_1) + P(A_2) + \cdots + P(A_N) = 1$. Note that we are using several definitions from the chapter:

$$P(S) = 1 \quad \text{for complete sample space } S$$

$$P(A_1 \cup A_2 \cup \cdots A_N) = P(A_1) + P(A_1) + \cdots + P(A_N)$$

since the events are mutually exclusive.

Consider the following example: Let S, the sample space, be a deck of cards; let the events $A_1, A_2, \ldots, A_4$ be the event of drawing a particular suit; i.e., let A_1 be any spade, etc. If we draw only one card, events $A_1, A_2, \ldots, A_4$ are mutually exclusive. Consider a new event B of drawing any king, as shown in Figure 1.7. Event B is contained in S, and we can express the probability associated with event B as

$$P(B) = P(B \cap S) = P[B \cap (A_1 \cup A_2 \cup A_3 \cup A_4)] \tag{1.15}$$

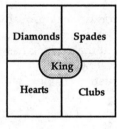

S

Figure 1.7 Venn diagram representing picking a king as the first card from a deck.

Since events A_1, A_2, A_3, and A_4 are mutually exclusive, we can express this last relationship as follows:

$$P(B) = P(B \cap A_1) + P(B \cap A_2) + P(B \cap A_3) + P(B \cap A_4) \tag{1.16}$$

We also know that

$$P(B \cap A_i) = P(B / A_i)P(A_i) \tag{1.17}$$

So we can express the probability of event B as follows:

$$P(B) = P(B/A_1)P(A_1) + P(B/A_2)P(A_2)$$
$$+ P(B/A_3)P(A_3) + P(B/A_4)P(A_4) \tag{1.18}$$

$$P(B) = \sum_i P(B/A_i)P(A_i) \tag{1.19}$$

This last equation is known as the *total probability* of event B.

The concept of total probability is now going to be utilized in the development of *Bayes' theorem*, or *theorem of inverse probability*. This theorem is actually very useful in the diagnosis of diseases based on symptoms. Physicians often use this technique to associate the likelihood that, given that a patient smokes, she or he will get lung cancer.

Consider now a deck of cards, and divide it into suits.

Let event A_1 be the event of picking a diamond.
Let event A_2 be the event of picking a spade.
Let event A_3 be the event of picking a club.
Let event A_4 be the event of picking a heart.
Let event B be the event of picking a king.

If we pick a spade, the conditional probability that it will be a king is $P(B/A_2)$. But consider now the inverse problem: What is the probability that a spade will be drawn, given that a king has been drawn? This time we are interested in the probability $P(A_2/B)$.

We now make use of Equation (1.14):

$$P(A_2/B) = \frac{P(A_2 \cap B)}{P(B)} = \frac{P(B \cap A_2)}{P(B)} \tag{1.20}$$

Note that $P(A_2 \cap B) = P(B \cap A_2)$.

Using Equation (1.19), we can expand this last equation as follows:

$$P(A_2/B) = \frac{P(B \cap A_2)}{P(B)} = \frac{P(B/A_2)P(A_2)}{P(B \cap A_1) + P(B \cap A_2) + P(B \cap A_3) + P(B \cap A_4)} \tag{1.21}$$

and each term in the denominator may be expanded as follows:

$$P(A_2/B) = \frac{P(B \cap A_2)}{P(B)}$$

$$= \frac{P(B/A_2)P(A_2)}{P(B/A_1)P(A_1) + P(B/A_2)P(A_2) + P(B/A_3)P(A_3) + P(B/A_4)P(A_4)} \quad (1.22)$$

And the more general form is

$$P(A_i/B) = \frac{P(B/A_i)\, P(A_i)}{\displaystyle\sum_j P(B/A_j)\, P(A_j)} \quad (1.23)$$

This is known as *Bayes' theorem*.

Example 1.5 A well-known national corporation XYZ produces workstations and test equipment. A recent survey by the site manager revealed that over the last 5 years, 35 percent of the electrical engineers hired had a master's of science in electrical engineering (MSEE). Of these, 80 percent are supervising special projects within the company. Of those electrical engineers who were hired without an MSEE, 10 percent also supervise special projects. Suppose that the site manager holds a special projects supervisor conference involving supervisors who have been with the company no longer than 5 years. What is the probability that a randomly selected person at the conference will hold an MSEE?

　　Let the events be as follows:

　　A:　The person selected is a special projects supervisor.
　　B:　The person selected holds an MSEE.
　　C:　The person selected does not hold an MSEE.

We are searching for $P(B/A)$.
　　By definition of conditional probability, this expression equals

$$P(B/A) = \frac{P(B \cap A)}{P(A)}$$

By use of Bayes' rule, we obtain

$$P(B/A) = \frac{P(A/B)P(A)}{P(A)}$$

Note that the denominator is the *total probability* that the person selected is a special projects supervisor. This probability equals

$$P(A) = P(B)P(A/B) + P(C)P(A/C)$$

Therefore, the preceding equation becomes

$$P(B/A) = \frac{P(A/B)P(B)}{P(B)P(A/B) + P(C)P(A/C)}$$

For this particular problem,

$$P(B) = P(\text{person selected holds an MSEE}) = 35\% = .35$$
$$P(C) = P(\text{person selected does not hold an MSEE})$$
$$= 1 - P(B) = .65$$
$$P(A/B) = P(\text{special projects supervisor/MSEE}) = .80$$
$$P(A/C) = P(\text{special projects supervisor/no MSEE}) = .10$$

Replacing these numbers yields

$$P(B/A) = \frac{P(A/B)P(B)}{P(B)P(A/B) + P(C)P(A/C)} = \frac{(.80)(.35)}{(.35)(.80) + (.65)(.1)} = .81$$

Therefore the probability that a randomly selected person at the conference will hold an MSEE is 81 percent.

1.6 INDEPENDENCE OF EVENTS

We now consider the concept of the *independence of events*. Two events are independent when the occurrence of one does not at all influence whether the other occurs. Analytically, we say that two events are independent *if and only if*

$$P(A \cap B) = P(A)P(B) \qquad (1.24)$$

There are some interesting implications to this definition. Equation (1.7) computes the probability of the union of two events A and B:

$$P(A \cup B) = P(A) + P(B) - P(A \cap B) \qquad (1.25)$$

If events A and B are independent events, then this equation may be expressed as follows:

$$P(A \cup B) = P(A) + P(B) - P(A)P(B) \tag{1.26}$$

We can see from the above equation that if A and B are mutually exclusive, then they can be independent only if either $P(A)$ or $P(B)$ is 0.

The conditional probability given by Equation (1.14),

$$P(A/B) = \frac{P(A \cap B)}{P(B)} \quad P(B) > 0 \tag{1.27}$$

may now be expressed by using Equation (1.24) as

$$P(A/B) = \frac{P(A)P(B)}{P(B)} = P(A) \tag{1.28}$$

Example 1.6 A pair of fair dice is rolled, the values of their faces are added, and the following events are defined:

$$A = \{\text{roll an odd sum}\}$$
$$B = \{\text{roll a sum that is an integer multiple of 3}\}$$

(a) Are events A and B statistically independent? Prove your answer.

Let us first compute the probability of event A:

One way of finding the probability of event A is by counting all possible pairs obtained when a pair of dice are rolled and adding the value of the faces.

						Sum of faces	Probability
{1,1}						2	1/36
{1,2},	{2,1}					3	2/36
{1,3},	{2,2},	{3,1}				4	3/36
{1,4},	{2,3},	{3,2},	{4,1}			5	4/36
{1,5},	{2,4},	{3,3},	{4,2},	{5,1}		6	5/36
{1,6},	{2,5},	{3,4},	{4,3},	{5,2},	{6,1}	7	6/36
	{2,6},	{3,5},	{4,4},	{5,3},	{6,2}	8	5/36
		{3,6},	{4,5},	{5,4},	{6,3}	9	4/36
			{4,6},	{5,5},	{6,4}	10	3/36
				{5,6},	{6,5}	11	2/36
					{6,6}	12	1/36
							36/36

Note that since each of these 36 possible events is mutually exclusive, the sum of the probabilities equals 1. To find the probability of event A, we notice that there are 18 possible ways in which an odd number may be obtained. Since all the events are mutually exclusive,

$$P(A) = \{\text{roll an odd number}\} = 18/36 = 1/2$$

Using the same logic, we now compute the probability of event B:

$$P(B) = \{\text{roll a number that is an integer multiple of 3}\} = 12/36$$

To determine if events A and B are independent, we need to compute the probability of the intersection of the two events $P(A \cap B)$. Therefore, we need to compute the probability of the event that the number rolled is both odd *and* an integer multiple of 3. This happens when the sum of the faces is 3 and 9, that is, each of these numbers is odd *and* an integer multiple of 3. Out of the 36 possible pairs, this combination occurs in 6 pairs. Therefore, $P(A \cap B) = 6/36$.

We now need to find whether

$$P(A \cap B) = P(A)P(B)$$
$$6/36 = (1/2)(12/36) = 6/36$$

Therefore, events A and B are independent.

(b) What is the probability of event $A \cup B$?

We now need to compute the probability of the event $A \cup B$. We have already computed all the necessary components:

$$P(A \cup B) = P(A) + P(B) - P(AB)$$

Note that since the events are independent, we can rewrite this equation as follows:

$$P(A \cup B) = P(A) + P(B) - P(A)P(B)$$

And this equals

$$P(A \cup B) = 1/2 + 12/36 - 6/36 = 24/36 = 2/3$$

Therefore, the probability of rolling an odd number or rolling a number that is an integer multiple of 3 equals 2/3.

1.7 BERNOULLI TRIALS

Suppose that an experiment is repeated n times and that it is desired to find the probability that a particular event occurs exactly m times. Assume that the experiment only has two possible outcomes:

$$\text{Success} = S \quad \text{probability of success } = p$$
$$\text{Failure } = F \quad \text{probability of failure } = q$$

If we perform the experiment n times, what is the probability that we will get m successes? We assume that these are independent trials. A typical experiment is one where a single coin is tossed successively and we consider the appearance of a head to be a success. The tosses are independent, and the outcomes are mutually exclusive; i.e., if a head appears, then a tail cannot. A typical sequence might look as follows:

$$\leftarrow \text{Independent} \rightarrow$$
$$S\,F\,F\,S\,S\,S\,F\,S\,F\,F\,F\cdots$$

where S implies a head or success and F implies a tail or failure.

If we obtain m successes, then we will have $n - m$ failures. The probability of any *one* given sequence (as the above) where the trials are independent having exactly m successes and $n - m$ failures is

$$P = P(S \cap F \cap F \cap S \cap S \cap S \cap F \cap S \cdots)$$
$$= P(S)P(F)P(F)P(S)P(S)P(S)P(F)P(S) \cdots \qquad (1.29)$$

and

$$P = p \cdot q \cdot q \cdot p \cdot p \cdot p \cdot q \cdot p \cdot q \cdot q \cdot q \cdots = p^m q^{n-m} \qquad (1.30)$$

But this is just the probability of *one* sequence. The sequence has been arranged in *one* particular way, so that we must now find the total number of possible ways in which we can arrange the total number of S's and F's. Therefore, given that we have m successes and $n - m$ failures, find the total number of possible sequences having exactly this number of successes and failures. This question may be formulated as follows:

Probability of arrangement shown

$$= \frac{1}{\text{total no. of possible arrangements}} \qquad (1.31)$$

or

$$P = \frac{1}{T} \tag{1.32}$$

Therefore,

$$T = \frac{1}{P} \tag{1.33}$$

We now find P by computing the probability of finding the particular arrangement shown:

Probability of a success in the first position: $\dfrac{m}{n}$

Probability of a failure in the second position: $\dfrac{n-m}{n-1}$

Probability of a failure in the third position: $\dfrac{n-m-1}{n-2}$

etc.

Therefore,

$$P = \frac{m(n-m)(n-m-1)(m-1)(n-m-2)\cdots}{n(n-1)(n-2)(n-3)(n-4)\cdots} \tag{1.34}$$

which can be reformulated as

$$P = \frac{m!\,(n-m)!}{n!} \tag{1.35}$$

where $n! = n(n-1)(n-2)\cdots$, and by definition $0! = 1$. Therefore,

$$T = \frac{n!}{m!\,(n-m)!} \tag{1.36}$$

Since each of the possible arrangements has the same probability and is independent from each other, the probability of obtaining exactly m successes out of n trials, when the probability of success is p and the probability of failure is q, equals

$$P_{m,n} = \frac{n!}{m!\,(n-m)!}\,p^m q^{n-m} \tag{1.37}$$

This is known as the *binomial distribution*.

Example 1.7 If a coin is tossed 6 times, what is the probability of obtaining at least 4 heads?

For this case $p = q = 1/2$. The probability of getting *exactly* 4 heads out of 6 tosses is calculated as follows:

$$P_{m,n} = \frac{n!}{m!(n-m)!}p^m q^{n-m} = \frac{6!}{4!(6-4)!}\left(\frac{1}{2}\right)^4\left(\frac{1}{2}\right)^{6-4} = \frac{15}{64}$$

Similarly, we can obtain the probability of obtaining *exactly* 5 heads out of 6 tosses

$$P_{5,6} = 6/64$$

and *exactly* 6 heads out of 6 tosses

$$P_{6,6} = 1/64$$

Therefore, the probability of obtaining at least 4 heads equals

$$P\{\text{at least 4 heads}\} = P\{\text{exactly 4 heads}\}$$
$$+ P\{\text{exactly 5 heads}\} + P\{\text{exactly 6 heads}\}$$

and

$$P\{\text{at least 4 heads}\} = 15/64 + 6/64 + 1/64 = 22/64$$

$$P(A) = \frac{365 \cdot 364 \cdot \cdot \cdot 336}{365^{30}}$$

In problems of this type we must often deal with factorials that may be cumbersome to compute. There is, however, an approximation that may be used for problems of this type. For large numbers we can make the following approximation (Stirling's approximation):

$$N! \cong (2\pi N)^{1/2}\left(\frac{N}{e}\right)^N \tag{1.38}$$

where $e = 2.718 \cdot \cdot \cdot$.

In addition, we can make the following approximation to the binomial distribution under specific conditions:

1. The number of successes m is small compared to the number of trials.
2. The number of trials is n.
3. The probability of success is small, that is, $p \ll 1$.

Then the following approximation applies:

$$P_{m,n} = \frac{(np)^m e^{-np}}{m!} \qquad (1.39)$$

This is also known as the *Poisson approximation* to $P_{m,n}$.

Example 1.8   The Binary Transmission Channel. The binary transmission channel problem is presented now because of its importance in communication systems. It also helps illustrate some of the concepts that we have just considered. Assume we have a source that transmits two symbols (m_0, m_1) and that two symbols are received. Let the received symbols be r_0 and r_1. A problem of this type may be represented using Figure 1.8. Let $P(m_0)$ and $P(m_1)$ be the a priori transmission probabilities. This implies that the symbol m_0 is sent with a probability $P(m_0)$. Since the source can transmit only two symbols, the sum of the two probabilities must add to 1; that is, $P(m_0) + P(m_1) = 1$. The forward transmission probabilities are $P(r_i/m_j)$. This is the probability that symbol r_i was received *given that* symbol m_j was sent. Then according to Equation (1.9),

$$P(r_j m_i) = P(r_j / m_i)P(m_i)$$

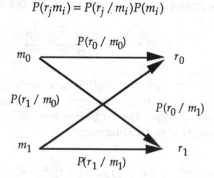

Figure 1.8 Binary transmission channel.

The probability of receiving one symbol may be found as follows:

$$P(r_j) = \sum_{i=0}^{1} P(m_i)P(r_j / m_i) = \sum_{i=0}^{1} P(m_i \cap r_j) \qquad (1.40)$$

The reasoning behind Equation (1.40) is that the symbol r_i can only be generated in two ways. The first way is through the $m_0 \rightarrow r_0$ path, the other is through the $m_1 \rightarrow r_0$ path. Another way of stating Equation (1.40) is to say that the sum of the probabilities of the intersections of r_i with m_j equals the total probability of symbol r_i.

We are perhaps more interested in the "backward probability," that is, $P(m_i/r_j)$. This is the probability that a symbol m_j was sent, *given that* a symbol r_i was received. In other words, we are measuring the output of the receiver, and we do not know which symbol was sent. Therefore, if we calculate $P(m_i/r_j)$, we will find the probability that if we receive symbol r_0, then symbol m_0 was sent. It should be obvious that the paths $m_1 \rightarrow r_0$ and $m_0 \rightarrow r_1$ are paths that will lead to erroneous results in reception.

$$P(m_i / r_j) = \frac{P(m_i \cap r_j)}{P(r_j)}$$

Multiply and divide by $P(m_i)$:

$$P(m_i / r_j) = \frac{P(m_i \cap r_j)}{P(r_j)} \cdot \frac{P(m_i)}{P(m_i)}$$

Rearranging gives

$$P(m_i / r_j) = \frac{P(r_j \cap m_i)}{P(m_i)} \cdot \frac{P(m_i)}{P(r_j)}$$

and finally

$$P(m_i / r_j) = \frac{P(r_j / m_i)P(m_i)}{P(r_j)} \qquad (1.41)$$

Example 1.9 Consider now the following problem: Given is a binary channel with two possible input message symbols $m_0 = 0$ and $m_1 = 1$ and two possible outcome symbols $r_0 = 0$ and $r_1 = 1$. Assume the a priori probabilities are $P(m_0) = .75$ and $P(m_1) = .25$ and that transmission probabilities are $P(r_0/m_0) = .67$ and $P(r_0/m_1) = .1$. This problem may be interpreted as follows: We have a transmitter that sends either 0s or 1s where P(sending a 0) = .75 and P(sending a 1) = .25. Whenever the source sends a 0, a 0 is received by the receiver with a probability equal to .667. On the other hand, whenever a 1 is sent, the receiver (in error!) receives a 0 with a probability equal to .1.

(a) Find $P(r_0)$ and $P(r_1)$.

We are given that $P(r_0/m_0) = .67$; therefore, we find that $P(r_1/m_0) = .33$. Notice that, by definition, $P(r_0/m_0) + P(r_1/m_0) = 1$. This is due to the fact that the symbol m_0 (or a 0) can only be sent through two possible paths. We are also given that $P(r_0/m_1) = .1$; therefore, by a similar argument, $P(r_1/m_1) = .9$.

The first figure (Figure 1.9 below) illustrates the original information supplied with the problem, and the second figure illustrates the information found after considering the different arguments used to solve part (a) of the problem.

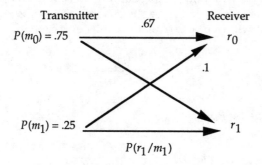

Figure 1.9 Original information supplied with the problem.

From Equation (1.40), we find

$$P(r_0) = P(m_0)P(r_0 / m_0) + P(m_1)P(r_0 / m_1)$$

and

$$P(r_0) = (.75)(.67) + (.25)(.1) = .525$$

The new information about the binary transmission channel is represented in Figure 1.10.

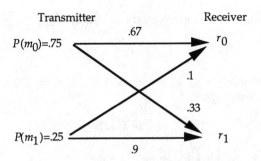

Figure 1.10 New information about binary transmission channel.

Similarly,

$$P(r_1) = P(m_0)P(r_1/m_0) + P(m_1)P(r_1/m_1)$$

and

$$P(r_1) = (.75)(.33) + (.25)(.1) = .475$$

Note that we could just as easily have reasoned that since $P(r_0) + P(r_1) = 1$, then $P(r_1) = 1 - P(r_0) = .475$.

(b) Find $P(m_0/r_0)$, $P(m_0/r_1)$, $P(m_1/r_0)$, and $P(m_1/r_1)$.
     These are the backward probabilities. Let us compute just one:

$$P(m_0/r_0) = \frac{P(r_0/m_0)P(m_0)}{P(r_0)} = \frac{(2/3)(3/4)}{21/40} = 20/21$$

(c) Find $P(m_0 \cap r_0)$.

$$P(m_0 \cap r_0) = P(r_0/m_0)P(m_0) = (2/3)(3/4) = 1/2$$

1.8 CHAPTER SUMMARY

We summarize this important chapter. For any event A,

$$0 \le P(A) \le 1$$

The union of events A and B is given by the equation

$$P(A \cup B) = P(A) + P(B) - P(A \cap B)$$

where $P(A \cap B)$ is the intersection of events A and B. Given that events A and B are mutually exclusive, then

$$P(A \cup B) = P(A) + P(B)$$

Given that events A and B are independent, then

$$P(A \cup B) = P(A) + P(B) - P(A)P(B)$$

Bayes' theorem is represented by the following formula:

$$P(A_i/B) = \frac{P(B/A_i)\, P(A_i)}{\sum_j P(B/A_j)\, P(A_j)}$$

The probability of obtaining exactly m successes in n independent trials where the probability of success equals p and the probability of failure equals q is given by

$$P_{m,n} = \frac{n!}{m!\,(n-m)!}\, p^m q^{n-m}$$

1:9 PROBLEMS

1. A card is drawn from a standard deck of cards. Find the probability that this card is a 2, or a 5, or a 10.

2. A dodecahedron is a solid object with 12 equal faces. It is frequently used as a calendar paperweight with one month placed on each face. If such a calendar is randomly placed on a desk, the outcome is taken to be the month of the upper face.

    (a) What is the probability of the outcome's being February?

    (b) What is the probability of the outcome's being January or April or August?

    (c) What is the probability that the outcome will be a month with 31 days?

(d) What is the probability that the outcome will be in the third quarter of the year?

3. Two fair, six-sided dice are thrown. Find the probability of

(a) Throwing a sum of 11

(b) Throwing two 7s

(c) Throwing a pair

4. In the network of switches shown below, each switch operates independently of the others and each has a probability of .5 of being closed. What is the probability of a complete path through the network?

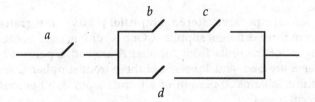

5. In a roulette wheel there are 38 slots in which a ball may land. There are numbers 0, 00, and 1 through 36. The odd numbers are red, the even numbers are black, and the 0s are green. A ball is thrown randomly into a slot.

(a) What is the probability it is red?

(b) What is the probability it is number 27?

(c) What is the probability it is either red or between 10 and 15, but not both?

(d) What is the probability it is not green?

6. A technician has three resistors, with values of 100, 300, and 900 Ω. If she connects them randomly in the configuration shown:

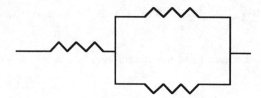

(a) What is the probability that the equivalent resistance of this network will be 390 Ω?

(b) What is the probability that the equivalent resistance will be greater than 390 Ω?

(c) What is the probability that the equivalent resistance will be less than 1000 Ω?

7. A manufacturer of inexpensive stereo amplifiers buys integrated circuit (IC) units from three different suppliers. One IC unit is used in each amplifier. Ten percent of the units from supplier A are bad, 5 percent of those from supplier B are bad, and 1 percent of those from supplier C are bad. The manufacturer obtains 50 percent of all units from A, 30 percent from B, and 20 percent from C.

(a) What is the probability that a randomly selected amplifier will contain a faulty IC?

(b) If an amplifier is found to have a faulty IC, what is the probability that it was obtained from supplier C?

(c) If the manufacturer buys one-third of his needs from each supplier, what is the probability that a randomly selected amplifier will contain a faulty IC?

8. A fair, six-sided die is thrown. If the outcome is odd, the experiment is terminated. If the outcome is even, the die is tossed a second time. Find the probability of

(a) Throwing a sum of 7

(b) Throwing a sum of 2

9. Given two events A and B with probabilities $P(A) = .25$, $P(B/A) = .5$, and $P(B/A) = .25$, clearly explain whether events A and B are mutually

exclusive.

10. A box contains a large number of 100-Ω resistors and an equal number of 400-Ω resistors. Three resistors are selected at random from the box and connected as shown below:

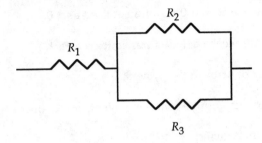

(a) What is the probability that the equivalent resistance of this network will be less than 200 Ω?

(b) What is the probability that the equivalent resistance will be greater than 500 Ω?

11. Two solid-state diodes are connected in parallel. Each diode has a probability of .05 that it will fail as a short circuit and a probability of .10 that it will fail as an open circuit. If the diodes are independent, what is the probability that the parallel combination will function? (Note that one short circuit causes the combination to fail, but one open circuit does not.)

12. A traffic control system directs cars, at random, into one of three alternate routes leading to the same destination. For cars on route A, the probability of having an accident is .1, for route B the probability is .15, while for route C it is .06.

(a) If one-half the cars are directed into route A, one-third into route B, and one-sixth into route C, what is the total probability of an accident?

(b) If an accident occurs, what is the probability that it is on route B?

(c) Suppose the assignment of cars into routes is changed so that each route carries one-third of the traffic. Now what is the total probability of an accident?

13. In a certain binary communication system, the messages are coded into 0 and 1. After coding, the probability of a 0 being transmitted is .55, while the probability of a 1 is .45. In the communication channel the probability that a transmitted 1 is erroneously received as a 0 is .1, while the probability of a 0 being received as a 1 is .2.

(a) Find the probability that a received 0 was transmitted as a 0.

(b) Find the probability that a received 1 was transmitted as a 1.

(c) Find the total probability of error in the system.

14. A pair of six-sided dice is rolled, and the following events are defined:

$$A = \{\text{rolling an odd number}\}$$
$$B = \{\text{rolling a number that is an integer multiple of 3}\}$$

(a) Are events A and B statistically independent? Prove your answer.

(b) What is the probability of the event $A \cup B$?

15. In playing an opponent of equal ability, what is more probable?

(a) To win 3 games out of 4, or to win 5 games out of 8.

(b) To win at least 3 games out of 4, or to win at least 5 games out of 8.

16. A small business firm has eight telephones, each of which is in use 10 percent of the time, independently of the others. How many trunk lines into the firm are required in order to meet its needs 90 percent of the time?

17. A five-stage register is constructed from five IC flip-flops. The probability that any one IC is faulty is .1, independent of any other IC. What is the probability that the shift register will work?

18. The probability that each relay in the figure below is closed is .5. The state of any relay is independent of the state of all other relays. Find the probability that a continuous path exists between terminals 1 and 2.

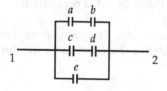

19. Three hunters, A, B, and C, fire into a flock of Canadian geese. Based on their previous hunting records, the probabilities of a hit are 1/2, 1/3, and 1/4, respectively. Assume that the firing of any one hunter has no effect on the performance of the others.

(a) Find the probability of at least one hit by the three hunters.

(b) Find the probability of at least two hits by the three hunters.

(c) One of the hunters is selected at random. She fires and makes a hit. Find the probability that the most accurate shooter was selected.

20. A satellite communication system has eight ground stations, each of which uses a channel through the satellite repeater 10 percent of the time. How many channels are needed in the repeater in order to make it available 90 percent of the time to all the ground stations desiring it?

1.10 COMPUTER EXAMPLES

MATHCAD® Computer Example 1.1*
COMPUTATION OF PROBABILITIES . . . Relative Frequency Approach

As an example of the relative frequency approach we will calculate the probability that a head will occur in a coin toss. We know, of course, that the probability of getting a head or a tail in a fair coin toss is .5; however, if we were to conduct an experiment to determine this, as we will, then we would see that the calculated probability approaches .5 as the number of trials in the experiment increases. We know that the formula for calculating the probability of an event by the relative frequency approach is

Pr(head) = no. of times a head occurs / total no. of coin tosses

Now we will define the following variables:

Heads = number of times a head occurs in a given experiment
Trials = total number of coin tosses performed in the experiment
Prob = Pr(head) = heads/trials = probability that a head will occur

There are 10 trials in the experiment; each consists of a single toss of the coin.

Trials := **100**   index of summation used below

k := **1 .. Trials**   number of heads per experiment

---

* Mathcad is a registered trademark of MathSoft, Inc., 101 Main Street, Cambridge, Massachusetts, 02142. Printed by permission of MathSoft, Inc.

$$\text{Heads} := \sum_k \text{flip}(1)$$

$$\text{Prob} := \frac{\text{Heads}}{\text{Trials}}$$

The experimentally calculated probability of getting a head for this experiment is

Prob = 0.55

To see the results for another experimental run with the same number of trials, put the cursor in the equation below and press the F9 key.

$$\text{flip}(\text{coin}) \equiv \Phi\,(\text{rnd}(1) - .5)$$

To see the results for more or fewer trials, change the value of Trials defined above to a number other than 10; try increasing it, try decreasing it. What happens?

MATHCAD Computer Example 1.2
DICE EXPERIMENT . . . Relative Frequency Approach

In this experiment we will imitate the roll of a single die a number of times and compute several different probabilities based on the relative frequency approach. We will compare these experimental values with those predicted theoretically. We start by defining a function to simulate the rolling of a die. This can be done by generating a sequence of uniformly distributed random variables between 0 and 6. Then, by forming a histogram with six equal intervals, we can count the number of occurrences within each interval and assign each interval a number from 1 to 6. The intervals are defined as follows:

If x is between 0 and < 1 → assign value of 1
If x is between 1 and < 2 → assign value of 2
If x is between 2 and < 3 → assign value of 3
If x is between 3 and < 4 → assign value of 4
If x is between 4 and < 5 → assign value of 5
If x is between 5 and < 6 → assign value of 6

The range variable to generate NTRIALS of the experiment (defined below) is:

$i := 0 .. \text{NTRIALS} - 1$

The function rnd(x) is a built-in function that generates uniform random values between 0 and the number within parenthesis.

data$_i$:= **rnd(6)**   counter to generate histogram intervals

k := **0 .. 7**

inter$_k$:= **k**   counter to generate histogram plot

j := **0 .. 5**

freq := **hist(inter , data)** built-in function that generates histogram

NTRIALS= **60**   number of times to roll the die

Table of the results:

freq

13
7
9
10
13
8

The probability of getting each of the above numbers can be calculated by the relative frequency approach:

$$\mathbf{Pr}_j := \frac{\mathbf{freq}_j}{\mathbf{NTRIAL}}$$

Pr$_j$

0.217
0.117
0.15
0.167
0.217
0.133

Try changing the value of NTRIALS above, and observe the changes. Increase NTRIALS to 600. What happens? The theoretical value of the probability of getting any single number on a fair die roll is, of course, 1/6 (or .166666 . . .).

$$A := \frac{1}{6}$$

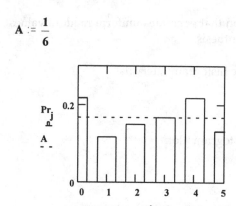

The dotted line is the theoretical value, or $1/6$.

MATLAB® Computer Example 1.1*
COMPUTATION OF PROBABILITIES . . . Relative Frequency Approach

As an example of the relative frequency approach, we will calculate the probability that a head will occur in a coin toss. We know, of course, that the theoretical probability of getting a head or a tail in a fair coin toss is .5. If we were to conduct an experiment to determine this number, the calculated probability would probably be different from .5.

   We know that the formula for calculating the probability of an event by the relative frequency approach is as follows:

Pr(head) = number of times a head occurs / total number of coin tosses

Now we will define the following variables:

   Heads = number of times head occurs in given experiment
   Trials = total number of coin tosses performed in experiment
   Prob = Pr(head) = heads/trials = probability that head will occur

In the experiment, 100 trials, each consists of a single toss of the coin.

```
%=============================================
% Input parameter
```

---

* MATLAB is a registered trademark of the MathWorks, Inc., 24 Prime Park Way, Natick, MA 01760-1500, Phone: 508-647-7000, WWW: http://www.mathworks.com

```
Trials = 100;


flip=rand(Trials);
Heads=0;
for i=1:1:Trials
 if flip(i) < 0.5
  Heads=Heads+1;
 end;
end;
Prob=Heads/Trials;
disp(['Prob: ',num2str(Prob)]),clear;
%============================================
```

Prob: 0.47

1. Rerun the experiment. Did you obtain the same answer? Why or why not?
2. Rerun the experiment making Trials = 1000.
Comment on your results.

MATLAB Computer Example 1.2
DICE EXPERIMENT . . . Relative Frequency Approach

In this experiment we will roll a single die a number of times and compute several different probabilities based on the relative frequency approach. We will compare these experimental values with those predicted theoretically. We start by defining a function to simulate the rolling of a die. This can be done by generating a sequence of uniformly distributed random variables between 0 and 6. Then, by forming a histogram with six equal intervals, we can count the number of occurrences within each interval and assign to each interval a number from 1 to 6. The intervals will be defined as follows:

$$
\begin{aligned}
&\text{If } x \text{ is between 0 and } < 1 \quad \rightarrow \text{ assign value of 1} \\
&\text{If } x \text{ is between 1 and } < 2 \quad \rightarrow \text{ assign value of 2} \\
&\text{If } x \text{ is between 2 and } < 3 \quad \rightarrow \text{ assign value of 3} \\
&\text{If } x \text{ is between 3 and } < 4 \quad \rightarrow \text{ assign value of 4} \\
&\text{If } x \text{ is between 4 and } < 5 \quad \rightarrow \text{ assign value of 5} \\
&\text{If } x \text{ is between 5 and } < 6 \quad \rightarrow \text{ assign value of 6}
\end{aligned}
$$

This is the range variable to generate NTRIALS of the experiment (defined below).

```
%============================================
% Input parameter
NTRIALS = 60;
```

This is the range variable to generate NTRIALS of the experiment (defined below).

```
%===========================================
% Input parameter
NTRIALS = 60;
data_i=6*rand(NTRIALS,1);
inter=6;
[freq_j,bins]=hist(data_i,inter);
clg;s1=subplot(2,2,1);
bar(0:1:inter-1,freq_j);
xlabel('j','Fontsize',[10]);
ylabel('freq_j','Fontsize',[10]);axisn(0:1:inter-1);
Pr_j=freq_j/NTRIALS;
set(s1,'Fontsize',[10])
s2=subplot(2,2,2);
bar(0:1:inter-1,Pr_j); axisn(0:1:inter-1);
xlabel('blue line : theoretical value ( 1/6 )' ,'Fontsize',[10]);
ylabel('Pr_j','Fontsize',[10]);
z=1/6*ones(size(0:0.01:6));
set(line(0:0.01:6,z),'Color',[0 0 1],'LineStyle','.')
set(s2,'Fontsize',[10])
s3=subplot(2,2,3);
axis('off'),
text(0.3,0.9,'freq_j : ','Fontsize',[10]);
for j=0:inter-1
 set(text(0,1/(inter+1)*(inter-1-
j),num2str(freq_j(j+1)),'Fontsize',[10]),'HorizontalAlignment','right');
 text(0,1/(inter+1)*(inter-1-j),['...number of times a ',num2str(j+1),'
occurred'],'Fontsize',[10]);
end;
s4=subplot(2,2,4);
axis('off')
text(0.1,0.9,'probability of occurrence :','Fontsize',[10])
for j=0:inter-1,
 set(text(0.1,1/(inter+1)*(inter-1-
j),'...probability','Fontsize',[10]),'HorizontalAlignment','left');
 set(text(0.6,1/(inter+1)*(inter-1-
j),num2str(Pr_j(j+1)),'Fontsize',[10]),'HorizontalAlignment','left');
end;
clear
%===========================================
```

1. Rerun the experiment. Did you obtain the same answer?
2. Rerun the experiment, making Trials = 600.

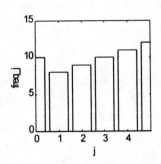 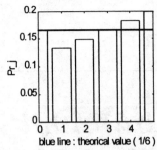

freq_j :	probability of occurrence :
10...number of times a 1 occurred	...probability 0.1667
8...number of times a 2 occurred	...probability 0.1333
9...number of times a 3 occurred	...probability 0.15
10...number of times a 4 occurred	...probability 0.1667
11...number of times a 5 occurred	...probability 0.1833
12...number of times a 6 occurred	...probability 0.2

CHAPTER

2 *Random Variables*

2.0 INTRODUCTION

The outcome of an experiment is completely described by the sample space S, where an event is a subset of the sample space. The outcome of the experiment may or may not be numerical in nature, such as the outcomes of the toss of a die or toss of a coin. It is convenient to map all outcomes, whether numerical or not, onto the set of real numbers. This mapping describes the random variable, which is very useful in solving many practical problems. In this chapter we begin the description of characteristics of both continuous and discrete random variables.

2.1 CONCEPT OF A RANDOM VARIABLE

In the first chapter we described a number of simple experiments involving the tossing of a die or a coin, and picking a card from a deck of cards. In these cases the outcome was always a random occurrence of an event. We said that the result was random since we were not able to predict what the outcome of each repetition of the experiment would be. In one case (coin toss), the outcome was a letter (*H* or *T*). In another (toss of a die), the outcome was a set of numbers (1 to 6); in yet another (draw a card from a deck of cards), the outcome was a combination of letters and numbers. These were all discrete cases in that the outcome could only be one of a number of finite possibilities. In addition, we knew ahead of time what the entire sample space *S* was. A *random variable* is simply a characterization of the outcome of a probabilistic experiment which assigns a numerical value to that outcome. In the case of an experiment which is not numeric in nature (toss of a coin), one can always code the output into a numeric scheme. We will assign to the output of a specific experiment the letter *x*, a specific value of the random variable *X*. The *range* of *X* depends on the particular experiment we are performing. For example, in the case of the single toss of a die, the range of *X* is from 1 to 6. In the case of a single toss of a coin, the outcome is a head or a tail to which we could arbitrarily assign a numerical value such as a 1 or a 2. This definition implies that the range of *X* is the set of all possible outcomes of the experiment we are performing, or, alternatively, the set of all possible outcomes of the random variable.

If the random variable takes only discrete values (where the range of values is countable), then we are dealing with a *discrete random variable*. For example, in the case of the single toss of a die, there are only six possible outcomes; consequently, we are dealing with a discrete random variable. Likewise, the single toss of a fair coin only has two possible outcomes. Consequently, the random variable is discrete. If the random variable takes a continuous range of values, then we are dealing with a *continuous random variable*, such as the output voltage of a seismograph as it registers earthquake events. Even if we assume that this output is limited to between −5 V and +5 V, there will be an infinite number of (uncountable) values between these two numbers that may be measured. Note that the assumption of continuity is adopted for mathematical convenience even if the output can only be measured with finite precision, resulting in a finite number of possible outcomes.

It is important that we emphasize the difference between a finite set of possible realizations and a finite range. In the case of the output of a seismometer, we have a finite range [−5, +5], and the random variable is continuous, since it can assume any value in the range. Conversely, in the case of the single toss of a fair die the range is [1, 6], and the random variable is discrete, since it can assume only the six possible discrete values of the output.

2.2 CUMULATIVE DISTRIBUTION FUNCTION

A random variable cannot be predicted exactly. Given a specific random experiment, we cannot predict with certainty the outcome of the next realization of the experiment. It is possible, however, to know certain rules which pertain to the specific experiment and which govern, in a general way, the behavior of the random variable X. For example, we know that when we toss a fair die, each of the faces has an equal chance of appearing. We can then say that the probability associated with a particular outcome x of X is 1/6. So even if we cannot predict the outcome of a specific toss, we can associate a probability with any face.

We now define the concept of a distribution function. A cumulative *distribution function* uniquely defines the probability that the random variable X is less than or equal to some allowed value x. Clearly, the shape and form of these functions differ depending on whether the random variable is discrete or continuous. Sometimes, the random variable exhibits a mixture of discrete and continuous behavior, and in these cases, the distribution function behaves accordingly. The distribution function is defined by the following equation:

$$F_X(x) = P(X \le x) \tag{2.1}$$

Now $F_X(x)$ specifies the probability that the random variable X assumes a value that is less than or equal to x. For example, consider the single toss of a die. We can ask the simple question, What is the probability that the outcome of a single toss will be less than or equal to 3? From results obtained in Chapter 1 (and by intuition), the answer is 1/2. Suppose that we now ask, What is the probability that the outcome of a single toss will be less than or equal to 6? Likewise, the answer is 1. Using this simple experiment as an example, we can construct a table of the values of $F_X(x)$ for this specific case:

$$F_X(0) = P(X \le 0) = 0$$
$$F_X(1) = P(X \le 1) = 1/6$$
$$F_X(2) = P(X \le 2) = 2/6$$
$$F_X(3) = P(X \le 3) = 3/6$$
$$F_X(4) = P(X \le 4) = 4/6$$
$$F_X(5) = P(X \le 5) = 5/6$$
$$F_X(6) = P(X \le 6) = 6/6$$

The values of the probability distribution function $F_X(x)$ may now be plotted as a function of the outcome x. Several properties of $F_X(x)$ are apparent once we examine Figure 2.1. For this particular case the random variable is *discrete*. Consequently, the distribution function $F_X(x)$ takes discrete steps. We also note that the distribution function has a maximum value equal to 1 and a minimum value equal to 0. Also note that $F_X(x)$ is nondecreasing as x increases. This indicates that the value of $F_X(x)$ at $-\infty$ equals 0 whereas the value of $F_X(x)$ at $+\infty$ equals 1.

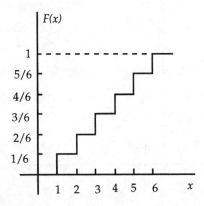

Figure 2.1 Values of the discrete distribution function $F_X(x)$ for the single toss of a die.

For the case of a discrete random variable, the value of $F_X(x)$ for a specific value of x is the value at the upper part of the step. The probability value associated with a specific discrete random variable x equals the size of the step.

Based on the observations made above, we conclude that the distribution function $F_X(x)$ has the following properties:

1. $0 \le F_X(x) \le 1$
2. $\lim_{x \to \infty} F_X(x) = 1$
3. $\lim_{x \to -\infty} F_X(x) = 0$
4. $F_X(x)$ is a nondecreasing function of x
5. $P(X = x) = F_X(x) - \lim_{\varepsilon \to 0} F_X(x - \varepsilon)$
6. $F_X(x) = P(X \le x) = 1 - P(X > x)$

The *first* property is really a statement of the fact that $F_X(x)$ is a probability function and therefore, has all of the properties associated with the concept of probability. The *second* and *third* properties are a consequence of the preceding statement. The *fourth* property is due to the fact that probability is a nonnegative quantity. The *fifth* property indicates that when the random variable is discrete, such as in the present case, the value of the probability associated with a specific value of the random variable is a finite number. When the random variable is continuous, the fifth property

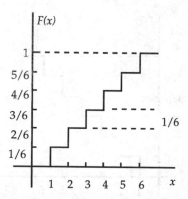

Figure 2.2 Values of the distribution function $F_X(x)$ for the single toss of a die. The value of $F_X(x)$ for $x = 3$ is shown as the step and equals $1/6$. The value of $F_X(x)$ for $x = 2$ is the value at the upper part of the step and equals $2/6$.

conveys the notion that the property associated with a specific value of the random variable is 0. This will be examined in greater detail later. The *sixth* property is a consequence of the fact that the events $X \leq x$ and $X > x$ are mutually exclusive and that $(X \leq x) \cup (X > x) = S$. Figure 2.2 shows the value of a discrete step.

Example 2.1 Consider Example 1.6. First draw the distribution function for the experiment of rolling a pair of fair dice.

(a) Find the probability that the sum of the faces will be 6 or less.
    By definition, we know that $F_X(6) = P(X \leq 6)$. Based on the comments made earlier, the value of $F(6)$ equals the upper part of the step. From the graph, we read this value to be $15/36$, which corresponds to the results of Example 1.6.

(b) Find the probability that the sum of the faces will be exactly equal to 10.
    The value of $P(X = 10)$ from the graph equals $F_X(10^+) - F_X(10^-)$, which equals $3/36$, or the jump in the discrete function.

(c) What is the probability that the sum of the faces will be less than or equal to 11?
    From property 6, we can solve this problem as follows:

$$F_X(11) = P(X \leq 11) = 1 - P(X > 11)$$

Therefore,

$$F_X(11) = 1 - 1/36 = 35/36$$

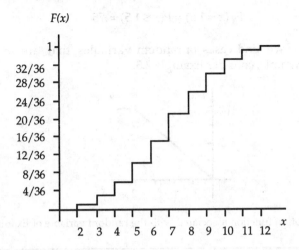

Figure 2.3 Distribution function associated with the experiment of rolling a pair of fair dice and adding the values of their faces. (The distribution function is drawn approximately to scale.)

$$F_X(11) = 1 - 1/36 = 35/36$$

Figure 2.3 shows the distribution function for this example.

When we deal with random variables that are strictly *continuous* the distribution function has no discontinuities.

Example 2.2 A random variable has a distribution function given by

$$F_X(x) = \begin{cases} 0 & -\infty < x \le 0 \\ 0.5x & 0 < x < 2 \\ 1 & 2 \le x < \infty \end{cases}$$

The distribution function associated with this random variable is shown in Figure 2.4. As can be observed, the cumulative distribution function does not have any discontinuities. If we examine the properties given earlier, we observe that the *fifth* property implies that in the case of a continuous random variable the probability value associated with a *specific value* of a continuous random variable is 0.

What is the significance of the value of $F_X(x)$ at $x = 1.5$? From the graph, we find that the value of $F_X(x)$ at $x = 1.5$ equals .75. This value is the probability that the random variable X is less than or equal to 1.5, that is,

$$F_X(x = 1.5) = P(x \le 1.5) = .75$$

Occasionally, we find cases of random variables that are of a mixed mode. For example, consider Example 2.3.

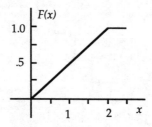

Figure 2.4 Distribution function associated with the random variable of Example 2.2.

Example 2.3 A random variable has a distribution function given by

$$F_X(x) = \begin{cases} 0 & -\infty < x \le -10 \\ \dfrac{1}{6} & -10 < x \le -5 \\ \dfrac{1}{15}x + \dfrac{1}{2} & -5 < x < 5 \\ \dfrac{5}{6} & 5 \le x < 10 \\ 1 & 10 < x < \infty \end{cases}$$

Figure 2.5 shows the distribution function associated with the random variable of Example 2.3.

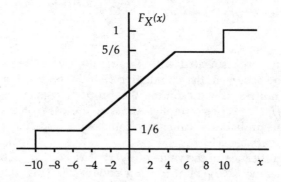

Figure 2.5 Distribution function associated with the random variable of Example 2.3.

This could be an example associated with a specific circuit where the output is such that anytime the voltage exceeds +5 V it is clamped at +10 V. This happens with a probability of 1/6. Likewise, anytime the output voltage is below −5 V the output is clamped at −10 V. This also happens with a probability of 1/6. We can also say that the output is between −5 V and +5 V with a probability of 4/6.

2.3 PROBABILITY DENSITY FUNCTION

The *probability density function* is defined as the *derivative* of the distribution function for continuous $F_X(x)$. Therefore,

$$f_X(x) = \frac{dF_X(x)}{dx} \tag{2.2}$$

However, if the random variable is discrete, then the density function is defined in terms of the values of the probability at the distinct values taken by the random variable (or probability mass functions):

$$f_X(x) = \begin{cases} P_X(x = x_i) & \text{if } x = x_i, \ i = 1, \cdots, n \\ 0 & \text{if } x \neq x_i \end{cases} \tag{2.3}$$

The interpretation of the information conveyed by this representation must be made carefully, taking into account the comments made earlier concerning the mixed-mode distribution function. Consider, for example, the density function shown by Figure 2.6.

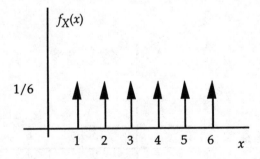

Figure 2.6 Probability density function associated with the random variable of Figure 2.1.

At each of the discrete steps associated with the distribution function, the density function is represented by a point mass function of strength equal to the jump, as shown in Figure 2.6. Therefore, for a discrete random variable, one could read directly off the graph the probability associated with the particular value of the random variable X. In this particular case, the probability of occurrence of any of the faces of the die is 1/6; and, moreover, all the faces have an equal probability of occurring. Before we consider the general properties of the probability density function, consider the probability density function associated with Example 2.3. As expected, at –10 V and +10 V there is a delta function of strength equal to the jump, or 1/6. From –10 to –5 V, the probability density function has a value of 0, since the random voltage cannot take on any of these values. From –5 to +5 V, the probability density function is a straight line. The area under this line can be found by answering the question, What is the probability that the random voltage lies between +5 and –5 V? In terms of the distribution function, this is equivalent to finding the value of the following expression:

$$F_X(5) - F_X(-5) = P_X(X \le 5) - P_X(X \le -5)$$

and this expression equals (directly off the graph) 2/3.

   In order to evaluate the height of the line, we compute

$$\int_{-5}^{5} a \, dx = \frac{2}{3} \tag{2.4}$$

From which $a = 1/15$.

The probability density function associated with this example is shown in Figure 2.7. Note that the probability that the random voltage lies between –5 and +5 V *plus* the probability that it lies at exactly –10 and +10 V equals 1.

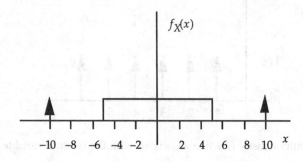

Figure 2.7 Density function associated with the random variable of Example 2.3.

The general properties associated with the probability density function are as follows. First,

$$f_X(x) \geq 0 \tag{2.5}$$

This equation states that the probability values are nonnegative, and it is a restatement of the fact that $F_X(x)$ is a nondecreasing function of the random variable X.

Second,

$$\int_a^b f_X(x)\, dx = P(a < X \leq b) \tag{2.6}$$

This is a most important property of the probability density function. It actually interprets for us the concept associated with the function $f_X(x)$ of the random variable X. This equation specifies the probability that the random variable X falls between the limits (a, b]. Obviously, this value must be greater than or equal to 0.

Third,

$$\int_{-\infty}^{\infty} f_X(x)\, dx = 1 \tag{2.7}$$

Based on Equation (2.6), if we let the limits of the integral approach $\pm\infty$, we see that the value of this integral must be identically equal to 1.

To illustrate the above properties, consider the case shown in Figure 2.8. In this case, the continuous random variable X has a probability density function given by a straight line within the limits [0,1]. This probability density function satisfies Equation (2.7), as the integral under the function is equal to 1.

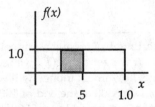

Figure 2.8 Probability density function associated with the random variable X. For this random variable, $f_X(x)$ is uniformly distributed between [0,1].

Suppose now that we wish to find the probability that the random variable X falls in [.25,.5], as depicted in the figure. To find this probability, the integral is evaluated as follows:

$$\int_{.25}^{.5} f_X(x)\, dx = \int_{.25}^{.5} 1 \cdot dx \qquad\qquad (2.8)$$

2.4 BASIC STATISTICAL PROPERTIES OF A RANDOM VARIABLE AND EXPECTED VALUES

There are a number of statistical properties of a random variable that will help us gain knowledge about the behavior of a random variable. Note that we can never describe the exact behavior of a random variable; otherwise we would no longer be dealing with a random variable. To obtain a complete statistical description of a random variable, the probability density function or distribution function is required. Often, and especially when dealing with measured data, we are only able to measure or estimate some of these properties. Although this gives us an incomplete characterization of the behavior of the random variable, often it is the only recourse available to us.

Averaging is a commonly used procedure performed on data samples. The data samples are realizations of a random variable. Therefore, we can extend the concept of averages to random variables through the introduction of expectation.

The *expected value E[X]*, or *mean value*, of a random variable is defined as

$$\mu = E[X] = \int_{-\infty}^{\infty} x f_X(x)\, dx \qquad\qquad (2.9)$$

In the case of a discrete random variable that takes discrete values x_i with associated probabilities $P(x_i)$, the expected value of X can be written as

$$\mu = E[X] = \sum_i x_i P(x_i) \qquad\qquad (2.10)$$

Example 2.4 Consider the experiment of tossing a fair coin. Assume that if we get a head (H), we receive \$1; if we get a tail (T), we receive \$0. We now toss the coin 1000 times. On average, how much money will we receive?

Intuitively, we expect to obtain \$500 at the end of 1000 tosses. How is such a number arrived at? On average, we expect to toss $500H$ and $500T$. Therefore, we quickly compute the following:

$$500(H) \times \$1 + 500(T) \times \$0 = \$500$$

We can also arrive at a formal answer to this problem by using Equation (2.10). Consider now the probability density function associated with this example, shown in Figure 2.9.

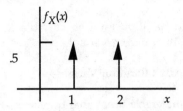

Figure 2.9 Probability density function associated with Example 2.4. Note that we have arbitrarily associated the number 1 with getting a head and the number 2 with getting a tail.

First note that this is a valid probability density (mass) function, i.e.,

$$\sum_i f_X(x_i) = 1 = .5 + .5$$

We now make use of Equation (2.10). First let us compute the average earnings in one throw:

$$\mu = E[X] = \sum_i x_i f_X(x_i) = \$1 \text{ times } .5 + \$0 \text{ times } 0.5 = \$.50$$

We multiply this value by the number of tosses and obtain the desired answer:

Average earnings per toss x number of tosses = $.50 times 1000 = $500

It is important to note that μ, the expected value of the random variable (also called the *population mean*), is *not* a random variable. This important concept will become clearer when we estimate the mean value of a population.

Example 2.5 Note that the mean of the random variable does not have to belong to the set of values of X. For example, consider the experiment of rolling a fair die. Figure 2.6 shows the probability density function of the random variable associated with the probability that a particular face of the die will occur. Compute the mean value of X.

Let us use Equation (2.10):

$$\mu = 1(\tfrac{1}{6}) + 2(\tfrac{1}{6}) + 3(\tfrac{1}{6}) + 4(\tfrac{1}{6}) + 5(\tfrac{1}{6}) + 6(\tfrac{1}{6}) = 3.5$$

As we can see, the mean value μ of the random variable X is not a member of the set of allowed values of the random variable.

Expected Value of a Function of a Random Variable

The random variable x can undergo a functional transformation as an input to a system. The functional transformation can be expressed as

$$Y = G_X(x) \tag{2.11}$$

The expected value of Y can be written based on Equation (2.9) as

$$E[Y] = \int_{-\infty}^{\infty} y f_X(y)\, dy \tag{2.12}$$

Since $Y = G_X(x)$, the expected value of $G_X(x)$ gives the same result. Therefore $E[Y]$ can be obtained as the expected value of $G_X(x)$:

$$E[Y] = \int_{-\infty}^{\infty} G(x) f_X(x)\, dx \tag{2.13}$$

For a discrete random variable Equation (2.13) reduces to

$$E[Y] = \sum_i G(x_i)\, P(x_i) \tag{2.14}$$

2.5 MOMENTS

Consider now the function of X, (X^n). We can find the mean value of this function by using Equation (2.13):

$$m_n = E\left[X^n\right] = \int_{-\infty}^{\infty} x^n f_X(x)\, dx \tag{2.15}$$

For different values of n, m_n is called the nth *moment* of the random variable X. When $n = 1$ we obtain the first moment, or the mean value, of the random variable. We defined this moment with Equation (2.13). For $n = 2$, we obtain the *mean square value* of the random variable X:

$$m_2 = E[X^2] = \int_{-\infty}^{\infty} x^2 f_X(x)\, dx \tag{2.16}$$

The mean square value is often associated with quantities such as the average power dissipated in a resistor, and its square root is equal to the *rms*, or *effective*, value of the voltage. The quantity $E[X^n]$ is called the nth moment of the random variable X.

The *central moments* of a random variable are defined as

$$m_1^C = E[(X-\mu)^n] = \int_{-\infty}^{\infty} (X-\mu)^n f_X(x)\, dx \tag{2.17}$$

The quantities m_n^C are called *central* moments of order n, because they are defined with reference to the expected value of the random variable. For $n = 1$, the first central moment is equal to 0:

$$m_1^C = E[(X-\mu)]$$
$$= \int_{-\infty}^{\infty} (x-\mu) f_X(x) dx = \int_{-\infty}^{\infty} x f_X(x)\, dx - \mu \int_{-\infty}^{\infty} f_X(x)\, dx \tag{2.18}$$

Note that since the quantity μ is a *constant*, it may be taken outside the integral. Therefore,

$$E[(X-\mu)] = \mu - \mu = 0$$

The value of the first integral in Equation (2.18) is equal to μ, and the last integral is, by definition, identically equal to 1. Therefore, the first central moment is equal to 0. Note that it is also convenient to solve this type of problem as follows:

$$E[(X-\mu)] = E[X] - E[\mu] = \mu - \mu = 0 \tag{2.19}$$

The second-order central moment is called the *variance* of the random variable X.

$$m_2^C = E[(X-\mu)^2]$$
$$= \int_{-\infty}^{\infty} x^2 f_X(x)\, dx - 2\mu \int_{-\infty}^{\infty} x f_X(x)\, dx + \mu^2 \int_{-\infty}^{\infty} f_X(x)\, dx \tag{2.20}$$

which equals

$$E\left[(X - \mu)^2\right] = E\left[X^2\right] - 2\mu^2 + \mu^2 = E\left[X^2\right] - \mu^2 \tag{2.21}$$

This is because

$$\int_{-\infty}^{\infty} x f_X(x)\, dx = \mu$$

$$\int_{-\infty}^{\infty} f_X(x)\, dx = 1$$

The variance of the random variable is denoted by the special symbol σ^2. Just like the mean value μ, σ^2 is a constant. The variance of a random variable is a very important quantity because, as we shall see later, it will give us an indication of the spread of the random variable about its mean. The square root of the variance is called the *standard deviation* of the random variable X. Another way of solving Equation (2.21) is as follows:

$$E\left[(X - \mu)^2\right] = E\left[X^2 - 2\mu X + \mu^2\right]$$
$$= E\left[X^2\right] - 2\mu^2 + \mu^2 = E\left[X^2\right] - \mu^2 = \sigma^2 \tag{2.22}$$

where we use the linearity of the expectation operation.

Example 2.6 Consider a random voltage V having a probability density function with the following shape (Figure 2.10):

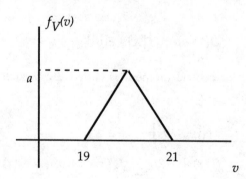

Figure 2.10 Probability density function associated with Example 2.6.

(a) Find the value of a.

In order for $f_V(v)$ to be a valid probability density function, the area under $f_V(v)$ must be equal to 1. Since the area of a triangle equals (base times height)/2, we have

$$\frac{(21-19)a}{2} = 1 \quad \therefore a = 1$$

(b) Find the mean value of the voltage.

Here we will directly apply Equation (2.13):

$$\mu = E[V] = \int_{-\infty}^{\infty} v f_V(v) \, dv = \int_{19}^{21} v f_V(v) \, dv$$

The equation of the probability density function is easily calculated and equals:

$$f_V(v) = x - 19 \quad \text{for } 19 \leq x \leq 20$$
$$f_V(v) = 21 - x \quad \text{for } 20 \leq x \leq 21$$

Consequently, we perform the integration as follows:

$$\mu = \int_{19}^{21} v f_V(v) \, dv = \int_{19}^{20} v(v-19) \, dv + \int_{20}^{21} v(21-v) \, dv = 20$$

Note that this value coincides exactly with the midpoint of the interval [19,21]. This is a fact well worth remembering: Anytime that the probability density function is symmetric about a value, this value is always the mean value of the random variable.

(c) Find the variance of the voltage.

In this case, we make use of Equation (2.22):

$$E\left[(X - \mu)^2\right] = E\left[X^2\right] - \mu^2$$

Since we already know the value of μ, we only need to compute $E[X^2]$:

$$E\left[X^2\right] = \int_{19}^{21} v^2 f_V(v) \, dv = \int_{19}^{20} v^2(v-19) \, dv + \int_{20}^{21} v^2(21-v) \, dv = 400.16$$

Therefore, the variance of the voltage equals

$$\sigma^2 = E\left[X^2\right] - \mu^2 = 400.1633 - 400 = .1633 \ V^2$$

2.6 EXAMPLES OF DIFFERENT DISTRIBUTIONS

Now we will describe a number of probability density functions commonly used in engineering problems.

Uniform Random Variables

A uniformly distributed random variable is equiprobable in some given range $[a,b]$. The probability density function associated with a uniformly distributed random variable is shown in Figure 2.11.

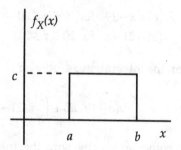

Figure 2.11 Probability density function of a uniformly distributed random variable x.

The amplitude c can easily be found by using Equation (2.7):

$$\int_{-\infty}^{\infty} f_X(x)\, dx = 1$$

and for the case

$$\int_a^b c\, dx = c \int_a^b dx = 1$$

which yields

$$c = \frac{1}{b-a} \quad a \leq x \leq b$$

The expected value is found as

$$E[X] = \int_{-\infty}^{\infty} x f_X(x)\, dx$$

$$E[X] = \int_a^b x \left(\frac{1}{b-a}\right) dx = \left(\frac{1}{b-a}\right) \frac{x^2}{2}\Big|_a^b = \frac{a+b}{2}$$

Likewise, the variance is found as follows:

$$E[X^2] = \int_a^b x^2 f_X(x)\, dx = \int_a^b x^2 \left(\frac{1}{b-a}\right) dx$$

$$= \left(\frac{1}{b-a}\right) \frac{x^3}{3}\Big|_a^b = \frac{b^3 - a^3}{3(b-a)} = \frac{b^2 + a^2 + ab}{3}$$

Therefore,

$$\sigma^2 = E[X^2] - \mu^2 = \frac{b^2 + a^2 + ab}{3} - \frac{(a+b)^2}{4} = \frac{(b-a)^2}{12}$$

Gaussian Random Variables

The probability density function associated with a gaussian distributed random variable is given by the following equation:

$$f_X(x) = \frac{1}{\sigma\sqrt{2\pi}} \exp\left[\frac{-(x-\mu)^2}{2\sigma^2}\right] \qquad -\infty < x < \infty \tag{2.23}$$

$$\sigma > 0 \quad \text{and} \quad -\infty < \mu < \infty$$

As can be observed from Equation (2.23), the gaussian probability density function can be completely specified through the parameters μ and σ. Also note that for the two constants which are parameters of the gaussian density, we have used the symbols μ and σ that are commonly used to indicate the mean and standard deviation, respectively, of a random variable. It will be shown later that the mean and standard deviation of the gaussian random variable are indeed the given constants μ and σ, respectively, in Equation (2.23). The importance of the gaussian distributed random variable is due to the fact many physical situations conform to this type of distribution. Also observe that the range of the random variable is $(-\infty, +\infty)$. The moments of a gaussian distributed random variable *with zero mean* are given by the following relationship:

$$m_n = E[x^n] = \begin{cases} 1 \cdot 3 \cdots (n-1)\sigma^n & \text{for } n \text{ even} \\ 0 & \text{for } n \text{ odd} \end{cases}$$

This states that all odd moments of a gaussian distributed random variable with zero mean are zero.

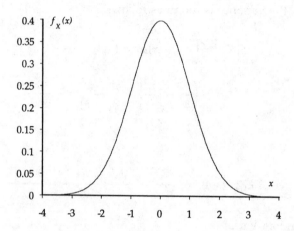

Figure 2.12 Probability density function of a gaussian distributed random variable x.

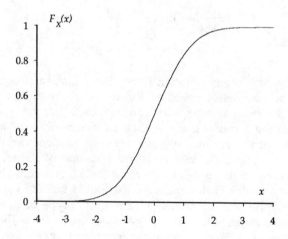

Figure 2.13 Distribution function of a gaussian distributed random variable x.

The probability density function associated with a gaussian distributed random variable with $\mu = 0$ and $\sigma = 1$ is shown by Figure 2.12. The distribution function of the gaussian distributed random variable is shown in Figure 2.13. Note that the probability density function is symmetric about its mean value, and that its maximum value is inversely proportional to the standard deviation of the random variable σ. This may be easily seen by setting $x = \mu$ in Equation (2.23).

Example 2.7

(a) Show, using MATHCAD, that the area under the gaussian probability density function equals 1.

   This is one of those exercises which is best proved by using MATHCAD. It could also be proved by mathematical manipulation; however, the advent of tools such as MATHCAD has made this type of problem easier to solve.

   Under MATHCAD, the following steps are taken:

   i.   Invoke the formula for the gaussian probability density function:

$$\frac{1}{\sigma\sqrt{2\pi}} \int_{-\infty}^{\infty} \exp\left(\frac{-x^2}{2\sigma^2}\right)$$

   ii. Load the symbolic processor.
   iii. Request to "Evaluate Symbolically" the expression.
   iv. The answer will appear as 1.

(b) Find the first moment of the gaussian probability density function given by the following equation:

$$f_X(x) = \frac{1}{\sigma\sqrt{2\pi}} \exp\left[\frac{-(x-a)^2}{2\sigma^2}\right] \qquad -\infty < x < \infty$$

In this case, we will prove that $E[X] = a$. Let us use Equation (2.9):

$$\mu = E[X] = \int_{-\infty}^{\infty} x f_X(x)\,dx$$

Therefore,

$$\mu = \frac{1}{\sigma\sqrt{2\pi}} \int_{-\infty}^{\infty} x \exp\left[\frac{-(x-a)^2}{2\sigma^2}\right] dx$$

Let $y = \dfrac{x-a}{\sqrt{2\sigma^2}}$. Then, $x = \sigma y\sqrt{2} + a$, and $dx = \sigma\sqrt{2}\, dy$. This yields

$$\mu = \frac{1}{\sigma\sqrt{2\pi}} \int_{-\infty}^{\infty} \left(\sigma y\sqrt{2} + a\right)\exp\left(-y^2\right)\sigma\sqrt{2}\, dy$$

$$= \int_{-\infty}^{\infty} \frac{\sigma\sqrt{2\pi} + a}{\sqrt{\pi}} \exp\left(-y^2\right) dy$$

$$= \frac{\sigma\sqrt{2\pi}}{\sqrt{\pi}} \int_{-\infty}^{\infty} y\exp\left(-y^2\right) dy + \frac{a}{\sqrt{\pi}} \int_{-\infty}^{\infty} \exp\left(-y^2\right) dy$$

From integral tables we obtain

$$\frac{\sigma\sqrt{2\pi}}{\sqrt{\pi}} \int_{-\infty}^{\infty} y\exp\left(-y^2\right) dy = 0$$

and

$$\frac{a}{\sqrt{\pi}} \int_{-\infty}^{\infty} \exp\left(-y^2\right) dy = \frac{a}{\sqrt{\pi}}\sqrt{\pi} = a$$

Modification of parameters μ and σ in Equation (2.23) has predictable effects on the shape of the probability density function $f_X(x)$. Changing the expected value μ has the effect of translating the curve $f_X(x)$ along the x axis, as shown in Figure 2.14. Increasing the value of the standard deviation σ has the effect of "flattening out" $f_X(x)$. As σ goes up in value, the probability density function becomes broader. This is shown in Figure 2.15. The gaussian probability density function is used frequently in statistical analysis. One can find an extensive number of tables containing the area under the gaussian probability density function. Obviously, it is impossible to have tables available that will take into account all possible combinations of mean values and standard deviations. It is therefore necessary to *normalize* the specific values or regions of the probability density function at hand in order

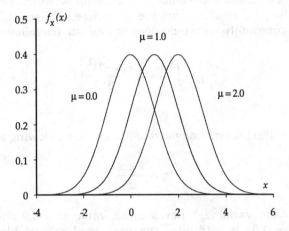

Figure 2.14 Effect of changing the expected value in a gaussian probability density function.

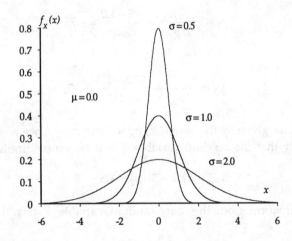

Figure 2.15 Effect of changing the standard deviation σ in a gaussian probability density function.

that we may use the commonly available tables. Let us illustrate the normalization procedure by means of an example.

Example 2.8 The normalization of a random variable is accomplished in the following manner. Let X be a gaussian distributed random variable with mean value $\mu = 1.0$ and standard deviation $\sigma = 2.0$. Suppose we need to compute the probability that the random variable is between $-.5$ and 1.5. In order to compute this probability we would need to evaluate the following integral:

$$\frac{1}{2\sqrt{2\pi}} \int_{-.5}^{1.5} \exp\left[\frac{-(x-1)^2}{8}\right] dx \qquad (2.24)$$

We now normalize the random variable X, using the following equation:

$$y = \frac{x-\mu}{\sigma}$$

The new random variable Y has a mean value $\mu = 0.0$ and a standard deviation $\sigma = 1.0$. In addition, the new random variable is gaussian distributed, which is a result that will be proved later in the chapter when we discuss the transformation of random variables. We also need to transform the values of the limits of the integral in the problem:

$$y_1 = \frac{x_1 - \mu_X}{\sigma_X} = \frac{-.5 - 1.0}{2.0} = -.75$$

$$y_2 = \frac{x_2 - \mu_X}{\sigma_X} = \frac{1.5 - 1.0}{2.0} = .25$$

Note that as given by the problem, $\mu_X = 1$ and $\sigma_X = 2$, we wish to compute the probability that the random variable X is between the limits

$$x_1 = -.5 \quad x_2 = 1.5$$

After the transformation, the new random variable Y has the following properties:

$$\mu_Y = 0 \quad \sigma_X = 1.0$$

The equivalent area under the new random variable Y is found by integrating under the limits:

$$y_1 = -.75 \quad y_2 = .25$$

Finding the area under the probability density function for the random variable X between $x_1 = -.5$ and $x_2 = 1.5$ when $\mu_X = 1.0$ and $\sigma_X = 2.0$ is equivalent to finding the area under the probability density function for the transformed random variable X, which we now call Y, between $y_1 = -.75$ and $y_2 = .25$. Compute now the required area by using MATHCAD.

$$\frac{1}{\sqrt{2\pi}} \int_{-.75}^{.25} \exp\left(\frac{-x^2}{2}\right) dx$$

$$= \frac{1}{\sqrt{2\pi}} \int_{-\infty}^{.25} \exp\left(\frac{-x^2}{2}\right) dx - \frac{1}{\sqrt{2\pi}} \int_{-\infty}^{-.75} \exp\left(\frac{-x^2}{2}\right) dx \qquad (2.25)$$

Using MATHCAD we find the following numbers (please note that a direct use of MATHCAD would have produced the desired results. This example is used merely as an illustration):

$$\frac{1}{\sqrt{2\pi}} \int_{-\infty}^{.25} \exp\left(\frac{-x^2}{2}\right) dx = .5987$$

and

$$\frac{1}{\sqrt{2\pi}} \int_{-\infty}^{-.75} \exp\left(\frac{-x^2}{2}\right) dx = .2237$$

Therefore, the probability that the random variable X is between $-.5$ and 1.5 equals $.5987 - .2237 = .375$, or 37.5 percent. The integrals can also be evaluated by tables of error function erf(x), defined as

$$\text{erf}(x) = \frac{1}{\sqrt{2\pi}} \int_0^x \exp\left(-\frac{1}{2}t^2\right) dt$$

Rayleigh Random Variables

The Rayleigh probability density function arises in connection with a variety of physical problems. Suppose that a rifle is aimed at a target and that the errors in the xy coordinates are gaussian distributed random variables. It may then be assumed that the total miss distance r is

$$r = \sqrt{x^2 + y^2}$$

where x and y are the miss distances in the X and Y directions. The concept of independence of random events can be extended to random variables, and it is more formally introduced in the next chapter.

If we further assume that the miss distances in the X direction and Y direction are independent of each other with zero mean and equal variances, then the probability density function of the miss distance r is said to be *Rayleigh distributed*. The probability density function of a Rayleigh distributed random variable is given by

$$f(x) = \frac{x}{\alpha^2} \exp\left(\frac{-x^2}{2\alpha^2}\right) u(x) \quad \alpha > 0 \tag{2.26}$$

where

$$u(x) = \begin{cases} 1 & \text{for } 0 \le x \le \infty \\ 0 & \text{elsewhere} \end{cases}$$

The probability density function and the distribution function of a Rayleigh distributed random variable with an arbitrary variance equal to 2 are given in Figures 2.16 and 2.17.

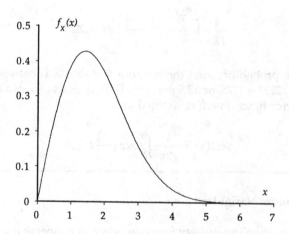

Figure 2.16 Probability density function of a Rayleigh distributed random variable X with an arbitrary variance equal to 2.

The mean value of the Rayleigh distributed random variable is found as follows:

$$\mu = E[X] = \int_0^\infty x^2 \exp\left(\frac{-x^2}{2\alpha^2}\right) dx = \sqrt{\frac{\pi}{2}}\,\alpha \qquad (2.27)$$

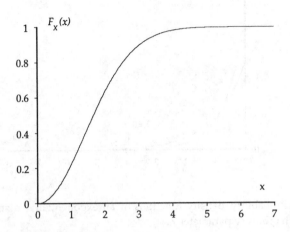

Figure 2.17 Distribution function of a Rayleigh distributed random variable X with an arbitrary variance equal to 2.

Exponential Random Variables

The exponential distribution function arises in several problems involving waiting time or time between events as well as in the description of the power fluctuation of noiselike signals. The probability density function associated with an exponentially distributed random variable with parameter λ is given by

$$f_X(x) = \lambda \exp(-\lambda x)\, u(x) \quad \lambda > 0 \qquad (2.28)$$

The probability density functions corresponding to an exponentially distributed random variable with arbitrary values of λ equal to .5, 1.0, and 1.5 are given in Figure 2.18.

The mean value is found as follows:

$$\mu = E[X] = \int_0^\infty x\lambda \exp(-\lambda x)\, dx = \frac{1}{\lambda}$$

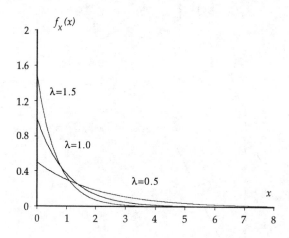

Figure 2.18 Probability density functions of an exponentially distributed random variable X with arbitrary parameter λ.

Example 2.9 A 486/66 personal computer (PC) clone has components which fail independently and uniformly with an average time between failure of 730 days (2 years). As you depart for college, your parents give you as a graduation present one of these computers. They, of course, hope that it will last the 4 years of your undergraduate academic career. What is the probability that it will last 1460 days without having to be fixed due to a component failure?

The equivalent question asks for the probability that the time to the first failure is *greater* than 1460 days. We now compute

$$1 - \frac{1}{730} \int_0^{1460} e^{\frac{-\tau}{730}} \, d\tau = .135 \tag{2.29}$$

In other words, there is a 13.5 percent probability that your computer will fail before you graduate.

Poisson Random Variables

Bernoulli trials were considered in Section 1.7. For this type of trial, the probability of success of the independent trials was given by the value p and the probability of failure by the value q. There are certain cases when the probability of success in a single trial approaches 0 while the number n of trials is very large. When this happens, the number of successes m in n trials is well approximated by a Poisson random variable. Therefore,

$$P_{m,n} = \frac{n!}{m!\,(n-m)!} p^m q^{n-m}$$

which can be approximated as

$$P_{m,n} \approx \frac{(np)^m e^{-np}}{m!} \tag{2.30}$$

where n = number of times an experiment is repeated

   m = number of times a particular outcome occurs

   p = probability of success

   q = probability of failure

To conform to more traditional ways of expressing the Poisson distribution, we will use the following notation:

$$P(N = k) = \frac{\alpha^k}{k!} e^{-\alpha} \quad k = 0, 1, 2, \ldots \tag{2.31}$$

where N is called a *Poisson random variable*.

The Poisson distribution has the property that its expected value equals the variance of the random variable which is the parameter α in the distribution.

Example 2.10 A random variable is Poisson distributed with parameter $\alpha = 2$. Compute the probability that the random variable is greater than or equal to its mean.

For a Poisson random variable, the parameter α equals its mean value. Consider the probability density function of a Poisson distributed random variable:

$$P(N = k) = \frac{\alpha^k}{k!} e^{-\alpha} \quad k = 0, 1, 2, \ldots$$

Since this is a discrete distribution with mean value equal to 2, we need to compute the following number:

$$P = \sum_{k=2}^{\infty} \frac{2^k}{k!} e^{-2}$$

Note that this is an infinite sum. However, the value of the terms in the sum becomes very small very fast. Using MATHCAD, we can compute the individual terms which yield the following values:

$k = 2$ $P_k = .271$

$k = 3$ $P_k = .18$

$k = 4$ $P_k = .09$

$k = 5$ $P_k = .036$

$k = 6$ $P_k = .012$

$k = 7$ $P_k = .003$

$k = 8$ $P_k = 8.593 \times 10^{-2}$

and the sum of these numbers over 10 values of k is given by MATHCAD Example 2.4 (page 98) as .594. Therefore, the probability that the Poisson distributed discrete random variable with mean value equal to 2 is greater than or equal to its mean is equal to .594.

The Poisson distribution is also very useful when we consider cases such as the arrival times of cars at tollbooths and radioactive decays. In cases such as these, the probability of success in any one interval is very small, and the events are independent.

2.7 MOMENT-GENERATING FUNCTIONS AND CHARACTERISTIC FUNCTIONS

The moments of a random variable play an important role in many applications. The moment-generating functions and characteristic functions can be used to obtain all the moments of a random variable. In addition, they provide a simple tool to obtain distributions of sums of random variables.

The moment-generating function $\vartheta(t)$ of a random variable X is defined as

$$\vartheta(t) = E\left[e^{tx}\right] \tag{2.32}$$

if the expected value exists for every value of t in some interval $-h < t < h$, $h > 0$. If the random variable is continuous, the moment-generating function can be obtained as

$$\vartheta(t) = \int_{-\infty}^{\infty} e^{tx} f_X(x)\, dx \tag{2.33}$$

and if the random variable is discrete, then $\vartheta(t)$ can be calculated by using the probability mass function:

$$\vartheta(t) = \sum_i e^{tx_i} P_X(x_i) \tag{2.34}$$

The moments of the random variable X can be obtained by successively differentiating $\vartheta(t)$ and then evaluating the result at $t = 0$. If we expand e^{tx} in (2.32) by its series expansion, we get

$$\vartheta(t) = E[1 + tx + \frac{t^2 x^2}{2!} + \cdots + \frac{t^n x^n}{n!} + \cdots] \tag{2.35}$$

Taking the expectation of each term, we get

$$\vartheta(t) = 1 + tm_1 + \frac{t^2}{2!} m_2 + \cdots + \frac{t^n}{n!} m_n + \cdots \tag{2.36}$$

where m_n is the nth moment of the random variable X. It can be seen from the above equation that if the nth-order moment exists, it can be evaluated by differentiating $\vartheta(t)$ n times as $t \to 0$.

$$m_n = \frac{d^n}{dt^n} \vartheta(0) \tag{2.37}$$

Thus the moment-generating function provides a convenient procedure to evaluate the moments. If the moment-generating function exists for a random variable, then this uniquely determines the corresponding distribution function.

The definition of characteristic functions $\Phi_X(\omega)$ is very similar to that of the moment-generating function and is given by

$$\Phi_X(\omega) = E[e^{j\omega x}] \tag{2.38}$$

where $j = \sqrt{-1}$. If the random variable is continuous, the characteristic function is given by

$$\Phi_X(\omega) = \int_{-\infty}^{\infty} f_X(x) e^{j\omega x} \, dx \tag{2.39}$$

and if the random variable is discrete, then $\Phi_X(\omega)$ can be obtained by the probability mass function as

$$\Phi_X(\omega) = \sum_i e^{j\omega x_i} P_X(x_i) \tag{2.40}$$

The expression for $\Phi_X(\omega)$ is very close to that of an inverse Fourier transform, except for a scaling constant. The main advantage of the characteristic function is that it always exists, unlike the moment-generating function. It can be observed from (2.38) that the value of $\Phi_X(0)$ is unity. In addition, it can be shown that

$$|\Phi_X(\omega)| \le 1 \tag{2.41}$$

The characteristic function can be used in a way similar to the moment-generating function for computing the moments, provided the moments exist. The nth-order moment m_n can be calculated from $\Phi_X(\omega)$ by

$$m_n = \frac{1}{j^n} \frac{d^n}{d\omega^n} \Phi_X(0) \tag{2.42}$$

The above result can be proved in a manner very similar to that of (2.37).

Characteristic Function of the Gaussian Random Variable

We now derive the characteristic function of a gaussian random variable X with zero mean and unit variance. The characteristic function of X can be written as

$$\Phi_X(\omega) = \int_{-\infty}^{\infty} \frac{1}{\sqrt{2\pi}} e^{-x^2/2} e^{j\omega x} \, dx \tag{2.43}$$

The exponent in (2.43) can be rewritten using the technique of completing the squares as

$$\Phi_X(\omega) = \int_{-\infty}^{\infty} \frac{1}{\sqrt{2\pi}} e^{-[x^2 - 2j\omega x + (j\omega)^2 - (j\omega)^2]/2} \, dx \tag{2.44}$$

which can be simplified to

$$\Phi_X(\omega) = e^{-\omega^2/2} \int_{-\infty}^{\infty} \frac{1}{\sqrt{2\pi}} e^{-(x-j\omega)^2/2} \, dx \qquad (2.45)$$

The integral in (2.45) is an integral of a gaussian probability density function, and, integrates to unity. Therefore $\Phi_X(\omega)$ of a gaussian random variable with zero mean and unit variance is given by

$$\Phi_X(\omega) = e^{-\omega^2/2} \qquad (2.46)$$

In a similar manner we can derive the characteristic function of a gaussian random variable with mean μ and variance σ^2 to be exp $[j\omega\mu - (1/2)\sigma^2\omega^2]$.

2.8 TRANSFORMATION OF A RANDOM VARIABLE

Often in the study of systems, a random variable is used as an input to a system. The easiest example to consider is the one where a measured signal is passed through a linear amplifier. In such a case the signal is just amplified without altering most of its characteristics. Suppose that the data being measured have a particular distribution. What would be the effect on the original distribution of such a linear amplification? This concept is illustrated by Figure 2.19.

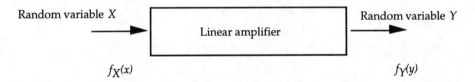

Figure 2.19 A random variable X with probability density functon $f_X(x)$ is linearly amplified. The output of the amplifier is the random variable Y with a probability density function $f_X(y)$.

Suppose now that we wish to determine the probability density function of the random variable Y which is obtained by transforming the random variable X. In terms of the random variable X we can express this transformation as follows:

$$Y = g(X) \ . \qquad (2.47)$$

In its simplest case, the linear case, the function $g(X)$ is

$$Y = g(X) = aX + b \quad a \neq 0 \tag{2.48}$$

This case will illustrate the methodology to be followed for the more general case. First, let us assume that $a > 0$. This may be illustrated by considering Figure 2.20.

Note that the linear relationship implies that whenever the random variable X extends from x to $x + dx$, the random variable Y lies between y and $y + dy$. Consider now the cumulative distribution function of Y (to make the distinction between the cumulative distribution functions of X and Y, we use appropriate subscripts X and Y:

$$F_Y(y) = P\{Y \leq y\} \tag{2.49}$$

We can substitute $Y = aX + b$ in the preceding equation:

$$F_Y(y) = P(aX + b \leq y) = P\left(X \leq \frac{y-b}{a}\right) = F_X\left(\frac{y-b}{a}\right) \tag{2.50}$$

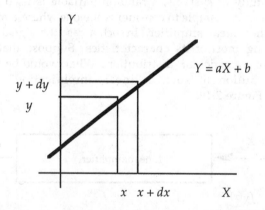

Figure 2.20 Illustration of the relationship $Y = aX + b$, for the case when $a > 0$.

This last result implies that for $a > 0$, and a linear transformation, we simply substitute in the probability distribution of X the value $(y - b)/a$. (Note that, as indicated earlier, $a \neq 0$.) For the case when $a < 0$, we obtain

$$F_Y(y) = P(aX + b \leq y) = P\left(X \geq \frac{y-b}{a}\right) \tag{2.51}$$

To obtain the probability density function, we take the derivative of the probability distribution as follows:

$$f(y) = \frac{d}{dy}[F_Y(y)] = \frac{d}{dy}\left[F_X\left(\frac{y-b}{a}\right)\right] \tag{2.52}$$

We solve this problem as follows:

We let $z = (y-b)/a$ and substitute in the preceding equation:

$$f(y) = \frac{d}{dy}[F_Z(z)] = \frac{dz}{dy}\frac{dF_Z(z)}{dz} = \frac{1}{a}f(z) = \frac{1}{a}f\left(\frac{y-b}{a}\right) \tag{2.53}$$

If we now incorporate the case when $a < 0$, we have

$$f(y) = \frac{1}{|a|}f\left(\frac{y-b}{a}\right) \tag{2.54}$$

We need to note here that the range of Y will be different from the range of X and the upper and lower limits of Y can be computed from those of X by using the transformation equation $g(X)$.

Example 2.11 Assume that X is a gaussian distributed random variable with $\mu = 1$ and $\sigma = 2$, and it undergoes the transformation shown in Figure 2.21.

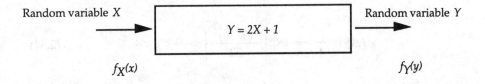

Random variable X $Y = 2X + 1$ Random variable Y

$f_X(x)$ $f_Y(y)$

Figure 2.21 A random variable X with gaussian probability density function $f(x)$ is linearly transformed as shown.

Let us now find $f(y)$. We know that

$$x = \frac{y-1}{2} \quad \text{and} \quad |a| = 2$$

Since X is gaussian distributed,

$$f_X(x) = \frac{1}{2\sqrt{2\pi}} \exp\left[\frac{-(x-1)^2}{8}\right] \quad -\infty < x < \infty \tag{2.55}$$

Using Equation (2.54)

$$f_Y(y) = \frac{1}{2} f_X\left(\frac{y-1}{2}\right) \tag{2.56}$$

Therefore, we can immediately write the expression for $f_Y(y)$:

$$f_Y(y) = \left(\frac{1}{2}\right) \frac{1}{2\sqrt{2\pi}} \exp\left[\frac{-\left(\frac{y-1}{2}-1\right)^2}{8}\right] \quad -\infty < y < \infty \tag{2.57}$$

Note that $x = (y-1)/2$. This yields:

$$f_Y(y) = \frac{1}{4\sqrt{2\pi}} \exp\left[\frac{-\left(\frac{y-3}{2}\right)^2}{8}\right] \quad -\infty < y < \infty \tag{2.58}$$

$$f_Y(y) = \frac{1}{4\sqrt{2\pi}} \exp\left[\frac{-(y-3)^2}{32}\right] \quad -\infty < y < \infty \tag{2.59}$$

Note that Equation (2.59) is again a gaussian probability density function with

$$\sigma_Y^2 = 4 \quad \text{and} \quad \mu_Y = 3$$

In other words, the linear transformation did not affect the type of probability density function. The random variable X is gaussian distributed, and after the linear transformation $Y = 2X + 1$, the random variable Y is also gaussian distributed. This important result extends to other types of density functions; i.e., a linear transformation of variables does not affect the type of probability density function. Using the notion that a linear transformation does not affect the type of probability density function, it would be possible to find the mean value and standard deviation of Y in an alternate way. In the present case, the resulting probability density function is going to be gaussian; therefore, we only need to find its mean value and standard deviation in order to totally describe $f_Y(y)$. Consider the following:

$$E[Y] = E[aX + b] = E[2X + 1] = 2E[X] + E[1] = 2 \cdot 1 + 1 = 3 \qquad (2.60)$$

This matches the result obtained by inspecting Equation (2.59). Likewise, consider the following:

$$\sigma^2 = E[Y^2] - \mu^2 \qquad (2.61)$$

We need only find the first term in the preceding equation:

$$E[Y^2] = E[(aX + b)^2] = E[(2X + 1)^2]$$
$$= E[4X^2 + 4X + 1] = 4E[X^2] + 4E[X] + 1 \qquad (2.62)$$

Since we are given that $\sigma_x = 2$ and $\mu_x = 1$, we can determine $E[X^2]$ as follows:

$$E[X^2] = \sigma_x^2 + \mu_x^2 = 4 + 1 = 5 \qquad (2.63)$$

Therefore,

$$E[Y^2] = 4 \cdot 5 + 4 \cdot 1 + 1 = 25 \qquad (2.64)$$

and

$$\sigma^2 = E[Y^2] - 2 = 25 - 9 = 16$$

which yields $\sigma_Y = 4$, which matches our earlier result. Consequently, knowing that $f(y)$ must be gaussian since $g(X)$ is linear, and having found the mean and standard deviation, we could have written the result of Equation (2.59) immediately.

The next case that we will consider is slightly more complicated. In this case the relationship between Y and X is not single-valued. For such a case the relationship between Y and X could be illustrated by Figure 2.24.

In a case such as this, the event $\{y \leq Y \leq y + dy\}$ has two possible equivalent events in the X domain: $\{x_1 \leq X \leq x_1 + dx_1\}$ or $\{x_2 \leq X \leq x_2 + dx_2\}$. In

Figure 2.22, these last two events have been denoted by dx_1 and dx_2, respectively. Since these two events are mutually exclusive, the probabilities may be found as follows:

$$P(y \leq Y \leq y + dy) = P[(x_1 \leq X \leq x_1 + dx_1) \cup (x_2 + dx_2 \leq X \leq x_2)] \qquad (2.65)$$

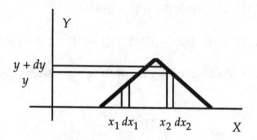

Figure 2.22 Illustration of a relationship between Y and X which is not single-valued.

which equals

$$P(y \leq Y \leq y + dy) = P(x_1 \leq X \leq x_1 + dx_1) + P(x_2 + dx_2 \leq X \leq x_2) \qquad (2.66)$$

For the present case $Y = g(X)$ is not a single-valued function of X, and in fact it has two roots, x_1 and x_2. Let us assume that $g(X)$ is a differential function. For sufficiently small dx, we can set

$$f_Y(y)\,dy = P(y \leq Y \leq y + dy) =$$
$$P(x_1 \leq X \leq x_1 + dx_1) + P(x_2 + dx_2 \leq X \leq x_2) \qquad (2.67)$$

which equals (when we do not have discontinuities in the cumulative distribution functions)

$$f_Y(y)\,dy = f_X(x_1)|dx_1| + f_X(x_2)|dx_2| \qquad (2.68)$$

We use the chain rule to obtain dy:

$$dy = g'(x_1)dx_1 \quad \text{and} \quad dy = g'(x_2)dx_2 \qquad (2.69)$$

Consequently

$$f_Y(y)\,dy = f_X(x_1)\frac{dy}{|g'(x_1)|} + f_X(x_2)\frac{dy}{|g'(x_2)|} \qquad (2.70)$$

which yields

$$f_Y(y) = \frac{f_Y(x_1)}{|g'(x_1)|} + \frac{f_Y(x_2)}{|g'(x_2)|} \tag{2.71}$$

For the more general case, when $Y = g(X)$ has n (finite number) real roots, we arrive at the generalized form of Equation (2.71):

$$f_Y(y) = \frac{f_X(x_1)}{|g'(x_1)|} + \cdots + \frac{f_X(x_n)}{|g'(x_n)|} \tag{2.72}$$

Example 2.12 Let us consider the case where $g(X) = aX^2 + b$, $a > 0$, and $f(x)$ is gaussian. Note that the function $g(X)$ has two roots.

First, we find the first derivative of $g(X)$:

$$g'(x) = 2ax \tag{2.73}$$

The roots of $g(X)$ are

$$x = \pm\sqrt{\frac{y-b}{a}}$$

which yields

$$x_1 = +\sqrt{\frac{y-b}{a}} \quad \text{and} \quad x_2 = -\sqrt{\frac{y-b}{a}}$$

for $y > b$; there are none for $y < b$.

We now calculate $g'(x_i)$. Since $g'(x_i) = 2ax_i$, we obtain

$$g'(x_1) = 2a\sqrt{\frac{y-b}{a}} \tag{2.74}$$

and

$$g'(x_2) = -2a\sqrt{\frac{y-b}{a}} \tag{2.75}$$

We actually need the absolute value of the first derivative:

$$|g'(x_{1,2})| = 2\sqrt{a(y-b)} \tag{2.76}$$

Therefore, using Equation (2.72) we are able to obtain an expression for the probability density function of Y:

$$f_Y(y) = \frac{1}{2\sqrt{a(y-b)}}\left[f_Y\left(\sqrt{\frac{y-b}{a}}\right) + f_Y\left(-\sqrt{\frac{y-b}{a}}\right)\right] \tag{2.77}$$

Consider now the case where the random variable X is gaussian distributed with $\mu = 1$ and $\sigma = 2$. Let the transforming equation be of the type $Y = X^2$, that is, $a = 1$ and $b = 0$. We can use Equation (2.77) as follows:

$$f_X(x) = \frac{1}{2\sqrt{2\pi}}\exp\left[\frac{-(x-1)^2}{8}\right] \quad -\infty < x < \infty \tag{2.78}$$

so that

$$f_X\left(\sqrt{\frac{y-b}{a}}\right) = \frac{1}{2\sqrt{2\pi}}\exp\left[\frac{-\left(\sqrt{\dfrac{y-b}{a}}-1\right)^2}{8}\right] \tag{2.79}$$

and

$$f_X\left(-\sqrt{\frac{y-b}{a}}\right) = \frac{1}{2\sqrt{2\pi}}\exp\left[\frac{-\left(-\sqrt{\dfrac{y-b}{a}}-1\right)^2}{8}\right] \tag{2.80}$$

Consequently,

$$f_Y(y) = \frac{1}{2\sqrt{a(y-b)}} \left\{ \frac{1}{2\sqrt{2\pi}} \exp\left[\frac{-\left(\sqrt{\frac{y-b}{a}}-1\right)^2}{8} \right] \right.$$

$$\left. + \frac{1}{2\sqrt{2\pi}} \exp\left[\frac{-\left(-\sqrt{\frac{y-b}{a}}-1\right)^2}{8} \right] \right\} \qquad (2.81)$$

which equals

$$f_Y(y) = \frac{1}{2\sqrt{y}} \left\{ \frac{1}{2\sqrt{2\pi}} \exp\left[\frac{-\left(\sqrt{y}-1\right)^2}{8} \right] + \frac{1}{2\sqrt{2\pi}} \exp\left[\frac{-\left(-\sqrt{y}-1\right)^2}{8} \right] \right\} \qquad (2.82)$$

and

$$f_Y(y) = \frac{1}{4\sqrt{2y\pi}} \left\{ \exp\left[\frac{-\left(\sqrt{y}-1\right)^2}{8} \right] + \exp\left[\frac{-\left(-\sqrt{y}-1\right)^2}{8} \right] \right\} \qquad (2.83)$$

Example 2.13 Now let us consider the case where the transformation is slightly more complicated. Let the random variable X be uniformly distributed from −1 to 3. Therefore $f_X(x)$ equals

$$f_X(x) = \begin{cases} 0.25 & -1 \le x \le 3 \\ 0 & \text{elsewhere} \end{cases}$$

Let the transformation $g(x)$ be given by the equation $Y = (|X|)^{.5}$. Find the probability density function of the random variable Y.

First, we find the derivative of $g(X)$ with respect to x:

$$g'(x) = \frac{1}{2\sqrt{|x|}} \qquad x \ne 0$$

Now we perform the transformation. We must consider two different cases which are valid for different portions of the interval of $f_X(x)$:

First case: $y = \sqrt{-x}$ valid for $-1 \leq x \leq 0$ and therefore $0 \leq y \leq 1$. In this case, the derivative of $g(x)$ equals

$$g'(x) = \frac{1}{2\sqrt{|x|}} = \frac{1}{2\sqrt{-x}} = \begin{cases} \dfrac{1}{2+\sqrt{-x}} = \dfrac{1}{2y} \\[2mm] \dfrac{1}{2-\sqrt{-x}} = \dfrac{1}{-2y} \end{cases}$$

Second case: $y = \sqrt{x}$ valid for $0 \leq x \leq 3$ and therefore $0 \leq y \leq \sqrt{3}$. In this case, the derivative of $g(x)$ equals

$$g'(x) = \frac{1}{2\sqrt{|x|}} = \frac{1}{2\sqrt{x}} = \begin{cases} \dfrac{1}{2+\sqrt{x}} = \dfrac{1}{2y} \\[2mm] \dfrac{1}{2-\sqrt{x}} = \dfrac{1}{-2y} \end{cases}$$

We are now ready to compute the probability density function $f(y)$:

$$f_Y(y) = \frac{f_X(x_1)}{|g'(x_1)|} + \frac{f_X(x_2)}{|g'(x_2)|}$$

This expression must be evaluated for each of the two separate ranges of y:
Consider the first case, when $0 \leq y \leq 1$:

$$f_Y(y) = \frac{.25}{|1/(2y)|} + \frac{.25}{|1/(-2y)|} = (.25)\,2y + (.25)\,2y = y$$

Consider the second case, when $0 \leq y \leq \sqrt{3}$:

$$f_Y(y) = \frac{.25}{|1/(2y)|} + \frac{.25}{|1/(-2y)|} = (.25)\,2y + (.25)\,2y = y$$

The resulting probability density function has the following form:

$$f_Y(y) = \begin{cases} 2y & 0 \le y \le 1 \\ y & 1 \le y \le \sqrt{3} \end{cases}$$

and is shown in Figure 2.23.

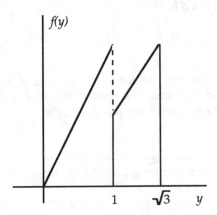

Figure 2.23 Resulting probability density function for example 2.13.

2.9 CONDITIONAL PROBABILITY DENSITY

In Chapter 1, we considered the cases where the probability of an event's occurring depended on whether another event had occurred. For example, consider the case of drawing from a deck of cards and considering the probability of the second card's being a spade, knowing that the first card was also a spade and was not replaced in the deck. For such a case, we used the equation

$$P(A/B) = \frac{P(AB)}{P(B)} \quad P(B) > 0 \tag{2.84}$$

where $P(A/B)$ is the probability of event A given that event B *has* occurred.

In the case of continuous random variables, we can arrive at similar definitions and concepts. In such cases, rather than computing the probability of an event given that an event B has occurred, we determine a *conditional cumulative distribution function* and a *conditional probability density function*. These functions retain all the properties of the conventional distribution and density functions explained earlier in this chapter.

The basic properties of the conditional probability density function are as follows:

1. $f_X(x/B) \geq 0$

2. $\int_a^b f_X(x/B)\, dx = P\big[(a \leq X \leq b)/B\big]$

3. $\int_{-\infty}^{\infty} f_X(x/B)\, dx = 1$

Note that these properties are similar to the properties given by Equations (2.5), (2.6), and (2.7). In this case, the probability density function is conditioned on some event B's having occurred. Let us offer some examples to illustrate these concepts.

Example 2.14 Consider the following probability density function:

$$f_X(x) = \begin{cases} 4x & 0 \leq x \leq 0.5 \\ -4x + 4 & 0.5 \leq x \leq 1 \\ 0 & \text{elsewhere} \end{cases}$$

It can be shown that

$$\int_0^1 f_X(x)\, dx = 1$$

The probability density function given by the above equation is illustrated in Figure 2.24.

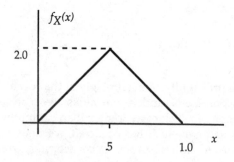

Figure 2.24 Probability density function for Example 2.14

Suppose that we now wish to find the statistics of the random variable X given that it is less than .4. In other words, we seek to find the probability density function $f_X(x/x \le .4)$. To solve this problem, we must consider first the conditional cumulative distribution function $F_X(x/B)$. We know that the definition of $F_X(x/B)$ is as follows:

$$F_X(x/B) = P[(X \le x)/B] = \frac{P[X \le x, B]}{P(B)} \quad P(B) > 0 \quad\quad (2.85)$$

For this example, event B is the event that the random variable X is less than or equal to the value .4. Consequently, for this example, the above equation changes to

$$F_X(x/x \le .4) = P[(X \le x)/X \le .4] = \frac{P(X \le x, X \le .4)}{P(X \le .4)} \quad\quad (2.86)$$

For this example, we wish to find the conditional probability density function of the random variable X given that it is less than or equal to .4.

Consider now the case when the random variable X is greater than .4. Then the expression in the numerator is equal to the expression in the denominator; i.e., the conditional distribution function has reached its maximum value of 1. When the random variable X is less than .4, then the expression equals

$$F_X(x/.4) = P[(X \le x)/.4] = \frac{P(X \le x)}{P(X \le .4)} = \frac{F(x)}{F(.4)} \qu\quad (2.87)$$

We can now find the expression of the conditional probability density function from this last expression by computing the derivative of Equation (2.87) with respect to X:

$$f_X(x/.4) = \begin{cases} \dfrac{f_X(x)}{\displaystyle\int_0^{.4} f_X(a)\,da} & 0 \le x \le .4 \\ 0 & \text{elsewhere} \end{cases} \quad\quad (2.88)$$

Note that the term in the denominator is just a constant, and therefore, the resulting expression is a scaled version of the original probability density function $f_X(x)$. The value of the denominator is equal to .32. Consequently, the expression for the conditional probability density function is

$$f_X(x/.4) = \begin{cases} 12.5x & 0 \le x \le .4 \\ 0 & \text{elsewhere} \end{cases} \quad\quad (2.89)$$

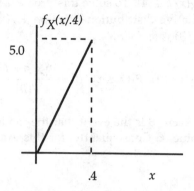

Figure 2.25 Conditional probability density function for Example 2.11.

The conditional probability density function we have just found is illustrated in Figure 2.25. Note that this is a scaled version of the original probability density function in the region of interest. We now prove that since the conditional probability density function is a valid density function, its integral must be equal to 1:

$$\int_0^{.4} 12.5x\, dx = 12.5\left(\left.\frac{x^2}{2}\right|_0^{.4}\right) = 12.5\left[(.4)^2\right] = 1 \qquad (2.90)$$

Suppose now that we ask a different question. What is the conditional probability density function for the case that the random variable X is greater than or equal to .6? The expression for the conditional probability distribution function becomes

$$F_X(x/B) = P\big[(X \le x)/B\big] = \frac{P(X \le x, B)}{P(B)}$$

$$= \frac{P(X \le x, X \ge .6)}{P(X \ge .6)} \qquad (2.91)$$

Note that now event B is the event that the random variable X is greater than or equal to .6. Just as before, consider the numerator of the last expression. We wish to find the conditional probability density function of the random variable X, given that it is greater than or equal to .6. Consider the case when the random variable X is less than .6. The expression in the numerator is equal to the null expression; i.e., the conditional distribution function equals 0. When the random variable X is greater than or equal to .6, the expression equals

$$F_X(x/B) = P[(X \le x)/B] = \frac{P(X \le x, B)}{P(B)}$$

$$= \frac{P(X \le x, X \ge .6)}{P(X \ge .6)} = \frac{P(X \le x)}{P(X \ge .6)} \qquad (2.92)$$

We can now find the expression of the conditional probability density function. First we evaluate the denominator:

$$P(X \ge .6) = \int_{.6}^{1} f(x) \, dx = \int_{.6}^{1} (-4x + 4) \, dx = .32 \qquad (2.93)$$

and the expression for the conditional probability density function becomes

$$f_X(x/x \ge .6) = \begin{cases} -12.5x + 12.5 & .6 \le x \le 1.0 \\ 0 & \text{elsewhere} \end{cases} \qquad (2.94)$$

2.10 CHAPTER SUMMARY

The cumulative distribution function of the random variable X is defined by

$$F_X(X) = P(X \le x)$$

The distribution function $F_X(x)$ has the following properties:

1. $0 \le F_X(x) \le 1$
2. $\lim_{x \to \infty} F_X(x) = 1$
3. $\lim_{x \to -\infty} F_X(x) = 0$
4. $F_X(x)$ is a nondecreasing function of x
5. $P(X = x) = F_X(x^+) - F_X(x^-)$
6. $F_X(x) = P(X \le x) = 1 - P(X > x)$

The probability density function is the derivative of the distribution function:

$$f_X(x) = \frac{dF_X(x)}{dx}$$

The properties associated with the probability density function are as follows:

1. $f_X(x) \geq 0$

2. $\int_a^b f_X(x)\,dx = P(a \leq X \leq b)$

3. $\int_{-\infty}^{\infty} f_X(x)\,dx = 1$

The expected value, or mean value, of a random variable is

$$\mu = E[X] = \int_{-\infty}^{\infty} x f_X(x)\,dx$$

The moments of a random variable are:

$$m_n = E[X^n] = \int_{-\infty}^{\infty} x^n f_X(x)\,dx$$

The central moments of a random variable X are

$$m_n^C = E[(x-\mu)^n] = \int_{-\infty}^{\infty} (x-\mu)^n f_X(x)\,dx$$

The variance of the random variable X is

$$E[(X-\mu)^2] = E[X^2] - \mu^2$$

The probability density of a uniformly distributed random variable is given by

$$f_X(x) = \begin{cases} \dfrac{1}{b-a} & b \leq x \leq a \\ 0 & \text{elsewhere} \end{cases}$$

The probability density of a gaussian distributed random variable is given by

$$f_X(x) = \frac{1}{\sigma\sqrt{2\pi}} \exp\left[\frac{-(x-\mu)^2}{2\sigma^2}\right] \quad -\infty < x < \infty$$

The probability density function of the random variable Y obtained by transforming the random variable X using the generalized transformation $Y = g(X)$ is given by

$$f_Y(y) = \frac{f_X(x_1)}{|g'(x_1)|} + \cdots + \frac{f_X(x_n)}{|g'(x_n)|}$$

where $g'(x_i)$ is the ith root of $g(X)$.

The properties of the conditional probability density function are as follows:

1. $f_X(x/B) \geq 0$

2. $\int_a^b f_X(x/B)\,dx = P\left[(a \leq X \leq b)/B\right]$

3. $\int_{-\infty}^{\infty} f_X(x/B)\,dx = 1$

2.11 PROBLEMS

1. A random variable X has the following probability density function:

$$f(x) = \begin{cases} e^{-1}e^{-x} & x > -1 \\ 0 & x \leq -1 \end{cases}$$

   (a) Find the mean value.

   (b) Find the mean square value.

   (c) Find the variance.

2. A gaussian random variable X has a mean value of 3 and a variance of 4.

   (a) Write the probability density function.

   (b) What is the value of the density function at $x = 0$, $x = 3$, and $x = 6$?

   (c) What is the probability that the random variable X will have a value greater than 0?

(d) What is the probability that the random variable X will lie in the range $1 < X \leq 5$?

3. A random variable X has the probability distribution function shown.

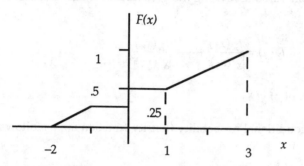

(a) Find $F_X(2)$.

(b) Find the probability that $-1 < X \leq 1$.

(c) Find the probability that $X > 0$.

(d) Find the probability that $X \leq 0$.

4. A random variable X has a probability density function given by

$$f_X(x) = Ae^{-|x|}$$

(a) Determine the value of the constant A.

(b) Find the probability that $X > 1$.

(c) Find the probability that $0 < X \leq 1$.

5. For the random variable of the preceding problem, find

(a) The mean value of the random variable X

(b) The mean square value of the random variable X

(c) The variance of the random variable X

6. An analog-to-digital converter samples a continuous random voltage V

and converts it into a discrete random variable X in accordance with the transformation shown in the sketch.

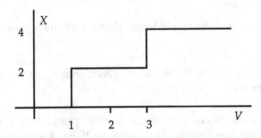

If V has an exponential probability density function with a mean value of 2.0, write the probability density function of X and find the mean value of X.

7. In a given telephone system, calls are placed with an average frequency of one call every 15 seconds (s). The times at which the calls are placed are statistically independent, and the time interval between successive calls is a random variable with an exponential probability density function.

   (a) What is the probability of no calls being placed in a time interval of 30 s?

   (b) If the last call was placed 10 s ago, what is the probability that the next call will be placed in the next 10 s?

8. A random variable has a probability density function given by the following equation:

$$f(x) = \begin{cases} .1 & -3 \le x \le 7 \\ 0 & \text{elsewhere} \end{cases}$$

   (a) Find the mean value.

   (b) Find the mean square value.

   (c) Find the variance.

9. Assume that vehicles on a highway have heights that are uniformly distributed between 4 and 14 feet. This highway has an underpass that has a clearance of only 12 ft.

(a) Write the conditional probability density function of the heights of vehicles, given that they have passed through the underpass.

(b) Find the conditional mean height of vehicles, given that they passed through the underpass.

(c) Find the conditional mean height of vehicles, given that they could not pass through the underpass.

(d) All vehicles over 8 ft tall must carry a special license plate. What is the probability that a vehicle bearing this plate will be able to pass through the underpass?

10. On a given limited-access highway, the minimum speed limit is 45 miles per hour (mi/h) and the maximum speed limit is 70 mi/h. After extensive observation it is determined that the probability density function of the actual speed S is as shown below:

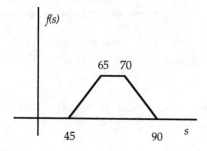

(a) What is the mean speed?

(b) What fraction of the vehicles are speeding?

(c) What is the conditional mean speed, given that a car is exceeding the speed limit?

(d) If the threshold of a police speed radar is set at 75 mi/h, what fraction of the speeders will be caught?

11. A gaussian random current has a mean value of 2 amperes (A) and a standard deviation of 3 A. It is flowing in a resistance of 5 Ω.

(a) Find the mean value of the power dissipated in the resistor.

(b) Find the variance of the power dissipated in the resistor.

12. Electrons in the beam of a cathode-ray tube (CRT) have two orthogonal components of velocity (perpendicular to the beam) that are independent gaussian random variables with zero mean and a standard deviation of 100 meters/second (m/s). The time required for the electrons to go from the electron gun to the screen is 10^{-5} s.

   (a) What is the most probable value of the distance from the center of the beam at which an electron will strike the screen?

   (b) What is the mean value of distance from the center at which the electron will strike the screen?

   (c) What is the probability of an electron's striking the screen at a distance greater than 2 millimeters (mm)?

13. A random variable X is gaussian distributed with zero mean and unit variance over $-\infty < x < \infty$. The random variable Y is related to X by

$$Y = \frac{1}{\sqrt{1 + X^2}}$$

   (a) Find the density function $f_Y(y)$.

   (b) Consider $M = \{0 < x \le 1\}$. Find the conditional density function $f(x/M)$.

14. Suppose X is a zero-mean, gaussian distributed random variable with variance σ^2. Suppose that we have a system which adds 1 to the random variable, squares the result, and subtracts the original random variable.

   (a) What is the expected value of the input of the system?

   (b) What is the expected value of the output of the system?

15. A gun is fired many times at a target, and it is found that one-half of all the shots fall inside a circle with a radius of 1 inch (in). What fraction of the shots fall outside a circle with a radius of 2 in?

16. A gun is fired many times at a target 1 ft in diameter, and one-sixteenth of the shots miss the target. For those shots that hit the target, find the

probability that they will hit within 3 in of the center.

17. A gaussian distributed random current has a mean value of 1 A and a standard deviation of 4 A. It is flowing in a resistance of 1 Ω.

(a) What is the mean value of the power dissipated in the resistor?

(b) What is the variance of the power dissipated in the resistor?

(c) How does the mean value of the current affect the mean and variance of the power?

18. A binary waveform has a probability density function of

$$f(x) = .6\delta(x) + .4\delta(x-1)$$

(a) Find the mean of the random variable X.

(b) Find the variance of the random variable X.

19. A random voltage V has a probability density function of the form

$$f_V(v) = \begin{cases} .02v & 0 \le v \le 10 \\ 0 & \text{elsewhere} \end{cases}$$

It is sampled and quantized into two levels in accordance with the rule

$$U = \begin{cases} 1 & V > v_1 \\ -1 & V \le v_1 \end{cases}$$

(a) What should be the value of v_1 so that the random variable U has zero mean?

(b) For this value of v_1, what is the variance of U?

20. The heights of the entering first-year engineering student at a typical university were found to be a Gaussian distributed random variable with mean value 68 in and standard deviation 10 in.

(a) What is the probability that a given person is taller than 68 in?

(b) What is the probability that the height's are between 60 and 70 in?

(c) What is the probability that the tallest 70 percent of the people are taller
      than that value?

2.12 COMPUTER EXAMPLES

MATHCAD Computer Example 2.1
UNIFORM PROBABILITY DENSITY FUNCTION ... Some Insights

This document will help you to understand the uniform probability density
function and give you some insight on what uniformly distributed random
variables are. We will start by generating 100 random variables which are
uniformly distributed between 0 and 6:

$i := 0..99$ range variable to generate 100 numbers

$data_i := rnd(6)$ rnd(x) is a built-in function that generates uniform random

                   variables between 0 and the number in parentheses, 6 here.

Now we will plot the numbers we just generated, one dot for each number:

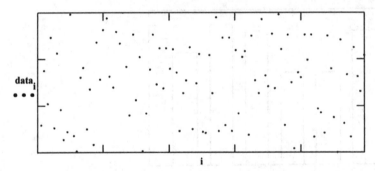

It will be easier to visualize the plot if we connect each of the points with a
straight line:

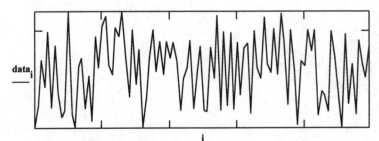

Note how the dots are distributed in the plot. To generate another set of 100
points, put the cursor on "rnd(6)" and hit [F9]. Try changing the range variable i
to go from 1 to 999 instead of 99; note again the distribution of points.

Now change the range variable i back to 99. Next we will create a histogram of the data points we just generated. The histogram will be displayed as a bar chart where the width of each bar represents a number interval and the height represents the number of times a variable within that interval was generated. For example, in the plot below, the first bar represents all the variables that fell between 0 and 1. The second bar represents those variables that fell between 1 and 2, and so on out to 6. The height of each bar is a "count" of the number of random variables that were generated in that interval.

$k := 0 .. 7$ counter to generate histogram intervals

$inter_k := k$ array that contains the intervals to be graphed:
                                        0 to 1, 1 to 2, 2 to 3, etc., out to 5 to 6

$j := 0 .. 6$ counter to generate histogram plot

$freq := hist(inter, data)$ built-in function that generates histogram

$max := max(freq) \cdot 1.1$ maximum value of plot

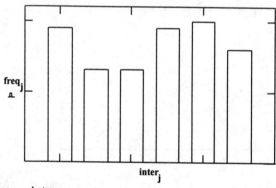

freq	inter
19	0
13	1
13	2
19	3
20	4
16	5
0	6

$max = 22$

Note the heights of the bars. The table below the graph with the title *freq$_j$* and index$_j$ shows the count for each interval, or the height of each bar. To see another 100 numbers and a new bar chart, in the Math window, depress the "Calculate Worksheet" option. This will recalculate everything using another set of 100 uniform random numbers. Try changing the range variable *i* in the top of the document to 999; then look at the bar chart again. In a uniform random distribution, all the numbers in the specified interval are supposed to be equally likely to occur. Why then are the bars not the same height?

We are drawing numbers from a computer-generated distribution which is not perfectly uniform. Even if the distribution were perfect, we would have to generate a very large number (approaching infinity) of variables before the bar heights became equal.

MATHCAD Computer Example 2.2
GAUSSIAN PROBABILITY DENSITY FUNCTION ... Some Insights

This document will help you to understand the gaussian probability density function and give you some insight on how the mean and variance are related to the position and shape of the gaussian or normal density function.
We start by defining the gaussian density function:

$$\text{inc} := \frac{(\mu + 3 \cdot \sigma) - (\mu - 3 \cdot \sigma)}{40} \qquad \text{define increments for plot}$$

$$x := \mu - 3 \cdot \sigma, \mu - 3 \cdot \sigma + \text{inc} .. \mu + 3 \cdot \sigma \qquad \text{range of values to plot}$$

$$f(x) := \frac{1}{\sigma \cdot \sqrt{2 \cdot \pi}} \cdot \exp\left[\frac{-(x - \mu)^2}{2 \cdot \sigma^2}\right] \qquad \text{gaussian density function}$$

$$\mu := -2 \qquad \text{value of the mean}$$

$$\sigma := 0.5 \qquad \text{value of standard deviation, equal to square root of variance}$$

Now try changing the value of the mean above (keep it between −20 and +20). Then try changing the standard deviation and observe the changes. As you can see, the mean value shifts the entire plot to the right or left, while the variance affects the width of the plot.

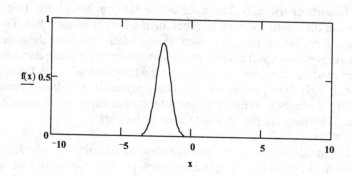

MATHCAD Computer Example 2.3
COMPUTATION OF AREAS UNDER THE GAUSSIAN PROBABILITY DENSITY FUNCTION

Let X be a gaussian distributed random variable with mean value equal to 1.0 and standard deviation equal to 2.0. We need to compute the probability that the random variable is between $-.5$ and 1.5.

Define the mean and the standard deviation:

$\mu \equiv 0.0$

$\sigma \equiv 2.0$

Define the increment of the random variable x:

$$\text{inc} := \frac{(\mu + 3 \cdot \sigma) - (\mu - 3 \cdot \sigma)}{40}$$

Define now the range of the random variable x:

$x := \mu - 3 \cdot \sigma, \mu - 3 \cdot \sigma + \text{inc} .. \mu + 3 \cdot \sigma$

Define the probability density function of x:

Limits:

$\text{upper} := 2.0$ $f(x) := \dfrac{1}{\sigma \cdot \sqrt{2 \cdot \pi}} \cdot \exp\left[\dfrac{-(x - \mu)^2}{2 \cdot \sigma^2}\right]$

$\text{lower} := -2.$ $y1 := 0.0, 0.01 .. f(\text{lower})$

$$\int_{lower}^{upper} f(x)\,dx = 0.683 \qquad y2 := 0.0, 0.01 .. f(upper)$$

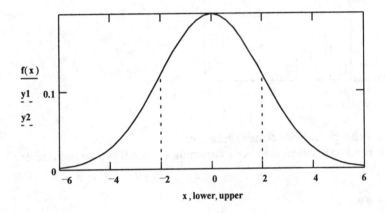

x , lower, upper

MATHCAD Computer Example 2.4
POISSON PROBABILITY DENSITY FUNCTION ... Examples

This computer example will help you to understand the Poisson probability density function. This is a discrete random variable, thus the probability density function is discrete. The Poisson distributed random variable has the unique property that the expectation equals the variance. The general form of the Poisson probability density function is as follows:

$$Pr(k) := \frac{\alpha^k}{k!} \cdot e^{-\alpha} \quad \square \quad k=0,1,2,...$$

where a is always greater than 0 and equals the expectation (and therefore the variance) of the random variable. Let us first plot the random variable as a function of parameter k:

$$k := 0, 1 .. 10$$

$$\alpha := 2.$$

$$f_k := \frac{\alpha^k}{k!} \cdot e^{-\alpha}$$

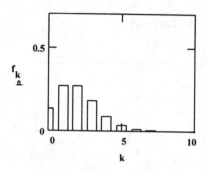

Is this a valid probability density function?

We need to perform the above summation from 0 to a large number:

$$\sum_{k} f_k = 0.9$$

Note that after just a few terms, the sum approaches 1.0 very rapidly. Change parameter a and see what effect this will have on the probability density function.

Consider now a Poisson distributed random variable with a mean value of 2, what is the probability that the random variable is greater than or equal to 2 be computed. We will set this problem up, using MATHCAD:

$h := 2, 3 .. 10$ This is the summation index.

$\alpha := 2$ This is the mean value of the random variable.

$$\sum_{h} f_h = 0.594 \qquad f_h := \frac{\alpha^h}{h!} \cdot e^{-\alpha}$$

The average number of impulses per hour is 8; therefore the probability of obtaining 3 impulses in 1 h is found as follows:

$\alpha := 8$

$h := 3$

$$f_h := \frac{\alpha^h}{h!} \cdot e^{-\alpha}$$

$$f_3 = 0.029$$

or about 3 percent.

MATLAB Computer Example 2.1
UNIFORM PROBABILITY DENSITY FUNCTION … Some Insights

This document will help you to understand the uniform probability density function and give you some insight on what uniformly distributed random variables are. We start by generating 100 random variables which are approximately uniformly distributed between 0 and 6:

```
%==========================================
% Input parameters
i = 100;
amplt_data = 6;
%==========================================
length_data=i;
inter=6;
data_i=amplt_data*rand(length_data,1);
[freq,bins]=hist(data_i,inter);
clg;s1=subplot(2,2,1);
plot(data_i,'r'); axisn(0:1:length_data-1);
xlabel('i','Fontsize',[10]);
ylabel('data_i','Fontsize',[10]);
title('Uniform random variable.','Fontsize',[10]);
set(s1,'Fontsize',[10])
s2=subplot(2,2,2);
plot(data_i,'.r'); axisn(0:1:length_data-1);
xlabel('i','Fontsize',[10]);
ylabel('data_i','Fontsize',[10]);
title('Uniform random variable.','Fontsize',[10]);
set(s2,'Fontsize',[10])
s3=subplot(2,2,3);
bar(0:1:inter-1,freq);axisn(-1:1:inter);
xlabel('inter_j','Fontsize',[10]);
ylabel('freq_j','Fontsize',[10]);
set(s3,'Fontsize',[10])
s4=subplot(2,2,4);
axis('off')
text(0.7,1,'freq_j : ','Fontsize',[10]);
```

```
text(0.2,1,'inter_j :','Fontsize',[10]);
for j=0:inter-1,
 set(text(0.85,1/(inter+1)*(inter-1-j),['
',num2str(freq(j+1))],'Fontsize',[10]),'HorizontalAlignment','right');
 set(text(0.3,1/(inter+1)*(inter-1-j),['
',num2str(j)],'Fontsize',[10]),'HorizontalAlignment','right');
end;
clear
```

Recalculate the entire document again. Did you get the same numbers? If not, why not? Change the range variable i in the top of the document to 999, then look at the bar chart again. In a uniform random distribution all the numbers in the specified interval are supposed to be equally likely to occur. Why are the bars not the same height?

MATLAB Computer Example 2.2
GAUSSIAN PROBABILITY DENSITY FUNCTION ... Some Insights

This document will help you to understand the gaussian probability density function and give you some insight on how the mean and variance are related to the position and shape of the gaussian or normal density function.

```
%===========================================
% Input parameters
mu = 3;
sigma = 5;
%===========================================
inc=((mu+3*sigma)-(mu-3*sigma))/40;
x=-20:inc:20;
mu=mu*ones(size(x));
f_x=(1/(sigma*sqrt(2*pi)))*exp(-(x-mu).^2/(2*sigma^2));
```

```
clg;s1=subplot(1,1,1);
plot(x,f_x,'r')
xlabel('x','Fontsize',[10]);
ylabel('f_x','Fontsize',[10]);
title('Gaussian probability density function','Fontsize',[10]);
set(s1,'Fontsize',[10])
clear
```

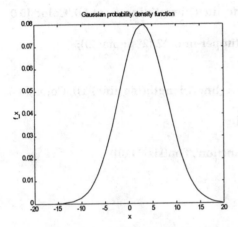

Change the value of the mean above (keep it between –20 and +20). Observe the effects on the plot. Try changing the standard deviation and observe the changes. As you can see, the mean value shifts the entire plot to the right or left, while the variance affects the width of the plot.

MATLAB Computer Example 2.3
COMPUTATION OF AREAS UNDER THE GAUSSIAN PROBABILITY
DENSITY FUNCTION

Let X be a gaussian distributed random variable with mean value equal to 1.0 and standard deviation equal to 2.0. We need to compute the probability that the random variable is between –.5 and 1.5.

```
%==========================================
% Input parameters
mu=0;
sigma=1;
lower=-1.;
upper=0.5;
%==========================================
inc=((mu+3*sigma)-(mu-3*sigma))/40;
x=(mu-3*sigma):inc:(mu+3*sigma);
x1=lower:0.01:upper;
```

```
f_x=(1/(sigma*sqrt(2*pi)))*exp(-(x-mu).^2/(2*sigma^2));
f_x1=(1/(sigma*sqrt(2*pi)))*exp(-(x1-mu).^2/(2*sigma^2));
P=trapz(x1,f_x1);
clg;s1=subplot(1,1,1);
plot(x,f_x,'r');axisn(x);
f_lower=(1/(sigma*sqrt(2*pi)))*exp(-(lower-mu).^2/(2*sigma^2));
inc_line=0:f_lower/100:f_lower;
lower=lower*ones(size(inc_line));
set(line(lower(1:length(inc_line)-1),inc_line(1:length(inc_line)-1)),'Color',[0 0
1],'LineStyle','.');
f_upper=(1/(sigma*sqrt(2*pi)))*exp(-(upper-mu).^2/(2*sigma^2));
inc_line=0:f_upper/100:f_upper;
upper=upper*ones(size(inc_line));
set(line(upper(1:length(inc_line)-1),inc_line(1:length(inc_line)-1)),'Color',[0 0
1],'LineStyle','.');
xlabel('x, lower, upper','Fontsize',[10]);
ylabel('f_x','Fontsize',[10]);
title('Gaussian probability density function','Fontsize',[10]);
set(s1,'Fontsize',[10])
disp(['Probability: ',num2str(P)]);
clear
```

Prob. 0.5328

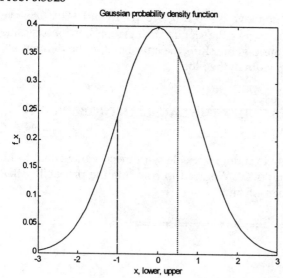

MATLAB Computer Example 2.4
POISSON PROBABILITY DENSITY FUNCTION ... Examples

This computer example will help you to understand the Poisson probability density function. This is a discrete random variable, thus the probability density function is discrete. The Poisson distributed random variable has the unique property that the expectation equals the variance. The general form of the Poisson probability density function is as follows:

```
%=========================================
% Parameters
alpha=2;
k=10;
min_sum=0;
max_sum=10;
%=========================================
fact=1;
f_k=[];
for n=0:1:k,
 if n~=0
  for ll=1:1:n
   fact=fact*ll;
  end
 end
 f_k(n+1)=((alpha^n)/fact)*exp(-alpha);
 fact=1;
end;
if(((min_sum)>=0)& ((max_sum)<length(f_k)))
 P=sum(f_k(min_sum+1:max_sum+1));
else
 disp('Define the correct range')
end;
disp(['<< sum(f): ',num2str(P)])
s1=subplot(1,1,1);
bar(0:1:k,f_k); axisn(0:1:k);
title('Poisson probability density function','Fontsize',[10]);
xlabel('k','Fontsize',[10]);
ylabel('f_k','Fontsize',[10]);
set(s1,'Fontsize',[10]);
clear
```

<< sum(f): 1

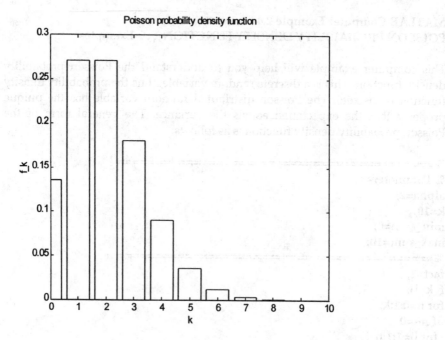

Change parameter alpha and see what effect this will have on the probability density function.

CHAPTER

Multiple Random Variables 3

3.0 INTRODUCTION

Many experiments involve the interaction of *several* random variables, and frequently the experimenter is interested in the joint behavior of these variables. For example, suppose that the Scholastic Assessment Test (SAT) scores, high school grade point average (GPA), and ending college GPA of college students are measured in a particular college. Let the SAT random variable be denoted by the random variable SAT(s), the high school GPA by the random variable GPAh(s), and the college GPA by the random variable GPAc(s), where s refers to a student. We may want to ask the following question: Is there any relationship between the first two random variables and the third? Suppose we add a fourth random variable, the age of the student $A(s)$, and we wish to explore the significance of this new random variable. We have already developed techniques that we can use to examine and estimate the statistical properties of any one of these random variables. We now need techniques that will help us determine the *joint* statistical properties of several random variables. In this chapter we will formalize the statistical concepts associated with more than one random variable. Once the tools to describe and analyze multiple random variables are developed, we introduce techniques associated with testing and analysis of measured data. This is perhaps a departure from traditional engineering textbooks in this topic. It is felt that an early introduction to the analysis of real cases coupled with the necessary theoretical background is essential in achieving a thorough understanding of this topic.

Theoretical analysis of hypothetical cases is an integral part of the understanding of the concepts explained in this book. The analysis of real data is the other part of the equation. For example, in a section of this chapter we analyze data obtained in a laboratory by testing their equivalence to a specified theoretical density function. First we test whether the data are indeed random (using the run test), then we test whether the data fit a specified theoretical density function, using two separate tests (the chi-square test and the Kolmogoroff-Smirnoff test). These tests were chosen because they are simple to use and they illustrate the examples given. They are not meant to be all-inclusive of the many different tests used for the given problem.

3.1 JOINT CUMULATIVE DISTRIBUTION FUNCTION OF TWO RANDOM VARIABLES

As with the case of a single random variable, we first define the concept of a *joint cumulative distribution function* for the case of two random variables. Later we will extend some of these concepts for more than two random variables. A joint distribution function of two random variables uniquely defines the probability of the joint events $\{X \leq x, Y \leq y\}$. The joint distribution function is defined by the following equation:

$$F_{XY}(x, y) = P(X \leq x, Y \leq y) \tag{3.1}$$

where $F_{XY}(x, y)$ specifies the probability that the random variable X assumes a value that is less than or equal to x *and* the random variable Y assumes a value that is less than or equal to y. For example, consider the toss of two dice. Let the random variable associated with the first die be designated X and the random variable associated with the second die Y. We can ask the questions: What is the probability that the value of the face of the first die will be less than or equal to 2, and what is the probability that the value of the face of the second die will be less than or equal to 2? In terms of Equation (3.1) we are trying to find the following value:

$$F_{XY}(2, 2) = P(X \leq 2, Y \leq 2) \tag{3.2}$$

We can now construct a table of the values of $F_{XY}(x, y)$ for this specific case. With this table we can read directly the value of $F_{XY}(x, y)$. For example, the value of the expression

$$F_{XY}(4,5) = P(X \leq 4, \ Y \leq 5) = 20/36 \tag{3.3}$$

$F_{XY}(x, y)$

$x =$	1	2	3	4	5	6
$y = $ 1	1/36	2/36	3/36	4/36	5/36	6/36
2	2/36	4/36	6/36	8/36	10/36	12/36
3	3/36	6/36	9/36	12/36	15/36	18/36
4	4/36	8/36	12/36	16/36	20/36	24/36
5	5/36	10/36	15/36	20/36	25/36	30/36
6	6/36	12/36	18/36	24/36	30/36	36/36

This table is constructed by considering all possible pairs of outcomes (36) and all possible ways in which a particular combination may appear out of the 36 possible pairs. For example,

$$F_{XY}(2,2) = P(X \leq 2, \ Y \leq 2)$$

is found by computing all possible ways in which the combinations (1,1), (1, 2), (2, 1), (2, 2) may occur. Since all these are mutually exclusive and there are a total of 36 such pairs, the probability is

$$P(X \leq 2, \ Y \leq 2) = 4/36$$

The properties of the joint cumulative distribution function are very similar to those for the distribution function of a single random variable (see Chapter 2). The joint cumulative distribution function of two random variables has the following properties:

1. $0 \leq F_{XY}(x, y) \leq 1$
2. $\lim F_{XY}(\infty, \infty) = 1$
3. $F_{XY}(x, y)$ is nondecreasing
4. $F_{XY}(-\infty, y) = F_{XY}(x, -\infty) = 0$
5. $F_X(x) = F_{XY}(x, \infty) = P(X \leq x, \ Y \leq \infty) = P(X \leq x)$
6. $F_Y(y) = F_{XY}(\infty, y) = P(X \leq \infty, \ Y \leq y) = P(Y \leq y)$

The *first* property is a statement of the fact that $F_{XY}(x, y)$ is a probability and, therefore, has all of the properties associated with the concept of probability. The *second* and *third* properties are a consequence of the preceding statement. The fourth property is a consequence of the limiting nature of the joint probability function. The *fifth* and *sixth* properties state that as one of the random variables reaches its upper limit, the joint probability distribution function equals the probability distribution function of the other random variable. For example, consider the table of $F_{XY}(x,y)$ for the case of throwing a pair of dice simultaneously. It would be complicated to draw this function as a surface, but it can be observed that as $x \to 6$ and $y \to 6$, the function $F_{XY}(x, y) \to 1$. When $x = 6$, however, $F_{XY}(6,y) = F_Y(y)$. Likewise, when $y = 6$, $F_{XY}(x, 6) = F_{XY}(x)$.

$$F_{XY}(x, y)$$

		$x=$ 1	2	3	4	5	$F_Y(y)$ $\downarrow$ 6
$y=$	1	1/36	2/36	3/36	4/36	5/36	6/36
	2	2/36	4/36	6/36	8/36	10/36	12/36
	3	3/36	6/36	9/36	12/36	15/36	18/36
	4	4/36	8/36	12/36	16/36	20/36	24/36
	5	5/36	10/36	15/36	20/36	25/36	30/36
$F_X(x) \to$	6	6/36	12/36	18/36	24/36	30/36	36/36

In the more general case, that is, $F_{XY}(\infty, y)$ and $F_{XY}(x, \infty)$, the cumulative distribution functions are called the *marginal cumulative distribution functions*.

As with the case of a single random variable, only when the random variables are discrete may the concept of the probability value associated with a specific value of the random variable be finite (nonzero).

3.2 JOINT PROBABILITY DENSITY FUNCTION OF TWO RANDOM VARIABLES

The *joint probability density function* is defined (if it exists) as the *derivative* of the joint cumulative distribution function. Note that since the joint cumulative distribution function has two variables, the derivative must be a partial derivative with respect to each of the variables. This derivative may be expressed by the following formula:

$$f_{XY}(x,y) = \frac{\partial^2 F_{XY}(x,y)}{\partial x \, \partial y} \tag{3.4}$$

The properties of the joint probability density function (for the case of two random variables) are very similar to those mentioned for the single-random-variable case:

Property 1:

$$f_{XY}(x,y) \geq 0 \qquad (3.5)$$

As in the case of a single random variable, the probability of the event $\{x < X < x + \Delta x; y < Y < \Delta y\}$ must be nonnegative. Consequently, $f_{XY}(x, y)$ must be nonnegative.

Property 2:

$$\iint\limits_{-\infty}^{\infty} f_{XY}(x,y) \, dx \, dy = 1 \qquad (3.6)$$

This equation states that the double integral under the joint probability density function must equal 1. This is, in fact, an extension of the second property of the probability density function of a single random variable.

Property 3:

$$f_X(x) = \int_{-\infty}^{\infty} f_{XY}(x,y) \, dy$$

$$(3.7)$$

$$f_Y(y) = \int_{-\infty}^{\infty} f_{XY}(x,y) \, dx$$

These equations state that the *marginal probability density function* of one of the random variables is found by integrating the joint probability density function over the other random variable.

Property 4:

$$P(x_a \leq X \leq x_b, \; y_c \leq Y \leq y_d) = \int_a^b \int_c^d f_{XY}(x,y) \, dx \, dy \qquad (3.8)$$

This equation states that the probability that the random variable X is between the limits x_a and x_b *and* the random variable Y is between the limits y_c and y_d is found by computing the value of the double integral of $f(x,y)$ over the specified limits. This property is also an extension of the second property of the probability density function of a single random variable.

Example 3.1 Consider the following joint probability density function:

$$f_{XY}(x,y) = \begin{cases} 9 \cdot \exp(-3x) \cdot \exp(-3y) & x, y \geq 0 \\ 0 & \text{elsewhere} \end{cases}$$

Find the probability that the random variable X is between the limits 0 and 1 and that the random variable Y is also between the limits 0 and 1.

First, we should confirm the fact that $f_{XY}(x,y)$, as given, is a valid joint probability density function. This we can do by testing property 2, that is,

$$\iint\limits_{-\infty}^{\infty} f_{XY}(x,y)\, dx\, dy = 1$$

For this example this means confirming that

$$\iint\limits_{-\infty}^{\infty} f_{XY}(x,y)\, dx\, dy = \int_0^\infty \int_0^\infty 9 \cdot \exp(-3x) \cdot \exp(-3y)\, dx\, dy = 1 \qquad (3.9)$$

This is proved as follows:

$$\iint\limits_{-\infty}^{\infty} 9 \cdot \exp(-3x) \cdot \exp(-3y)\, dx\, dy = 3 \cdot \int_0^\infty \exp(-3x)\, dx \cdot 3 \cdot \int_0^\infty \exp(-3y)\, dy \qquad (3.10)$$

The first integral is evaluated by

$$3 \cdot \int_0^\infty \exp(-3x)\, dx = 3 \left[-\frac{1}{3} \exp(-3x) \Big|_0^\infty \right] = 3 \left[-\frac{1}{3}(0-1) \right] = 1 \qquad (3.11)$$

with the second integral having an identical value.

We therefore conclude that this is indeed a valid joint probability density function. We now proceed to answer the question in the example by evaluating the following double integral:

$$9 \cdot \int_0^1 \int_0^1 \exp(-3x) \cdot \exp(-3y)\, dx\, dy =$$

$$9 \cdot \int_0^1 \exp(-3x)\, dx \cdot \int_0^1 \exp(-3y)\, dy = .92 \qquad (3.12)$$

This result implies that the probability that the random variable X is between the limits 0 and 1 and that the random variable Y is also between the limits 0 and 1 is .92.

Example 3.2 Consider the following joint probability density function:

$$f_{XY}(x,y) = \begin{cases} \dfrac{4}{225}xy & 0 \le x \le 3, \quad 0 \le y \le 5 \\ 0 & \text{elsewhere} \end{cases}$$

Find the probability that the random variable X is between the limits 0 and 2 and that the random variable Y is between the limits 0 and 4.

As with the previous example, the first thing that should be checked is whether the given joint probability density function is valid. We need to prove that

$$f_{XY}(x,y) = \begin{cases} \dfrac{4}{225}xy & 0 \le x \le 3, \quad 0 \le y \le 5 \\ 0 & \text{elsewhere} \end{cases}$$

Find the probability that the random variable X is between the limits 0 and 2 and that the random variable Y is between the limits 0 and 4.

As with the previous example, the first thing that should be checked is whether the given joint probability density function is valid. We need to prove that

$$\frac{4}{225}\int_0^3\int_0^5 xy \, dx \, dy = 1$$

We then proceed to solve the problem as follows:

$$\frac{4}{225}\int_0^2 x \, dx \cdot \int_0^4 y \, dy = \frac{4}{225}\left(\frac{x^2}{2}\Big|_0^2\right)\left(\frac{y^2}{2}\Big|_0^4\right) = \frac{4}{225}\left(\frac{4}{2}\right)\left(\frac{16}{2}\right) = .28 \qquad (3.13)$$

This result implies that 28 percent of the time the random variable X is between the limits 0 and 2 *and* that the random variable Y is between the limits 0 and 4.

3.3 STATISTICAL PROPERTIES OF JOINTLY DISTRIBUTED RANDOM VARIABLES: JOINT MOMENTS

Just as in the case of single random variables, there are a number of statistical properties of jointly distributed random variables that will help us gain knowledge about the "behavior" of the random variables. Two jointly distributed random variables X and Y are said to be *statistically independent* of each other if and only if the joint probability density function equals the product of the two marginal probability density functions. This is expressed by the following relationship:

$$f_{XY}(x,y) = f_X(x)f_Y(y) \tag{3.14}$$

The joint moments of two random variables yield information concerning their joint behavior. It is important to realize that since we are now dealing with two random variables, we need to specify the order of the moment *for each* of the two random variables. We are therefore required to specify, for example, the mth moment of the random variable X and the nth moment of the random variable Y as the mnth joint moment of the X and Y random variables. The physical significance of these joint moments will not be as easy to understand, except in some very specific cases. The mnth moment of the random variables X and Y is defined as follows:

$$E\left[X^m Y^n\right] = \int\int_{-\infty}^{\infty} x^m y^n f_{XY}(x,y)\, dx\, dy \tag{3.15}$$

When $m = 0$, we obtain the moments of Y, and when $n = 0$, we obtain the moments of X. Consider now the case when $m = 1$ and $n = 1$:

$$E[XY] = \int\int_{-\infty}^{\infty} xy f_{XY}(x,y)\, dx\, dy \tag{3.16}$$

This quantity is called the *correlation* between X and Y. If the correlation between X and Y is equal to the product of their means, then we say that they are *uncorrelated*. Uncorrelated random variables are usually described in terms of covariance, which is discussed later in this chapter. If the correlation between X and Y is equal to 0, then we say that the random variables are *orthogonal* to each other. We must be careful with the concept of correlation, as often this is confused with the concept of independence.

Example 3.3 Consider the random variable X with mean value 3 and variance 2. A second random variable Y is defined as $Y = 3X - 11$. Find the mean value of Y and the correlation of X and Y.

We first find the mean square value of the random variable X:

$$E\left[X^2\right] = \sigma_X^2 + \mu_X^2 = 11 \tag{3.17}$$

We now find the mean value of the random variable Y:

$$E[Y] = E[3X - 11] = 3E[X] - 11 = 9 - 11 = -2$$

The correlation of X and Y is

$$E[XY] = E[X(3X - 11)] = E\left[3X^2 - 11X\right] \tag{3.18}$$

which equals

$$E[XY] = 3E\left[X^2\right] - 11E[X] = 3 \cdot 11 - 11 \cdot 3 = 0$$

Based on this result, we say that X and Y are orthogonal. Are they uncorrelated? For X and Y to be uncorrelated, the correlation between X and Y must equal the product of their means:

$$E[XY] = \mu_X \mu_Y$$

In this case, this is obviously not true, since

$$E[XY] = 0 \neq \mu_X \mu_Y = (3)(-2) = -6$$

Therefore, although X and Y are orthogonal, they are correlated.

The two joint probability density functions considered above

$$f(x,y) = \begin{cases} 9 \cdot \exp(-3x) \cdot \exp(-3y) & x, y \geq 0 \\ 0 & \text{elsewhere} \end{cases}$$

and

$$f(x,y) = \begin{cases} \dfrac{4}{225}xy & 0 \le x \le 3, \quad 0 \le y \le 5 \\ 0 & \text{elsewhere} \end{cases}$$

are independent since they are separable into $f(x)f(y)$. Consider the second example:

$$f(x) = \int_{-\infty}^{\infty} f(x,y)\, dy = \int_{0}^{5} \frac{4}{225}xy\, dy = \frac{4}{225}x \int_{0}^{5} y\, dy = \frac{2}{9}x \qquad (3.19)$$

and likewise,

$$f(y) = \int_{-\infty}^{\infty} f(x,y)\, dx = \int_{0}^{3} \frac{4}{225}xy\, dx = \frac{4}{225}y \int_{0}^{3} x\, dx = \frac{2}{25}y \qquad (3.20)$$

which proves that

$$f_{XY}(x,y) = f_X(x)f_Y(y)$$

Example 3.4 Two random variables X and Y have a joint probability density function given by

$$f_{XY}(x,y) = \begin{cases} Ax^2(1-y) & 0 \le x \le 1, \quad 0 \le y \le 1 \\ 0 & \text{elsewhere} \end{cases} \qquad (3.21)$$

(a) Find the value of A.

In order to find A, we will use the fact that the integral over the joint probability density function must equal 1, that is,

$$\iint_{-\infty}^{\infty} f_{XY}(x,y)\, dx\, dy = 1 \qquad (3.22)$$

Therefore,

$$\int_{0}^{1}\int_{0}^{1} Ax^2(1-y)\, dx\, dy = 1 \qquad (3.23)$$

which equals

$$A\int_{0}^{1} x^2\, dx \int_{0}^{1}(1-y)\, dy = A\left[\frac{x^3}{3}\right]_{0}^{1} \cdot \left(y - \frac{y^2}{2}\right)\Big|_{0}^{1} = \frac{A}{6} = 1 \qquad (3.24)$$

Therefore, $A = 6$.

(b) Find $f_X(x)$ and $f_Y(y)$.
    The two marginal densities are found as follows:

$$f_X(x) = \int_0^1 6x^2(1-y)\, dy = 6x^2\left[y - \frac{y^2}{2}\right]_0^1 = 3x^2 \quad 0 \le x \le 1 \qquad (3.25)$$

$$f_Y(y) = \int_0^1 6x^2(1-y)\, dx = 2x^3(1-y)\Big|_0^1 = 2(1-y) \quad 0 \le y \le 1 \qquad (3.26)$$

The random variables are statistically independent, since $f_{XY}(x,y) = f_X(x)f_Y(y)$.

(c) Find the correlation between the two random variables.
    Here we seek to find $E[XY]$:

$$E[XY] = \int_0^1\int_0^1 xy\left[x^2(1-y)\right] dx\, dy \qquad (3.27)$$

which equals

$$E[XY] = 6\int_0^1 x^3\, dx \cdot \int_0^1 \left(y - y^2\right) dy = 6E[X]E[Y] = 6\mu_x\mu_y = \frac{1}{4} \qquad (3.28)$$

This states that the correlation of two statistically independent random variables equals the product of their mean values. If either one of the two random variables has a zero mean, then the result will equal zero. For the present case, we say that the two random variables are *uncorrelated*.

The *covariance* between the random variables X and Y is the $m = 1, n = 1$ joint central moments:

$$\text{cov}(X,Y) = \int\limits_{-\infty}^{\infty}\!\!\int (x - \mu_x)(y - \mu_y)f_{XY}(x,y)\, dx\, dy \qquad (3.29)$$

Example 3.5 Find the covariance of the random variables of Example 3.4.
First, find the expected value of each of the random variables:

$$E[X] = \int_0^1 x f(x)\, dx = \int_0^1 3x^3\, dx = \frac{3}{4} \tag{3.30}$$

and likewise,

$$E[Y] = \int_0^1 y f(y)\, dy = \int_0^1 y \cdot 2(1-y)\, dy = \frac{1}{3} \tag{3.31}$$

We now find the covariance as follows:

$$\mathrm{cov}(X,Y) = \int\!\!\int_{-\infty}^{\infty} (x - \mu_x)(y - \mu_y) f_{XY}(x,y)\, dx\, dy$$

$$= 6 \int_0^1\!\!\int_0^1 \left(x - \frac{3}{4}\right)\left(y - \frac{1}{3}\right) x^2(1-y)\, dx\, dy = 0 \tag{3.32}$$

The expression for the covariance could have been evaluated as follows:

$$\mathrm{cov}(X,Y) = \int\!\!\int_{-\infty}^{\infty} (x - \mu_x)\cdot(y - \mu_y)\cdot f_{XY}(x,y)\, dx\, dy = E\big[(X - \mu_x)\cdot(Y - \mu_y)\big] \tag{3.33}$$

which equals

$$\mathrm{cov}(X,Y) = E\big[XY - X\mu_y - Y\mu_x - \mu_x\mu_y\big]$$

$$= E[XY] - \mu_y E[X] - \mu_x E[Y] - \mu_x\mu_y \tag{3.34}$$

$$\mathrm{cov}(X,Y) = E[XY] - \mu_y\mu_x - \mu_x\mu_y + \mu_x\mu_y = E[XY] - \mu_x\mu_y \tag{3.35}$$

This form is easier to evaluate than the preceding integral. When $\mathrm{cov}(X,Y)$ is zero, then the random variables X and Y are *uncorrelated*. From the above equation we can see that when the random variables X and Y are uncorrelated the correlation $E[XY]$ is equal to the products of the means μ_x and μ_y.

A quantity which is also of great interest is the *correlation coefficient* between two random variables. The correlation coefficient between two random variables is defined as follows:

$$\rho_{XY} = \frac{\mathrm{cov}(X,Y)}{\sigma_x\sigma_y} = \frac{E[XY] - \mu_x\mu_y}{\sigma_x\sigma_y} \tag{3.36}$$

The correlation coefficient is a number that varies between -1 and $+1$. When the correlation coefficient $\rho = 1$, then the random variables are said to be perfectly correlated. This is typically achieved when there is a linear relationship between X and Y. When the correlation coefficient $\rho = -1$, then the random variables are also perfectly correlated but with a negative correlation. If the correlation coefficient $\rho = 0$, then the random variables are uncorrelated. Note that if the random variables are statistically independent, then they are also uncorrelated, since their covariance is zero.

Example 3.6 The random variables X and Y are defined by the relationship $Y = 2X + 3$. Find their correlation coefficient.

To calculate the correlation coefficient, we need to evaluate the following expression:

$$\rho = \frac{\text{cov}(X,Y)}{\sigma_x \sigma_y} = \frac{E[XY] - \mu_x \mu_y}{\sigma_x \sigma_y} = \frac{E[X(2X+3)] - \mu_x \mu_y}{\sigma_x \sigma_y} \tag{3.37}$$

which equals

$$\rho = \frac{E[2X^2 + 3X] - \mu_x \mu_y}{\sigma_x \sigma_y} = \frac{2E[X^2] + 3\mu_x - \mu_x(2\mu_x + 3)}{\sigma_x \sigma_y} = \frac{2\left(E\left[X^2\right] - \mu_x^2\right)}{\sigma_x \sigma_y}$$

This equation is simplified, since Y is a function of X, by noting that the expression for the variance of Y may be expressed in terms of the variance of X as follows:

$$\sigma_y^2 = E[Y^2] - \mu_y^2 = E[(2X+3)^2] - (2\mu_x + 3)^2 = 4\sigma_x^2 \tag{3.39}$$

Therefore, the correlation coefficient equals

$$\rho = \frac{2\sigma_x}{\sigma_y} = \frac{2\sigma_x}{2\sigma_x} = 1 \tag{3.40}$$

We have thus shown that if two random variables are linearly dependent, their correlation coefficient equals 1. Had the relationship been of the type $Y = -2X + 3$, then the correlation coefficient would be -1.

Let us consider again the definition of the correlation coefficient:

$$\rho_{XY} = \frac{\text{cov}(X,Y)}{\sigma_X \sigma_Y} \tag{3.41}$$

The numerator, or cov(X, Y), is given by [from Equation (3.35)]

$$\text{cov}(X,Y) = E[XY] - \mu_X \mu_Y$$

Note that if the random variables X and Y are uncorrelated, i.e., if $\rho_{XY} = 0$, then cov(X, Y) = 0 and

$$E[XY] = \mu_X \mu_Y \tag{3.42}$$

3.4 JOINTLY DISTRIBUTED GAUSSIAN RANDOM VARIABLES

Jointly distributed gaussian variables are commonly encountered in practical problems. The joint probability density function of gaussian distributed random variables is given by the following expression:

$$f_{XY}(x,y) = \frac{1}{2\pi\sigma_X \sigma_Y \sqrt{1-\rho^2}} \exp\left\{\frac{-1}{2(1-\rho^2)}\right.$$

$$\times \left. \left[\frac{(x-\mu_X)^2}{\sigma_X^2} - 2\rho\frac{(x-\mu_X)(y-\mu_Y)}{\sigma_X \sigma_Y} + \frac{(y-\mu_Y)^2}{\sigma_Y^2}\right]\right\} \tag{3.43}$$

The quantity ρ is the *correlation coefficient* between the two gaussian distributed random variables X and Y. And μ_X, μ_Y, σ_X, σ_Y, and ρ are the parameters of the distribution. The marginal densities of X and Y are also gaussian density functions. In fact, it can be proved that

$$f_X(x) = \frac{1}{\sigma_x \sqrt{2\pi}} \exp\left[\frac{-(x-\mu_x)^2}{2\sigma_x^2}\right] \tag{3.44}$$

i.e., the marginal probability density function of the random variable X is also a gaussian probability density function with parameters μ_X and σ_X. One could prove a similar result for the random variable Y.

Whenever two random variables are statistically independent, their correlation coefficient equals zero. The opposite, however, is not necessarily true; i.e., if the correlation coefficient between two random variables is zero, it does not necessarily imply that they are statistically independent. One exception to this rule is in the case of two jointly distributed gaussian random variables. Consider the expression of Equation (3.43):

$$f_{XY}(x,y) = \frac{1}{2\pi\sigma_X\sigma_Y\sqrt{1-\rho^2}} \exp\left\{\frac{-1}{2(1-\rho^2)}\right.$$

$$\left. \times \left[\frac{(x-\mu_X)^2}{\sigma_X^2} - 2\rho\frac{(x-\mu_X)(y-\mu_Y)}{\sigma_X\sigma_Y} + \frac{(y-\mu_Y)^2}{\sigma_Y^2}\right]\right\}$$

Let the correlation coefficient $\rho = 0$. Then the preceding expression reduces to

$$f_{XY}(x,y) = \frac{1}{2\pi\sigma_X\sigma_Y} \exp\left\{\frac{-1}{2}\left[\frac{(x-\mu_X)^2}{\sigma_X^2} + \frac{(y-\mu_Y)^2}{\sigma_Y^2}\right]\right\} \qquad (3.45)$$

By inspection of the preceding equation, we realize that $f_{XY}(x,y)=f_X(x)f_X(y)$. Therefore, for the case of jointly distributed gaussian random variables, if it can be shown that their correlation coefficient equals 0, the two random variables are statistically independent.

3.5 FUNCTIONS OF GAUSSIAN JOINTLY DISTRIBUTED RANDOM VARIABLES

There are a number of probability density functions related to the gaussian density. Consider the case of a shooter aiming at a target. Assume that the center of the target is at the center of the coordinate system (x,y). As the shooter aims, there is an error associated with each x and y. Assume now that the error in each of the directions is gaussian distributed and that the errors are independent of each other. This situation is illustrated by Figure 3.1. The X and Y "miss" amounts are considered to be independent gaussian distributed amounts.

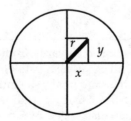

Figure 3.1 This figure illustrates a target and the miss amounts in the X and Y directions.

The miss distance r is found to be

$$r = \sqrt{x^2 + y^2} \tag{3.46}$$

where X and Y are identically distributed independent gaussian random variables. If X and Y are gaussian distributed with variance σ^2, the probability density function of r can be shown to be Rayleigh and has the following form:

$$f_R(r) = \frac{r}{\sigma^2} \exp\left(\frac{-r^2}{2\sigma^2}\right) \cdot u(r) \tag{3.47}$$

Figure 3.2 shows a plot of $f_R(r)$ for the case when the standard deviation σ equals 2.

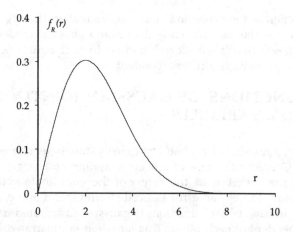

Figure 3.2 Rayleigh probability density function. For this case the standard deviation of both independent random variables x and y is $\sigma = 2$. Note that the maximum of $f_R(r)$ is at σ and the maximum value of $f_R(r)$ is .606/σ.

Example 3.7 A shooter is positioned 50 yards from a target in the first round of the state qualifier match. She is using a caliber 45 Model 1911 U.S. Army pistol. In order to qualify for the state finals, she must not exceed a 2-in radius from the target. Assume that the x and y coordinates from the center of the target are independent gaussian distributed random variables with zero mean and variance equal to 1.0 in. Calculate the probability that she will qualify for the state finals.

The distance r from the center of the target is Rayleigh distributed. Consequently, in order to find the probability that she will not go to the state finals, we must compute

$$P = 1 - \int_0^2 \frac{r}{\sigma^2} \exp\left(\frac{-r^2}{2\sigma^2}\right) dr \qquad (3.48)$$

where the variance equals 2.2. The value of the integral is found to be .865. Consequently, the probability that she will not go to the state finals equals 13.5 percent.

3.6 CONDITIONAL PROBABILITY DENSITY

In Chapter 2 we saw that the conditional cumulative distribution function of a random variable *given that* a certain event B has occurred is given by

$$F_X(x/B) = P\big[(X \leq x)/B\big] = \frac{P(X \leq x, B)}{P(B)} \quad P(B) > 0 \qquad (3.49)$$

and the corresponding conditional probability density function is

$$f_X(x/B) = \frac{f_X(x)}{P(B)} \qquad (3.50)$$

where $P(B)$ is the probability of event B. When more than one random variable is involved, the condition can often be expressed in terms of one of the random variables. For example, we seek to find the probability of occurrence of the random variable X given that the random variable Y has a given value. If both of these are continuous random variables, the probability that the random variable Y attains a specific value is identically equal to zero. Therefore, we confine the condition not to a specific value, but to an interval of given width Δy. We then extend the concept of Equation (3.50), as condition B involves the random variable Y. Equation (3.50) is therefore rewritten conditioned on the event $\{y < Y \leq y + \Delta y\}$:

$$F_X(x/y < Y \leq y + \Delta y) = \frac{P(X \leq x; y < Y \leq y + \Delta y)}{P(y < Y \leq y + \Delta y)} \qquad (3.51)$$

Without loss of generality, taking the derivative of the above expression and letting the interval Δy approach zero yield the following result:

$$f_X(x/y) = \frac{f_{XY}(x,y)}{f_Y(y)} \tag{3.52}$$

Likewise, the conditional density of Y may be expressed as follows:

$$f_Y(y/x) = \frac{f_{XY}(x,y)}{f_X(x)} \tag{3.53}$$

For those cases where X and Y are statistically independent, the above expressions reduce to the following trivial cases:

$$f_X(x/y) = \frac{f_{XY}(x,y)}{f_Y(y)} = \frac{f_X(x)f_Y(y)}{f_Y(y)} = f_X(x) \tag{3.54}$$

$$f_Y(y/x) = \frac{f_{XY}(x,y)}{f_X(x)} = \frac{f_X(x) \cdot f_Y(y)}{f_X(x)} = f_Y(y) \tag{3.55}$$

Example 3.8 Let X and Y be the random variables introduced in Example 3.3, whose joint probability density function is defined as follows:

$$f_{XY}(x,y) = \begin{cases} \dfrac{4}{225} xy & 0 \le x \le 3, \quad 0 \le y \le 5 \\ 0 & \text{elsewhere} \end{cases} \tag{3.56}$$

Find $f_X(x/y)$ and $f_Y(y/x)$.

We found the marginal probability density functions were

$$f_X(x) = \frac{2}{9}x \quad \text{and} \quad f_Y(y) = \frac{2}{25}y$$

Therefore,

$$f_X(x/y) = \frac{f_{XY}(x,y)}{f_Y(y)} = \frac{\dfrac{4}{225}xy}{\dfrac{2}{25}y} = \frac{4}{225} \cdot \frac{25}{2}x = \frac{2}{9}x \tag{3.57}$$

In this particular case, we know that the random variables, X and Y are independent. Consequently,

$$f_{XY}(x,y) = f_X(x)f_Y(y) \tag{3.58}$$

Therefore, according to Equation (3.57),

$$f_X(x/y) = f_X(x) \tag{3.59}$$

which is confirmed by the result just obtained.
    Likewise, we find that

$$f_X(y/x) = \frac{f_{XY}(x,y)}{f_X(x)} = \frac{\frac{4}{225}xy}{\frac{2}{9}x} = \frac{4}{225} \cdot \frac{9}{2}y = \frac{2}{25}y = f_Y(y) \tag{3.60}$$

Example 3.9 Two random variables X and Y have a joint probability density function of the form

$$f_{XY}(x,y) = \begin{cases} A(x+y) & 0 \le x \le 1,\, 0 \le y \le 1 \\ 0 & \text{elsewhere} \end{cases}$$

(a) Find the value of A.
    We find the value of A by solving the following equation:

$$\int_0^1 \int_0^1 f_{XY}(x,y)\, dx\, dy = \int_0^1 \int_0^1 A(x+y)\, dx\, dy = 1 \tag{3.61}$$

This equals

$$\int_0^1 A(\frac{1}{2}+y)\, dy = A(\frac{y}{2}+\frac{y^2}{2})\Big|_0^1 = 1 \tag{3.62}$$

Therefore $A = 1$.

(b) Find the marginal probability density functions $f_X(x)$, and $f_Y(y)$.

$$f_X(x) = \int_{-\infty}^{\infty} f_{XY}(x,y)\, dy = \int_0^1 (x+y)\, dy = \begin{cases} x+\frac{1}{2} & 0 \le x \le 1 \\ 0 & \text{elsewhere} \end{cases} \tag{3.63}$$

Likewise,

$$f_Y(y) = \begin{cases} y + \dfrac{1}{2} & 0 \le y \le 1 \\ 0 & \text{elsewhere} \end{cases} \tag{3.64}$$

(c) Find the conditional probability density functions $f(x/y)$ and $f(y/x)$:

$$f_X(x/y) = \frac{f_{XY}(x,y)}{f_Y(y)} = \frac{x+y}{y + \dfrac{1}{2}} \tag{3.65}$$

and

$$f_Y(y/x) = \frac{f_{XY}(x,y)}{f_X(x)} = \frac{x+y}{x + \dfrac{1}{2}} \tag{3.66}$$

(d) Are X and Y statistically independent? Why or why not?

$$f_X(x) \cdot f_Y(y) = (x + \frac{1}{2}) \cdot (y + \frac{1}{2}) = xy + \frac{1}{2}x + \frac{1}{2}y + \frac{1}{4} \ne (x+y) = f_{XY}(x,y)$$

Therefore, X and Y are *not* statistically independent.

The *conditional expectation* of random variables is defined in the same manner that expectations were defined earlier in Chapter 2. The conditional expectation of X given that $Y = y$ is

$$E[X/Y = y] = \int_{-\infty}^{\infty} x f_X(x/y)\, dx \tag{3.67}$$

Other moments are likewise defined.

3.7 PROBABILITY DENSITY FUNCTION OF SUM OF TWO RANDOM VARIABLES

This is an important case since it is frequently found in the analysis of physical systems. Consider the case of the random variable X and the random variable Y, and consider the sum of the two random variables yielding the new random variable Z:

$$Z = X + Y \tag{3.68}$$

Suppose we know the joint probability density function $f_{XY}(x,y)$, and we wish to determine the probability density function of Z. We can analyze this problem by considering the probability distribution function of Z and then taking the derivative of $F_Z(z)$ to find the probability density function $f_Z(z)$:

$$F_Z(z) = P(Z \le z) \tag{3.69}$$

Since we know the relationship among the three random variables, we can make the following substitution:

$$F_Z(z) = P(Z \le z) = P(X + Y \le z) \tag{3.70}$$

The probability distribution function defined by Equation (3.70) may now be found by integrating the joint probability density function over the region to the left of the line $X + Y$, as shown in Figure 3.3.

Therefore $F(z)$ is found as follows:

$$F_Z(z) = \int_{-\infty}^{\infty} \int_{-\infty}^{z-y} f_{XY}(x,y) \, dx \, dy \tag{3.71}$$

Taking the derivative with respect to the random variable Z, we obtain

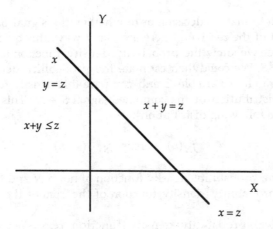

Figure 3.3 Representation of the region of integration for the case of the sum of two random variables $Z = X + Y$.

$$f_Z(z) = \frac{d[F_Z(z)]}{dz} = \int_{-\infty}^{\infty} f_{XY}(z-y,y)\, dy \tag{3.72}$$

A case that is of particular interest results when the random variables X and Y are statistically independent. In such a case we can simplify Equation (3.72) as follows:

$$f_Z(z) = \int_{-\infty}^{\infty} f_{XY}(z-y,y)\, dy = \int_{-\infty}^{\infty} f_X(z-y) f_Y(y)\, dy \tag{3.73}$$

This last integral is the convolution of the two probability density functions $f_X(x)$ and $f_Y(y)$. We therefore have found the following basic theorem:

Theorem 3.1. Given that two random variables X and Y with probability density functions $f_X(x)$ and $f_Y(y)$ are statistically independent such that $f_{XY}(x,y) = f_X(x)f_Y(y)$, the new random variable $Z = X + Y$ will have a probability density function $f_Z(z)$ equal to the convolution of the two probability density functions $f_X(x)$ and $f_Y(y)$.

Suppose we are measuring the output of a receiver Z. The input to the receiver may be a random noise N, or random noise N with an additive independent random signal S. This may be illustrated as follows:

$$Z = N \quad \text{or} \quad Z = S + N$$

We may wish to make a decision as to whether the signal S is present or not in the output of the receiver. There are many ways this question may be answered. Suppose we knew the probability density function of the noise N and of the signal S. We could then estimate the probability density function of the measured random variable Z and compare this density function with the theoretical distribution of the combined signal $S + N$. This implies the comparison of the following distributions:

$$f_N(n) \quad \text{versus} \quad f_{S+N}(s+n)$$

where $f_N(n)$ is the probability density function of noise N and $f_{S+N}(s + n)$ is the theoretical probability density function of the sum of the two random variables s and n.

The theoretical probability density function $f_{S+N}(s + n)$ equals the convolution of $f_S(s)$ and $f_N(n)$ since it was stipulated that s and n were independent random variables. Therefore, comparison of the two distributions may lead one to conclude the presence or absence of the signal at the output of the receiver. In practice, one may want to use a more robust procedure for the detection of a signal in the presence of additive independent noise. Nevertheless this simple example illustrates the usefulness of the basic theorem which specifies the probability density function of the sum of two independent random variables.

The convolution of two functions is often calculated using Fourier transforms. However, in the context of obtaining the probability density function of the sum of two random variables, the Fourier transform approach to computing the convolution of the individual probability density functions can be interpreted in terms of characteristic functions. It was shown in Chapter 2 that the expression to obtain the characteristic function of a random variable was nearly a Fourier transform operation. Therefore many of the properties of the Fourier transform can be applied to characteristic functions. One particular property of interest here is that the Fourier transform of a convolution is the product of the individual Fourier transforms. Therefore the characteristic function of the random variable Z obtained as the sum of two random variables X and Y can be written as

$$\Phi_Z(\omega) = \Phi_X(\omega)\Phi_Y(\omega) \tag{3.74}$$

where $\Phi_X(\omega)$, $\Phi_Y(\omega)$, and $\Phi_Z(\omega)$ are characteristic functions of random variables X, Y, and Z, respectively. This result can be easily extended to the sum of n random variables. If a random variable Z is given as a sum of the random variables $X_1 + X_2 + \cdots + X_n$ such that

$$Z = X_1 + X_2 + \cdots + X_n \tag{3.75}$$

then

$$\Phi_Z(\omega) = \Phi_{X_1}(\omega)\Phi_{X_2}(\omega) \cdots \Phi_{X_n}(\omega) \tag{3.76}$$

This property is useful in computing the probability density functions of sums of random variables.

Example 3.10 Let us further consider the preceding case with the following example: Let the noise term be a random variable uniformly distributed between −2 and +2 V. Let the signal be a random variable independent of the noise and uniformly distributed between +3 and +4 V. Determine the theoretical probability density function of $f(s + n)$. The probability density functions of the two random variables are square functions as in Figure 3.4.

By inspection, we determine that the amplitude of $f_N(n)$ is $1/4$, and the amplitude of $f_S(s)$ equals 1. The probability density functions have the following equations:

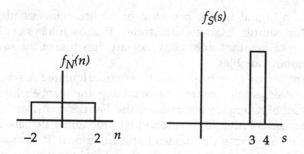

Figure 3.4 Probability density functions of the two random variables for Example 3.10.

$$f_N(n) = \begin{cases} 1/4 & -2 \le n \le 2 \\ 0 & \text{elsewhere} \end{cases}$$

and

$$f_S(s) = \begin{cases} 1 & 3 \le s \le 4 \\ 0 & \text{elsewhere} \end{cases}$$

Since the two random variables are statistically independent, the probability density function of their sum $f(s + n)$ is found by obtaining the convolution of the respective probability density functions. This convolution is illustrated by Figure 3.5.

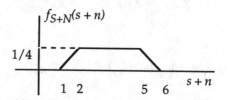

Figure 3.5 Illustration of the convolution of the respective probability density functions for Example 3.10.

We may conclude that

$$\int_1^6 f_{S+N}(s+n)\, d(s+n) = 1$$

Example 3.11 We now consider the case of the sum of two independent random variables which are gaussian distributed, and we prove that the resulting probability density function is also gaussian. Let X be a gaussian distributed random variable with zero mean and standard deviation equal to 1. Let Y be a gaussian distributed random variable independent of X, with zero mean and standard deviation also equal to 1. Consider now a new random variable formed by the sum of the two random variables X and Y. The respective probability functions are the following:

$$f_X(x) = \frac{1}{\sqrt{2\pi}} \exp\left(\frac{-x^2}{2}\right) \quad -\infty < x < \infty \tag{3.77}$$

and

$$f_Y(y) = \frac{1}{\sqrt{2\pi}} \exp\left(\frac{-y^2}{2}\right) \quad -\infty < y < \infty \tag{3.78}$$

Since the random variables are independent, the probability density function of their sum is the convolution of the two respective probability functions:

$$f_{X+Y}(x+y) = f_Z(z) = \int_{-\infty}^{\infty} f_X(z-y)f_Y(y)\, dy \tag{3.79}$$

We now evaluate this convolution integral:

$$f_Z(z) = \frac{1}{2\pi} \int_{-\infty}^{\infty} \exp\left(\frac{-(z-y)^2}{2}\right) \exp\left(\frac{-y^2}{2}\right) dy \tag{3.80}$$

which equals

$$f_Z(z) = \frac{1}{2\pi} \int_{-\infty}^{\infty} \exp\left[\frac{-\left(z^2 - 2zy + 2y^2\right)}{2}\right] dy \tag{3.81}$$

In order to solve this integral, we need to complete the square inside the exponential function:

$$f_Z(z) = \frac{1}{2\pi} \int_{-\infty}^{\infty} \exp\left[\frac{-\left(\dfrac{z^2}{2} + \dfrac{z^2}{2} - 2zx + 2y^2\right)}{2}\right] dy \tag{3.82}$$

and this expression equals:

$$f_Z(z) = \frac{1}{2\pi} \int_{-\infty}^{\infty} \exp\left\{ \frac{-\left[\frac{z^2}{2} + \left(\frac{z}{\sqrt{2}} \right)^2 - 2zx + \left(y\sqrt{2} \right)^2 \right]}{2} \right\} dy \qquad (3.83)$$

We now complete the square to obtain:

$$f_Z(z) = \frac{1}{2\pi} \int_{-\infty}^{\infty} \exp\left(\frac{-z^2}{4} \right) \exp\left\{ \frac{-\left[\left(\frac{z}{\sqrt{2}} \right)^2 - 2zx + \left(y\sqrt{2} \right)^2 \right]}{2} \right\} dy \qquad (3.84)$$

The function of the variable z may be taken outside the integral:

$$f_Z(z) = \frac{1}{2\pi} \exp\left(\frac{-z^2}{4} \right) \int_{-\infty}^{\infty} \exp\left\{ \frac{-\left[\left(\frac{z}{\sqrt{2}} \right)^2 - 2zx + \left(y\sqrt{2} \right)^2 \right]}{2} \right\} dy \qquad (3.85)$$

The expression inside the exponential function now forms the following perfect square:

$$f_Z(z) = \frac{1}{2\pi} \exp\left(\frac{-z^2}{4} \right) \int_{-\infty}^{\infty} \exp\left[\frac{-\left(\frac{z}{\sqrt{2}} - y\sqrt{2} \right)^2}{2} \right] dy \qquad (3.86)$$

which equals

$$f_Z(z) = \frac{1}{\sqrt{2\pi}} \exp\left(\frac{-z^2}{4}\right) \frac{1}{\sqrt{2\pi}} \int_{-\infty}^{\infty} \exp\left[\frac{-\left(\frac{z}{\sqrt{2}} - y\sqrt{2}\right)^2}{2}\right] dy \qquad (3.87)$$

Using a change of variable, we can evaluate the integral.
Let

$$\omega = y\sqrt{2} - \frac{z}{\sqrt{2}}$$

Then

$$dy = \frac{1}{\sqrt{2}} d\omega$$

and

$$f_Z(z) = \frac{1}{\sqrt{2\pi}} \exp\left(\frac{-z^2}{4}\right) \frac{1}{\sqrt{2}} \left[\frac{1}{\sqrt{2\pi}} \int_{-\infty}^{\infty} \exp\left(\frac{-\omega^2}{2}\right) d\omega\right] \qquad (3.88)$$

The expression within the brackets equals 1. Consequently, the probability density function of the random variable z equals

$$f_Z(z) = \frac{1}{\sqrt{2\pi}\sqrt{2}} \exp\left(\frac{-z^2}{4}\right) \qquad (3.89)$$

The probability density function $f_Z(z)$ is gaussian distributed. Note that

$$\mu_z = 0 \quad \text{and} \quad \sigma = \sqrt{2}$$

The above result can be obtained in a simpler way by using characteristic functions. It is shown in this section that the characteristic function of the sum of two random variables is given by the product of the characteristic function of the individual random variables. Therefore we can write

$$\Phi_Z(\omega) = \Phi_X(\omega)\Phi_Y(\omega) \qquad (3.90)$$

where $\Phi_X(\omega) = \Phi_Y(\omega) = e^{-\omega^2/2}$.

Substituting these in (3.90), we obtain

$$\Phi_Z(\omega) = e^{-2\omega^2/2} \tag{3.91}$$

From the results of Section 2.7 we know that the above characteristic function corresponds to a gaussian random variable with mean 0 and variance 2, which concurs with the result obtained from direct convolution.

For those cases where the two independent random variables X and Y have means and standard deviations different from zero, it can be proved that the resulting random variable Z, given by the sum of the two random variables, will have a mean equal to the sum of the two means and a variance equal to the sum of the two variances. This proof is left for the reader as a (simple?) exercise. In the preceding example, the convolution integral was handled in a fairly simple manner. In typical physical situations the convolution integral is not handled that easily, and one must resort to more traditional methods of computing the convolution between two probability density functions.

Example 3.12 Consider now the case of two independent random variables X and Y with probability density functions of the following type:

$$f_X(x) = \begin{cases} a\exp(-ax) & 0 \le x \le \infty \\ 0 & \text{elsewhere} \end{cases}$$

and

$$f_Y(y) = \begin{cases} b\exp(-by) & 0 \le y \le \infty \\ 0 & \text{elsewhere} \end{cases}$$

Consider now the random variable Z given by the sum of the two random variables X and Y:

$$Z = X + Y$$

As noted before, the probability density function of the random variable Z is given by the convolution of the two probability density functions $f_X(x)$ and $f_Y(y)$. We will use a graphical technique to help us visualize the convolution of these two functions. The probability density functions $f_X(x)$ and $f_Y(y)$ are shown in Figure 3.6.

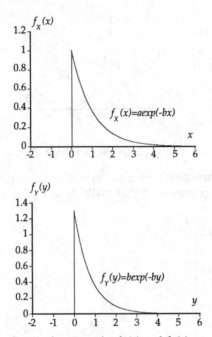

Figure 3.6 Probability density functions for $f_X(x)$ and $f_Y(y)$.

The probability density function of the random variable Z is given by the convolution of the two probability density functions $f_X(x)$ and $f_Y(y)$. The formula we wish to evaluate is given by Equation (3.79), and it is represented by the graph in Figure 3.7. Consider first the function $f_Y(-y)$:

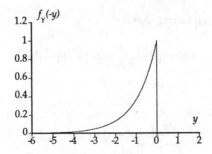

Figure 3.7 Probability density function for $f_Y(-y)$.

Figure 3.8 illustrates the evaluation of Equation (3.79) for the case when $Z = 0$. Since there is no overlap between $f_X(z - y)$ and $f_Y(y)$, the value of the convolution is 0 when $z \leq 0$.

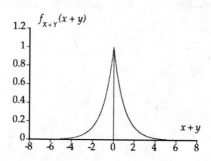

Figure 3.8 Evaluation of Equation (3.79) for the case when $Z = 0$.

The value of the convolution for the case when the random variable Z is greater than 0 may be represented pictorially as in Figure 3.9.

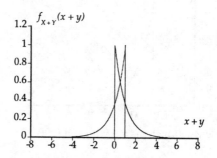

Figure 3.9 Evaluation of Equation (3.79) for the case when $Z = 1$.

For those cases when $Z > 0$, $f_Z(z)$ has the following value:

$$f_Z(z) = \int_0^z a\exp\left[-a(z - y)\right] b\exp(-by)\, dy$$

$$= ab\exp(-az)\int_0^z \exp\left[(a - b)y\right]\, dy \tag{3.92}$$

which equals

$$f_Z(z) = \frac{ab}{a - b}\left[\exp(-bz) - \exp(-az)\right] \quad a \neq b \tag{3.93}$$

Figure 3.10 is a graphical illustration of the probability density function $f_Z(z)$ for the case when $a = 1.0$ and $b = .75$.

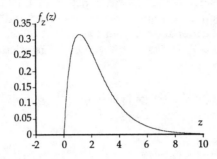

Figure 3.10 Graphical illustration of the probability density function $f_Z(z)$ for the case when $a = 1.0$ and $b = .75$.

When $a = b$, the problem has the following solution:

$$f_Z(z) = \int_0^z a \exp\left[-a(z-y)\right] a \exp(-ay)\, dy \tag{3.94}$$

which equals

$$f_Z(z) = a^2 \exp(-az) \int_0^z \exp(+ay) \exp(-ay)\, dy = a^2 \exp(-az)\, z \tag{3.95}$$

Figure 3.11 illustrates the probability density function of $f_Z(z)$ for the case when $a = b$ and $a = 1$.

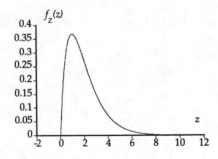

Figure 3.11 Probability density function of $f_Z(z)$ for the case when $a = b$ and $a = 1$.

3.8 EXPECTED VALUE OF SUMS OF RANDOM VARIABLES

In Section 3.7 we considered the density function of the sum of two random variables. In this section, we consider the expected value of the sums of random variables. Consider the random variable Z which is the sum of two random variables X and Y. The expected value of Z is calculated as follows:

$$E[Z] = E[X + Y] = \int\int_{-\infty}^{\infty} (x + y) f_{XY}(x, y)\, dx\, dy \tag{3.96}$$

which equals

$$E[Z] = \int\int_{-\infty}^{\infty} x f_{XY}(x, y)\, dx\, dy + \int\int_{-\infty}^{\infty} y f_{XY}(x, y)\, dx\, dy \tag{3.97}$$

These integrals are evaluated as follows:

$$E[Z] = \int_{-\infty}^{\infty} x f_X(x)\, dx + \int_{-\infty}^{\infty} y f_Y(y)\, dy = \mu_x + \mu_y \tag{3.98}$$

This result is easily extended for the case of more than two random variables. In such cases, the expected value of the sum of random variables equals the sum of the expected values of the respective random variables. If $X_1, X_2, \ldots, X_N$ are N random variables, then $E[X_1 + X_2 + \cdots + X_N] = E[X_1] + E[X_2] + \cdots + E[X_N]$. This is a direct consequence of the linearity property of the expectation operation.

3.9 VARIANCE OF SUM OF RANDOM VARIABLES

In this section we consider the variance of the sum of random variables. Consider the sum of two random variables X and Y:

$$Z = X + Y$$

The expected value of the sum equals

$$E[Z] = \mu_X + \mu_Y \tag{3.99}$$

Consider now the variance of the sum:

$$\sigma_Z^2 = E\left[(Z - \mu_z)^2\right] \tag{3.100}$$

We now substitute for $Z = X + Y$ and $\mu_Z = \mu_X + \mu_Y$:

$$\sigma_Z^2 = E\left[\left[(X+Y) - (\mu_X + \mu_Y)\right]^2\right] \tag{3.101}$$

$$\sigma_Z^2 = E\left[\left[(X - \mu_X) + (Y - \mu_Y)\right]^2\right] \tag{3.102}$$

which yields

$$\sigma_Z^2 = E\left[(X - \mu_X)^2\right] + E\left[(Y - \mu_Y)^2\right] + 2E[(X - \mu_X)(Y - \mu_Y)] \tag{3.103}$$

This can be written in terms of the variances of X *and* Y and the covariances as

$$\sigma_Z^2 = \sigma_X^2 + \sigma_Y^2 + 2\,\mathrm{cov}(X, Y) \tag{3.104}$$

By using the definition of correlation coefficient in (3.36), this can be further written as

$$\sigma_Z^2 = \sigma_X^2 + \sigma_Y^2 + 2\sigma_X \sigma_Y \rho_{XY} \tag{3.105}$$

If the random variables are uncorrelated, this simplifies to

$$\sigma_Z^2 = \sigma_X^2 + \sigma_Y^2$$

This result can be extended to the sum of more than two independent random variables. If $X_1, X_2, \ldots, X_N$ are uncorrelated random variables with variances given by

$$\sigma_{X_1}^2, \sigma_{X_2}^2, \ldots, \sigma_{X_N}^2$$

respectively, then the variance of the sum of the random variables is given by the sum of the variances of the individual random variables:

$$\mathrm{var}[X_1 + X_2 + \cdots + X_N] = \sigma_{X_1}^2 + \sigma_{X_2}^2 + \cdots + \sigma_{X_N}^2 \tag{3.106}$$

3.10 BRIEF INTRODUCTION TO CENTRAL LIMIT THEOREM

The central limit theorem is concerned with the probability density function of the sum of large numbers of independent random variables. Loosely described, the central limit theorem states that the probability density function of the sum of a large number of random variables tends toward the normal probability density function, under certain condition, called *Lindeberg conditions* (Papoulis, 1984). We state the theorem without proof and learn it to the extent that we use it for practical applications.

Consider the sequence of mutually independent random variables $X_1, \ldots, X_n$ with the corresponding means $\mu_1, \ldots, \mu_n$ and variances given by

$$\sigma_1^2 \cdots \sigma_n^2$$

Let us form a new random variable

$$Y_n = \frac{1}{\sqrt{n}}(X_1 + X_2 + \cdots + X_n) \tag{3.107}$$

Then

$$E[Y_n] = \frac{1}{\sqrt{n}}(E[X_1] + E[X_2] + \cdots + E[X_n]) \tag{3.108}$$

which yields

$$\mu_{Y_n} = \frac{1}{\sqrt{n}}(\mu_1 + \mu_2 + \cdots + \mu_n) \tag{3.109}$$

since the random variables are independent

$$\sigma_{Y_n}^2 = \frac{1}{n}\left(\sigma_1^2 + \sigma_2^2 + \cdots + \sigma_n^2\right) \tag{3.110}$$

The various σ_i satisfy the condition that

$$\sigma_i^2 \ll n\sigma_{y_n}^2 \tag{3.111}$$

Then the probability density function of the normalized sum Y_n converges to the gaussian distribution with mean μ_{Y_n} and variance $\sigma^2_{Y_n}$, for large n or

$$f_{Y_n}(y_n) \rightarrow \frac{1}{\sqrt{2\pi}\sigma_{y_n}} \exp\left[\frac{-(y_n - \mu_{y_n})^2}{2\sigma^2_{y_n}}\right] \quad -\infty < y_n < \infty \quad (3.112)$$

This theorem has also been shown to be valid for special cases of dependent random variables. However, the majority of the applications of the central limit theorem are related to independent random variables. This theorem can also be used to generate gaussian distributed random variables. The central limit theorem deals with a type of convergence called *convergence in distribution*. However, the convergence works better near the center of the distribution (near the mean) than near the tails of the distribution.

In the following we prove a special case of the central limit theorem, where we consider the sum of random variables with identical distributions with zero mean and unit variance. Let

$$Y_n = \frac{1}{\sqrt{n}} \sum_{i=1}^{n} X_i \qquad (3.113)$$

where the means and variances equal 0 and 1, respectively, for all the random variables.

The proof is done using characteristic functions where we show that in the limit as $n \rightarrow \infty$, the characteristic function of Y_n approaches that of a normal probability density function with mean of zero and variance of unity.

The characteristic function of Y_n can be written as

$$\Phi_{Y_n}(\omega) = E\left[e^{j\omega Y_n}\right] \qquad (3.114)$$

However, Y_n is the sum of $X_1, X_2, \ldots, X_n$, with a scaling factor of $1/\sqrt{n}$. Therefore,

$$\Phi_{Y_n}(\omega) = \left[\Phi_{\tilde{X}}(\omega)\right]^n \qquad (3.115)$$

where $\Phi_{\tilde{X}}(\omega)$ is the characteristic function of $X_i/\sqrt{n}$.

It is easy to see from the definition of characteristic function that

$$\Phi_{\tilde{X}}(\omega) = \Phi_X(\omega/\sqrt{n}) \qquad (3.116)$$

Substituting Equation (3.116) into (3.115), we obtain

$$\Phi_{Y_n}(\omega) = \left[\Phi_X(\omega/\sqrt{n})\right]^n \tag{3.117}$$

We now expand this term in a finite Taylor series about the origin ($\omega = 0$) and obtain

$$\Phi_X\left(\frac{\omega}{\sqrt{n}}\right) = 1 + \Phi_X^1(0)\omega + \frac{1}{2}\Phi_X^2(0)\omega^2 + R(\omega) \tag{3.118}$$

where $\Phi_X^1(0)$ and $\Phi_X^2(0)$ are the first- and second- order derivatives and $R(\omega)$ is the remainder term, given by

$$R(\omega) = \frac{1}{3!}\Phi_X^3\left(\frac{\xi}{\sqrt{n}}\right)\omega^3 \tag{3.119}$$

where ξ is some point in the interval $[0, \omega]$. TheTaylor expansion of $\Phi_X(\omega/\sqrt{n})$ can be further simplified by carrying out the differentiation of $\Phi_X(\omega/\sqrt{n})$, and using the information on mean and variance of X, to

$$\Phi_X(\omega/\sqrt{n}) = 1 - \frac{1}{2n}\omega^2 + \frac{\tilde{R}(\omega)}{n\sqrt{n}} \tag{3.120}$$

where $\tilde{R}(\omega)$ is the remainder term.

Taking the logarithm of both sides of (3.117), we obtain

$$\ln\left[\Phi_{Y_n}(\omega)\right] = n\left[\ln\Phi_X\left(\frac{\omega}{\sqrt{n}}\right)\right] \tag{3.121}$$

Substituting (3.120) into (3.121), we can write

$$\ln\left[\Phi_{Y_n}(\omega)\right] = n\ln\left[1 - \frac{1}{2n}\omega^2 + \frac{\tilde{R}(\omega)}{n\sqrt{n}}\right] \tag{3.122}$$

The value of n can be chosen large enough that

$$\left|\frac{-\omega^2}{2n} + \frac{\tilde{R}(\omega)}{n\sqrt{n}}\right| < 1$$

and we expand the right side of (3.122) as a series of the form $\ln(1 + \delta)$, when

$|\delta| < 1$, to get

$$\ln\left[\Phi_{Y_n}(\omega)\right] = \frac{-\omega^2}{2} + \text{terms involving } \frac{1}{\sqrt{n}}, \frac{1}{n}, \cdots \qquad (3.123)$$

Taking the limit as $n \to \infty$ on both sides of (3.123), we obtain

$$\lim_{n\to\infty} \ln\left[\Phi_{Y_n}(\omega)\right] = \frac{-\omega^2}{2} \qquad (3.124)$$

which can be written in the form

$$\lim_{n\to\infty} \Phi_{Y_n}(\omega) = e^{-\omega^2/2} \qquad (3.125)$$

which is the characteristic function of a normal probability density function with zero mean and unit variance.

Therefore, we have shown that the characteristic function of Y_n is the same as that of a normal probability density function. When we can exchange the limiting operation in Equation (3.124) and the Fourier integral operation that defines the characteristic function, we can extend this proof to say that the probability density function of Y_n converges to a normal density function.

3.11 ESTIMATE OF POPULATION MEAN, EXPECTED VALUE, AND VARIANCE OF ESTIMATE

When dealing with random data, we are often asked to determine statistical parameters associated with the process that is being observed to record the data. This can sometimes be an impossible task if we only have available a record of limited duration of the process. We may have to estimate, for example, the underlying probability density function of the process. This is dealt with in a later section of this chapter. At first we may simply have to estimate the mean of the data and decide whether this is a good estimate of the mean of the process. From Equation (2.9) we know that

$$\mu = E[X] = \int_{-\infty}^{\infty} xf_X(x)\, dx \qquad (3.126)$$

It would be practically impossible to evaluate the preceding equation for several reasons. First, we may not know $f_X(x)$. Second, we have finite samples of a process. We therefore can compute the average of the data as follows:

$$\hat{\mu} = \frac{1}{n}\sum_{i=1}^{n} X_i \tag{3.127}$$

where the caret in $\hat{\mu}$ implies that this is an estimate of the mean value of the random variable.

Intuitively, Equation (3.127) makes sense. Not knowing anything about the underlying population of the random variable X, we determine an *estimate* of its mean value by adding the n values and then dividing by the number of values. Note that $\hat{\mu}$ is an estimate. Since it is only an estimate, it is possible that if we were to measure the random variable X at a different time, or for a different n, the estimate of its mean value $\hat{\mu}$ would also change. Here $\hat{\mu}$ is obtained by adding several random variables and scaling the sum. Therefore, $\hat{\mu}$ is a *random variable*. Since it is a random variable, it has a probability density function, a mean value, variance, higher-order moments, etc.

It is interesting to note that if the n samples of the random variable X are independent, then the probability density function of

$$\frac{(\hat{\mu} - \mu)\sqrt{n}}{\sigma}$$

approaches a gaussian probability density function with $\mu = 0$ and $\sigma = 1$ as $n \to \infty$. This is a direct consequence of the central limit theorem explained earlier. We are going to require that the samples be independent from each other. Consider now the random variable $\hat{\mu}$. We find its mean value as follows:

$$E[\hat{\mu}] = E\left[\frac{1}{n}\sum_{i=1}^{n} X_i\right] = \frac{1}{n}\sum_{i=1}^{n} E[X_i] \tag{3.128}$$

and we recognize

$$\frac{1}{n}\sum_{i=1}^{n} E[X_i] = \frac{n\mu}{n} = \mu \tag{3.129}$$

which is the mean value of the process.

Equation (3.129) implies that our choice of estimating function, namely, Equation (3.127), is a good choice because our estimate has the same mean as the random variable. We also say that $\hat{\mu}$, as defined by Equation (3.114), is an *unbiased estimate* of μ, the mean value of the process. Any estimator whose expected value is the same as the quantity being estimated is said to be an unbiased estimator.

As stated earlier, $\hat{\mu}$ is a random variable. As we have just calculated, its expected value is μ, the mean value of the process. We will now proceed to find the *variance* of the estimate. The variance of $\hat{\mu}$ can be written as

$$\text{var}(\hat{\mu}) = \text{var}\left(\frac{1}{n}\sum_{i=1}^{n} X_i\right) \tag{3.130}$$

This can be rewritten as

$$\text{var}(\hat{\mu}) = \frac{1}{n^2}\text{var}\left(\sum_{i=1}^{n} X_i\right) \tag{3.131}$$

We know from Equation (3.106) that variance of the sum of N independent random variables is given by the sum of the variances of the random variables. Therefore,

$$\text{var}(\hat{\mu}) = \frac{1}{n^2}\sum_{i=1}^{n} \text{var}(X_i) \tag{3.132}$$

However, $\text{var}(X_i)$ is σ^2 which is the variance of the underlying distribution. Therefore, $\text{var}(\hat{\mu})$ can be simplified to

$$\text{var}(\hat{\mu}) = \frac{n}{n^2}\sigma^2 = \frac{\sigma^2}{n} \tag{3.133}$$

This important result relates the variance of the estimate of the mean of the random variable to the variance of the random variable and the number of data points. In other words, the variance of the estimate of the mean value of the random variable is directly proportional to the variance of the random variable and is inversely proportional to the number of points used to make the estimate. One can also say that the variance of the estimate goes up as the variance of the random variable goes up. However, the variance will decrease if more points are used in the estimate.

One question that may be asked relates to the confidence we place in an estimate such as $\hat{\mu}$. The confidence is based on the measure of error between the estimated value and the true value of μ. Based on the central limit theorem, $\hat{\mu}$ is approximately gaussian distributed. Assume now that we have n samples of the random variable X. Further assume that the variance of the population is known and equals σ^2. The estimate of the mean of the population is given by $\hat{\mu}$. We now ask, What is the probability that the interval

$$\left(\hat{\mu} - \frac{2\sigma}{\sqrt{n}}, \hat{\mu} + \frac{2\sigma}{\sqrt{n}}\right)$$

includes the population mean μ? The main assumption is that we know the population variance. In order for the above interval to include the population mean, the following event must occur:

$$\left(\hat{\mu} - \frac{2\sigma}{\sqrt{n}} < \mu < \hat{\mu} + \frac{2\sigma}{\sqrt{n}}\right)$$

This event may be written as follows:

$$-2 < \frac{\sqrt{n}(\hat{\mu} - \mu)}{\sigma} < 2$$

Note that

$$\frac{\sqrt{n}(\hat{\mu} - \mu)}{\sigma}$$

has mean equal to 0 and standard deviation equal to 1 (see Example 2.8). We are now able to evaluate the probability that the given interval includes the population mean by computing

$$\int_{-2}^{2} \frac{1}{\sqrt{2\pi}} \exp\left(\frac{-x^2}{2}\right) dx = .954$$

Therefore the probability that the population mean μ is in the prescribed interval is .954. This interval

$$\left(\hat{\mu} - \frac{2\sigma}{\sqrt{n}}, \ \hat{\mu} + \frac{2\sigma}{\sqrt{n}}\right)$$

is called the 95.4 percent confidence interval for the mean μ. The 95 percent confidence interval is given by

$$\left(\hat{\mu} - \frac{1.96\sigma}{\sqrt{n}}, \ \hat{\mu} + \frac{1.96\sigma}{\sqrt{n}}\right)$$

The length of the confidence interval is given by

$$\frac{4\sigma}{\sqrt{n}}$$

which is a constant for a given n. Once σ and n are known, then the length of the confidence interval will be known. Note that for a given population variance, the interval may be made as large or as small as desired by taking appropriate samples.

Example 3.13 The variance of a gaussian distributed random variance is known to be .9738. For $n = 10, 20, 30$, and 40, the mean value was estimated as .1554, −.38063, −.33492, −.33833, respectively. For the different values of n, define the lengths of the .95 confidence interval.

Based on the previous discussion, the .95 confidence interval equals

$$\left(\hat{\mu} - \frac{1.96\sigma}{\sqrt{n}} < \mu < \hat{\mu} + \frac{1.96\sigma}{\sqrt{n}} \right)$$

For $n = 10$ and $\hat{\mu} = .1554$, this formula equals

$$-.456 < \mu < .767$$

For $n = 20$ and $\hat{\mu} = -.38063$,

$$-.813 < \mu < .052$$

For $n = 30$ and $\hat{\mu} = -.33492$,

$$-.688 < \mu < .018$$

For $n = 40$ and $\hat{\mu} = -.33833$,

$$-.644 < \mu < .032$$

These intervals are called the 95 percent confidence interval for μ.

3.12 ESTIMATE OF POPULATION VARIANCE

Another parameter we can estimate from the measured data that is of equal importance to the estimate of the mean value is the estimate of the variance of the random variable. Proceeding in a similar way, we can postulate that an estimate of the variance of the random process is given by

$$\hat{\sigma}^2 = \frac{1}{n}\sum_{i=1}^{n}(X_i - \hat{\mu})^2 \tag{3.134}$$

where the $\hat{\sigma}^2$ symbol implies that this is an estimate of the variance of the process. The quantity $\hat{\sigma}^2$ is a random variable; let us calculate its expected value:

$$E\left[\hat{\sigma}^2\right] = E\left[\frac{1}{n}\sum_{i=1}^{n}(X_i - \hat{\mu})^2\right] = \frac{1}{n}E\left[\sum_{i=1}^{n}(X_i - \hat{\mu})^2\right] \tag{3.135}$$

We now proceed to evaluate this expression. Note that

$$\frac{1}{n}E\left[\sum_{i=1}^{n}(X_i - \hat{\mu})^2\right] = \frac{1}{n}E\left[\sum_{i=1}^{n}\left[(X_i - \mu) - (\hat{\mu} - \mu)\right]^2\right]$$

$$= \frac{1}{n}E\left[\sum_{i=1}^{n}\left[(X_i - \mu)^2 - 2(X_i - \mu)(\hat{\mu} - \mu) + (\hat{\mu} - \mu)^2\right]\right] \tag{3.136}$$

where we have added and subtracted μ. Now we note that

$$\frac{1}{n}E\left[\sum_{i=1}^{n}(X_i - \mu)^2\right] = \frac{1}{n}\sum_{i=1}^{n}E\left[(X_i - \mu)^2\right] = \frac{1}{n}\sum_{i=1}^{n}\sigma^2 = \frac{n\sigma^2}{n} = \sigma^2 \tag{3.137}$$

and

$$\frac{1}{n}E\left[\sum_{i=1}^{n}-2(X_i - \mu)\cdot(\hat{\mu} - \mu)\right] = \frac{-2}{n}E\left[(\hat{\mu} - \mu)\cdot\sum_{i=1}^{n}(X_i - \mu)\right] \tag{3.138}$$

which equals

$$= \frac{-2}{n}E\left[(\hat{\mu} - \mu)n(\hat{\mu} - \mu)\right] = \frac{-2n}{n}E\left[(\hat{\mu} - \mu)^2\right] = -2\frac{\sigma^2}{n} \tag{3.139}$$

and

$$\frac{1}{n}E\left[\sum_{i=1}^{n}(\hat{\mu}-\mu)^2\right]=\frac{1}{n}E\left[n(\hat{\mu}-\mu)^2\right]=E\left[(\hat{\mu}-\mu)^2\right]=\frac{\sigma^2}{n} \qquad (3.140)$$

Therefore,

$$\frac{1}{n}E\left[\sum_{i=1}^{n}(X_i-\hat{\mu})^2\right]=\sigma^2-2\frac{\sigma^2}{n}+\frac{\sigma^2}{n}=\sigma^2-\frac{\sigma^2}{n}$$

$$=\frac{n\sigma^2-\sigma^2}{n}=\frac{(n-1)\sigma^2}{n} \qquad (3.141)$$

which is different from σ^2.

This result implies that our estimator for the variance of the sample is *biased*. In other words, the estimator given by Equation (3.134) is not an unbiased estimator. It can be prove that if we change our estimator slightly in Equation (3.134), we will obtain an unbiased estimate of the variance of the sample. The unbiased estimator is given by the following equation:

$$\hat{\sigma}^2=\frac{1}{n-1}\sum_{i=1}^{n}(X_i-\hat{\mu})^2 \qquad (3.142)$$

As may be observed, the only difference is that the denominator of Equation (3.135) has the term $n-1$ rather than the term n. Obviously, for large n, $n>30$, the two estimates will yield nearly the same result.

3.13 COMPUTER GENERATION OF UNIFORM RANDOM VARIABLES

Most computer systems have subroutines in their systems called *pseudorandom number generators*. A typical name for such routines is RAN. The most commonly used method of generating random numbers makes use of the following recursion:

$$x_{n+1}=(ax_n+c) \ (\text{mod } m) \qquad (3.143)$$

where a, c = positive constants
mod m = modulus of m, usually taken to be word length of machine
x_n = present value of the number
x_{n+1} = next value of the number

The standard notation in number theory $x = a$ (mod m) means that $x - a$ is divisible by m. Algorithms like this have been around for a long time (see Hamming, 1962) and are called *linear congruential operators*. The recurrence of Equation (3.136) will eventually repeat itself, and, consequently, repeat the sequence it is producing. One major disadvantage of this method is that it is possible for the numbers to be sequentially correlated on successive calls. It is therefore imperative that the numbers generated by such a method be carefully inspected for problems such as trends.

The very popular International Mathematical and Statistical Library (IMSL) has a number of subroutines that produce random numbers, including ggubs, ggubt, and ggu. Also, the book *Numerical Recipes, The Art of Scientific Computing* by Press et al. (1987) has a good collection of routines for the generation of uniform deviates.

3.14 COMPUTER GENERATION OF GAUSSIAN DISTRIBUTED RANDOM VARIABLES USING THE CENTRAL LIMIT THEOREM

Now that we are able to generate a uniform distribution of numbers, we are in a position to generate other distributions. gaussian deviates will be fairly easy to generate if we use the central limit theorem. The sum of very few random numbers from a distribution usually gives a good approximation to a gaussian distribution (see Hamming, p. 389). It takes the sum of as few as 10 numbers, uniformly distributed, to generate a good approximation to a gaussian distributed number.

Since we need to invoke the central limit theorem, we need to add a series of numbers the sum of which will be gaussian distributed. If we add 12 uniformly distributed random numbers between 0 and 1, the mean of the sum will be 6. The variance of the sum will be $\sigma^2 = n/12$, and since $n = 12$, the variance will equal 1. As we mentioned in Section 3.10, this procedure does not work well near the tails of the gaussian distribution. Another commonly used algorithm to generate a gaussian distribution is the Box-Mueller technique. This technique is introduced in Computer Example 3.5.

3.15 TESTING THE EQUIVALENCE OF A PROBABILITY DENSITY FUNCTION OF EXPERIMENTAL DATA TO A THEORETICAL DENSITY FUNCTION

The Run Test

As engineers, we may be called upon to make measurements of a particular process. The data are often collected by a digital computer, and then we are asked to make inferences concerning their statistical properties. For example, suppose we have a vibration test bed, where instruments are tested for loose parts. The vibration bed has pneumatic actuators that provide the random vibration to the setup. Two three-axis accelerometers measure the output of the system, and they also serve to provide a point of reference for the closed-loop controller. The controller averages the output of the accelerometers, determines if the average is between specified limits, and closes the loop by controlling the input to the table of the actuators. It would be interesting to verify if the output of each of the accelerometers were within a specified range, and whether the output was random and fit an anticipated probability density function. This is a hypothetical case, but it illustrates the type of data we are likely to have as a basis for decisions.

One of the first questions we have to answer is whether the data that we collected from this experiment are independent of one another. If the collected data are not independent, the measurements may be considered suspect, and the measurement scheme has to be investigated to check why the data are not independent. The first test we use on the data will tell us whether the data, as collected, are random. This test, called the *run test*, is distribution-free, i.e., it does not depend on the distribution of the data it is testing. This test serves to determine the independence of sampled values from collected data.

The first step in performing the run test is to divide the measured data into two equal classes. One easy way to accomplish this is by using the *median* of the measured data. The median of *n* data values is calculated by ordering the set of data values by magnitude and then selecting the middle value for *n* odd, or the average of the two middle values for *n* even. The data are then scored as being above or below the median. The actual run test is illustrated via the following example.

Example 3.14 Consider the following 20 numbers selected from a table of random numbers. These numbers should pass the run test:

50, 55, 72, 78, 17, 86, 87, 90, 25, 64, 56, 37, 49, 41, 23, 76, 73, 48, 8, 28

The median of these numbers is 52.5. Since we have an even number of data values, we have taken the median to be the mean of the two middle values of the set of ordered data. We now score the data as being above or below the median, entering a − sign if the number is below the median and a + if the number is above the median. Scoring the data given above yields the following sequence:

50, 55, 72, 78, 17, 86, 87, 90, 25, 64, 56, 37, 49, 41, 23, 76, 73, 48, 8, 28
 −   +   +   +   −   +   +   +   −   +   +   −   −   −   −   +   +   −   −   −

We now define a run as a sequence of identical observations that is preceded or followed by other symbols. For this particular case, we identify the following as runs:

50, 55, 72, 78, 17, 86, 87, 90, 25, 64, 56, 37, 49, 41, 23, 76, 73, 48, 8, 28

   1   2   3   4   5   6   7   8   9

This set of data, as determined by the median, has 9 runs. The number of runs in a particular sequence gives an indication of whether the data are random. If there are a small number of runs, it may mean that similar observations are getting clustered. For example, suppose that we had only 2 runs. This may mean that there is a low-frequency phenomenon affecting the data. If there are too many runs, this may indicate a faster, perhaps periodic, change in the data being measured. Once we have the number of runs, the data are evaluated in comparison to prespecified run behavior of random data with a hypothesis that the data is from random samples. This process is called *hypothesis testing*. Suppose we want to make the statement that the data are part of random samples. This statement or hypothesis is tested experimentally. Similar to the 95 or 99 percent confidence interval for the estimation of the mean, the hypothesis testing is also done at different levels of significance. Suppose we wish to test at the 5 percent level of significance whether we can accept the hypothesis that the observations are independent. We find that for a 5 percent confidence interval the number of observations falling on either side of the median is 10, and that the number of observed runs must fall in the following interval:

$$r_{10;95\%} < r_{observed} < r_{10;5\%} = 6 < r_{observed} < 16$$

In our particular case, since the number of observed runs equals 9, we can assume that the observations are independent.

The Chi-Square Goodness-of-Fit Test

Suppose that we would like to test the numbers obtained from such a random variable to see whether we can state that they are gaussian distributed. One way to make such a test is to (1) determine whether these are indeed independent observations of the random variable and (2) use the chi-square test to determine if the numbers' deviations from some predetermined probability density function. We use the following process:

1. The data we have observed, N points long, have an unknown distri-bution

$G(x)$. We wish to determine whether $G(x)$ equals some predetermined distribution $F(x)$. In other words, we wish to conduct a test that will tell us whether we can accept the hypothesis that $G(x) = F(x)$. In the present case $G(x)$ is the unknown distribution óf the N data points, and $F(x)$ is a known distribution. Group the N data points into M bins. The number M depends on the number of samples we have obtained.

2. To perform a chi-square goodness-of-fit test, we compute a random variable which is a function of the observed random variables (without any unknown parameters), called the *test statistic*. This test statistic is defined by

$$\chi^2 = \sum_{i=1}^{M} \frac{(g_i - f_i)^2}{f_i} \qquad (3.144)$$

where g_i is the observed frequency of the data in the Mth data interval, and f_i is the expected frequency in the Mth interval.

This test statistic has a χ^2 distribution with n degrees of freedom as $N \rightarrow \infty$. The number of degrees of freedom is related to the sample size and the number of parameters estimated from the sample. For example, assume that $F(x)$ is gaussian and that we are dealing with M intervals of data. Although there are M intervals, once we have calculated $M - 1$ intervals, the observed frequencies on the Mth interval are fixed. We have therefore "lost" a degree of freedom. For a gaussian distributed random variable, two additional constraints are necessary, since we must estimate the mean and the variance of the observed data. Therefore, for a gaussian distributed random variable where the observed data have been divided into M bins

$$n = M - 3 \text{ degrees of freedom}$$

The obtained statistic is then compared with the critical values of a χ^2 distribution at a specified level. The calculations required to test this statistic are illustrated in the following examples.

Example 3.15 Suppose that we conduct an experiment with a fair die. This die is tossed a total of 120 times, and the following data are obtained:

Face number	1	2	·3	4	5	6
Observed frequency	18	20	22	19	21	20

Since we are dealing with a fair die, we wish to test the observed distribution against a uniform distribution. For a uniform distribution (and a fair die) the data should be equiprobable, meaning that the expected frequency at each face should be 20. We now compute the test statistic:

$$\chi^2 = \sum_{i=1}^{6} \frac{(g_i - f_i)^2}{f_i}$$

$$= \frac{(18-20)^2}{20} + \frac{(20-20)^2}{20} + \frac{(22-20)^2}{20} + \frac{(19-20)^2}{20} + \frac{(21-20)^2}{20} + \frac{(20-20)^2}{20}$$

$$= .5$$

The number of degrees of freedom n is in this case equal to 5 since there are 6 intervals. Note that we do not need to calculate the mean and the variance for this case. In a χ^2 table at the 5 percent level, the tabulated value is smaller than the required one. We are therefore able to accept the hypothesis that this is a fair die.

Example 3.16 In the case of testing for normality the procedure is similar, with the exception that one must choose whether to select class intervals that will have equal expected observations or to select class intervals of equal width (thereby yielding unequal expected observations in each interval). Several suggestions have been made in the literature concerning the number of class intervals if the equal expected number of observations procedure is used. The table given below shows one such suggestion:

N	200	400	600	800	1000
M	16	20	24	27	30

where N is the number of observations ($M \sim \sqrt{N}$) and M is the suggested number of bins to be used.

This is the minimum number of class intervals suggested for the given number of points when we use the 5 percent confidence level. We will now illustrate the use of the chi-square test when testing for normality. Suppose that we have 200 points which have been independently obtained and it is desired to test the hypothesis that these numbers are gaussian distributed.

The following numbers were generated using a RAN function which produces uniformly distributed random numbers between 0 and 1. Twelve such numbers were added, and the value 6.0 was subtracted from each value. Each value was then multiplied by 3 in order to increase the variance of the data.

Intervals of Z		Intervals of X
Interval 1 [$-\infty, -1.53$]	$\rightarrow$	[$-\infty, -4.26$]
Interval 2 [$-1.53, -1.15$]	$\rightarrow$	[$-4.26, -3.14$]
Interval 3 [$-1.15, -.89$]	$\rightarrow$	[$-3.14, -2.38$]
Interval 4 [$-.89, -.67$]	$\rightarrow$	[$-2.38, -1.74$]
Interval 5 [$-.67, -.49$]	$\rightarrow$	[$-1.74, -1.21$]
Interval 6 [$-.49, -.32$]	$\rightarrow$	[$-1.21, -.71$]
Interval 7 [$-.32, -.16$]	$\rightarrow$	[$-.71, -.24$]
Interval 8 [$-.16, 0$]	$\rightarrow$	[$-.24, .23$]
Interval 9 [$0, .16$]	$\rightarrow$	[$.23, .70$]
Interval 10 [$.16, .32$]	$\rightarrow$	[$.70, 1.17$]
Interval 11 [$.32, .49$]	$\rightarrow$	[$1.17, 1.67$]
Interval 12 [$.49, .67$]	$\rightarrow$	[$1.67, 2.20$]
Interval 13 [$.67, .89$]	$\rightarrow$	[$2.20, 2.84$]
Interval 14 [$.89, 1.15$]	$\rightarrow$	[$2.84, 3.60$]
Interval 15 [$1.15, 1.53$]	$\rightarrow$	[$3.60, 4.72$]
Interval 16 [$1.53, \infty$]	$\rightarrow$	[$4.72, \infty$]

Table 3.1 Table showing the limits for z, and the transformed limits.

The data were then tested using the run test to determine whether there was an underlying trend to the data. The *median* of the numbers was found to be equal to .229, the *mean* equal to .229, and the *standard deviation* equal to 2.936. The number of runs about the median was found to be 92; therefore, we conclude that the observations are independent. Note that for $N = 200$, with the observations divided equally about the median, the limits of the run test are 88 and 113. If the number of runs falls between these limits, we conclude that the observations are independent. Since we have a total of 200 observations, we will follow the suggestion made earlier and divide the interval into 16 classes with equal probability. The idea, then, is to find the limits of the 16 intervals within a gaussian probability density function having equal areas. The limits for these areas will be valid for a gaussian distributed random variable having $\mu = 0$ and $\sigma = 1$. In our particular case the random samples have a mean of .229 and a standard deviation of 2.9356. Consequently, the limits must take these numbers into account by performing the following transformation:

$$x = \hat{\sigma} z + \hat{\mu}$$

	Sample observations	Expected observations	$\lvert g - f\rvert$	$\lvert g - f\rvert^2/f$
Interval 1	14	12.5	1.5	.18
Interval 2	14	12.5	1.5	.18
Interval 3	9	12.5	3.5	.98
Interval 4	14	12.5	1.5	.18
Interval 5	17	12.5	4.5	1.6
Interval 6	6	12.5	6.5	3.38
Interval 7	15	12.5	2.5	.5
Interval 8	11	12.5	1.5	.18
Interval 9	10	12.5	2.5	.5
Interval 10	17	12.5	4.5	1.62
Interval 11	14	12.5	1.5	.18
Interval 12	8	12.5	4.5	1.62
Interval 13	14	12.5	1.5	.18
Interval 14	11	12.5	1.5	.18
Interval 15	13	12.5	.5	.02
Interval 16	13	12.5	.5	.02
Total	200	200.0		11.52

Table 3.2 Table showing the calculation of the chi-square value.

Here $\hat{\sigma}$ is the data standard deviation, and $\hat{\mu}$ is the mean of the data. The random variable z is a gaussian distributed random variable with $\mu = 0$ and $\sigma = 1$. First, the limits for z, then the transformed limits are given in Table 3.1.

The random numbers were sorted in ascending order. The number of observations that fall within each interval, or g_i in Equation (3.144), and the expected number of observations that should fall within each interval if the data were gaussian distributed are given below. Also given are the rest of the values needed to compute Equation (3.144).

We can see from Table 3.2 that the χ^2 value is 11.52. The number of degrees of freedom is now calculated as follows: There are a total of $K = 16$ intervals; we now subtract 2 degrees of freedom, since we had to calculate the mean and the variance from the data. We subtract another degree of freedom since, once we know the number of observations in the first 15 intervals, the number of observations in the 16th interval is fixed. Consequently, the number of degrees of freedom is equal to 13. At the 5 percent confidence level and with 13 degrees of freedom, $\chi^2 \leq 22.36$. Since the calculated value is 11.52, the hypothesis of normality is accepted at the 5 percent level of significance.

The Kolmogoroff-Smirnoff Test

The Kolmogorov-Smirnoff (K-S) test is another procedure which can be used to determine whether a set of observations is from a prespecified continuous distribution. The K-S test is often used to detect deviations from a prespecified distribution and is particularly useful when the sample size is small. Just as with the chi-square test, we seek to test whether the data we have measured, N points long, belong to an unknown distribution $G(x)$ and whether we can accept the hypothesis that $G(x)$ equals some prespecified distribution $F(x)$. In the present case $G(x)$ is the unknown distribution of the N data points, and $F(x)$ is a known distribution.

The procedure is simple. Given a sample of N observations, one determines the following statistic:

$$D = \max|F(x) - G(x)| \tag{3.145}$$

where $G(x)$ is the sample cumulative distribution and $F(x)$ is the predetermined cumulative distribution. One then tests whether the value D exceeds a critical value in the K-S table. Often, the mean and variance of $G(x)$ are not known and must be determined from the data at hand. When N is small, the chi-square test may yield questionable results. A classic paper by Lilliefors (1967) arrived at a table of critical values for D when one is testing the hypothesis that the measured data are normally distributed with unknown mean and variance. These must be estimated from the data. We now illustrate the use of the K-S test with the following two examples.

Example 3.17 A set of 30 numbers was constructed by calling a typical RAN function and then using the central limit theorem to construct a hypothetical set of numbers that are approximately gaussian distributed. The 30 numbers are

1.6375, −.62194, .45484, −1.0209, −.93947, .15868, 1.5446, −.080282, −.31020, .82311, .51504, −.71454, .13264, −.77693, −.92907, .14282, −2.2608, −2.7259, .54301, −.096875, .19121, 1.2848, .32185, 1.3082, .79608, −1.4439, −.29512, −1.5404, −1.5552, .46353

These numbers have a mean of −.166, and standard deviation of 1.065. The numbers were tested for independence using the run test, and a total of 17 runs were found; this passes the test for independence at the 5 percent level. The critical D is given by the equation $D = \max |F(x) - G(x)|$ and equals .0077. Comparing the result to the critical D, .161 for $n = 30$ at the 5 percent level, we cannot reject the hypothesis that the data are gaussian distributed. Figure 3.12 compares the theoretical (gaussian) distribution with the measured distribution.

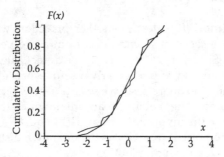

Figure 3.12 Comparison of the theoretical (gaussian) distribution function with the measured distribution.

Example 3.18 We again use the K-S test to test the hypothesis that a set of numbers generated by the RAN function is gaussian distributed. The set of numbers is given below:

1.1579, −2.7218, −.40407, .49773, 1.2581, −.40833, 1.5282, .026621, .36743, −1.4166, 1.0186, −.43683, −2.2177, 1.1813, .13648, 1.3459, 1.2890, −2.1638, −2.8879, .083694, −1.1094, −2.6381, −1.0328, 1.6660, −2.8362, −2.7645, −1.3033, −1.9428, −.94326, −1.2769

These numbers have a mean of −.5649 and standard deviation of 1.479. The numbers were tested for independence using the run test, and a total of 14 runs were found, which passes the test for independence at the 5 percent level. Figure 3.13 compares the theoretical (gaussian) distribution with the measured distribution.

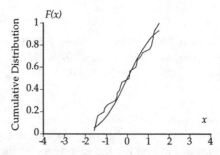

Figure 3.13 Comparison of the theoretical (gaussian) distribution function with the measured distribution.

The critical D was found to be .094 and, therefore, compared to the critical D in the K-S table, which is 0.161 at the 5 percent level, we accept the hypothesis that the data are gaussian distributed.

3.16 CHAPTER SUMMARY

The joint cumulative distribution function for the case of two random variables is given by

$$F_{XY}(x,y) = P(X \le x, \ Y \le y)$$

The joint cumulative distribution function of two random variables has the following properties:

1. $0 \le F_{XY}(x, y) \le 1$
2. $\lim F_{XY}(\infty, \infty) = 1$
3. $F_{XY}(x, y)$ is nondecreasing
4. $F_{XY}(-\infty, y) = F_{XY}(x, -\infty) = 0$
5. $F_X(x) = F_{XY}(x, \infty) = P(X \le x, \ Y \le \infty) = P(X \le x)$
6. $F_Y(y) = F_{XY}(\infty, y) = P(X \le \infty, \ Y \le y) = P(Y \le y)$

The joint probability density function is defined (if it exists) as follows:

$$f_{XY}(x, y) = \frac{\partial^2 F_{XY}(x, y)}{\partial x \, \partial y}$$

The properties associated with the joint probability density function are as follows:

Property 1:

$$f_{XY}(x, y) \ge 0$$

Property 2:

$$\int\!\!\int_{-\infty}^{\infty} f_{XY}(x, y) \, dx \, dy = 1$$

Property 3:

$$f_X(x) = \int_{-\infty}^{\infty} f_{XY}(x, y) \, dy$$

$$f_Y(y) = \int_{-\infty}^{\infty} f_{XY}(x, y) \, dx$$

The mnth moment of the X and Y random variables is defined as follows:

$$E\left[X^m Y^n\right] = \int\limits_{-\infty}^{\infty}\!\!\int x^m y^n f_{XY}(x,y)\, dx\, dy$$

The covariance between the random variables X and Y is defined as the $m = 1$, $n = 1$ joint central moments:

$$\text{cov}(X,Y) = \int\limits_{-\infty}^{\infty}\!\!\int (x - \mu_x)(y - \mu_y) f_{XY}(x,y)\, dx\, dy$$

The correlation coefficient between two random variables is defined as follows:

$$\rho_{XY} = \frac{\text{cov}(X,\,Y)}{\sigma_x \sigma_y} = \frac{E[XY] - \mu_x \mu_y}{\sigma_x \sigma_y}$$

The probability density function of the sum of two random variables when the random variables are statistically independent is calculated as follows:

$$f_Z(z) = \int_{-\infty}^{\infty} f_{XY}(z - y,\, y)\, dy = \int_{-\infty}^{\infty} f_X(z - y) f_Y(y)\, dy$$

The expected value of the sum of two random variables is calculated as follows:

$$E[Z] = E[X + Y] = \int\limits_{-\infty}^{\infty}\!\!\int (x + y) f_{XY}(x,y)\, dx\, dy$$

The variance of the sum of two random variables is calculated as follows:

$$\sigma_Z^2 = \sigma_X^2 + \sigma_Y^2 + 2\,\text{cov}\,(X,Y)$$

If the random variables are uncorrelated, this simplifies to

$$\sigma_Z^2 = \sigma_X^2 + \sigma_Y^2$$

The estimate of the population mean is

$$\hat{\mu} = \frac{1}{n}\sum_{i=1}^{n} X_i$$

and the variance of this estimate is given by

$$\text{var}(\hat{\mu}) = \frac{\sigma^2}{n}$$

The unbiased estimate of the population variance is calculated as follows:

$$\hat{\sigma}^2 = \frac{1}{n-1}\sum_{i=1}^{n}\left(X_i - \hat{\mu}\right)^2$$

3.17 PROBLEMS

1. Given the joint probability density function for two random variables X and Y:

$$f(x,y) = \begin{cases} kx(1+y) & 0 < x \le 2,\ 0 < y \le 1 \\ 0 & \text{elsewhere} \end{cases}$$

(a) What is the value of k?

(b) What is the joint distribution function?

(c) Find the marginal density function $f_Y(y)$.

(d) What is the probability that $y > 0$?

2. Two random variables X and Y have joint probability density function of the form

$$f(x,y) = \begin{cases} \frac{3}{2}(x^2 + y^2) & 0 \le x \le 1,\ 0 \le y \le 1 \\ 0 & \text{elsewhere} \end{cases}$$

(a) What is the probability that both x and y are larger than .5?

(b) Find the mean value of X.

(c) Find the standard deviation of X.

(d) Find the correlation $E[XY]$.

3. Two random variables X and Y have a joint probability density function of the form

$$f(x,y) = \begin{cases} Ke^{-(2x+3y)} & x > 0, \ y > 0 \\ 0 & \text{elsewhere} \end{cases}$$

(a) What is the value of K?

(b) Are X and Y statistically independent? Why?

(c) Find the conditional probability density function $f(x/y)$.

4. Given two random variables X and Y, each uniformly distributed between $+1$ and -1 with a joint probability density function

$$f(x,y) = \begin{cases} 0.25 & -1 \le x \le 1, \ -1 \le y \le 1 \\ 0 & \text{elsewhere} \end{cases}$$

define another random variable $Z = X + Y$

(a) X and Y are independent. Why? Find the marginal density functions $f_X(x)$ and $f_Y(y)$.

(b) Find the mean values, mean square values, and variances of X and Y. Find also the correlation coefficient between X and Y.

(c) Find the mean value of the random variable Z, its mean square value, and its variance.

(d) Find and sketch the probability density function $f_Z(z)$.

5. A random signal X can be observed only in the presence of independent, additive noise N. The observed quantity is $Y = X + N$. The joint probability density function of X and Y is

$$f(x,y) = \begin{cases} K(x^3+1)(y^4+2) & 0 \le x \le 1,\ 0 \le y \le 1 \\ 0 & \text{elsewhere} \end{cases}$$

If the observed value of Y is $y = .5$, what is the best estimate of X?

6. Two random variables X and Y are related by

$$Y = 2X - 1$$

and X is uniformly distributed from 0 to 1. Find the correlation coefficient for X and Y.

7. A random variable X has a mean value of 10 and a variance of 36. Another random variable Y has a mean value of -5 and a variance of 64. The correlation coefficient for X and Y is $-.25$. Find the variance of the random variable $Z = X + Y$.

8. Two random variables X and Y have a joint probability density function of

$$f(x,y) = \begin{cases} 2e^{-(x+2y)} & x \ge 0,\ y \ge 0 \\ 0 & \text{elsewhere} \end{cases}$$

Let $Z = X + Y$.

(a) Determine the probability density function $f_Z(z)$.

(b) Determine the probability distribution function $F_Z(z)$.

9. Two random variables X and Y have means and variances of

$$\mu_X = 1 \quad \sigma_X^2 = 1 \quad \mu_Y = 0 \quad \sigma_Y^2 = 2$$

The correlation coefficient of these two random variables is $\rho = .5$.

(a) Find the mean of $X + Y$.

(b) Find the variance of $X + Y$.

(c) Find $E[XY]$.

(d) Find $E[(X + Y)(X - Y)]$.

10. Variable X is a random variable that is uniformly distributed between -3 and $+3$. Variable Y is another random variable that is uniformly distributed between -6 and $+6$. Also, X and Y are statistically independent and are added to obtain $Z = X + Y$.

   (a) Sketch (with numerical values labeled) the probability density function for Z.

   (b) Find the correlation coefficient ρ associated with the random variables Z and X.

11. In the circuit below, X and Y may be considered to be statistically independent random variables.

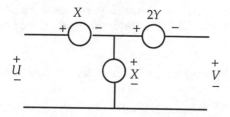

   Now X is a random variable with mean 1 and variance 9; Y is a random variable with mean 0 and variance 4. What is the correlation coefficient of U and V?

12. Generate a surface plot of the joint probability density function of X and Y where X and Y are jointly gaussian distributed. The standard deviations of X and Y are 1 and 3, respectively, and their mean values are -1 and 1. (*Hint*: Use MATHCAD or MATLAB.)

13. Compute the correlation coefficient between the file named *corr1* and a set of random numbers. Convince yourself that your program works correctly by first computing the correlation between corr1 *and* corr1. What should the correlation coefficient be? Place the cursor on the computing equation for the random numbers and recompute the worksheet. What happens to the correlation coefficient? Repeat for all the files named *corr**. (*Hint*: Use MATHCAD or MATLAB.)

14. Estimate the mean value and the standard deviation of the files named *corr**. (*Hint*: Use MATHCAD or MATLAB.)

15. Generate 100 gaussian distributed random variables. Use the central limit theorem approach for this problem. The mean of the numbers should be 1.5 and the standard deviation 1.0. Plot a histogram of the numbers generated. (*Hint*: Use MATHCAD or MATLAB)

16. Generate 100 uniformly distributed random numbers.

    (a) Determine whether these numbers pass the run test.

    (b) Regenerate the random numbers by recomputing the worksheet. Do they always pass the run test?

    (c) Establish some correlation between adjacent numbers and determine now whether they pass the run test.

    (*Hint*: Use MATHCAD or MATLAB.)

17. Generate 200 uniformly distributed random numbers. Adjust their mean value to 1.0 and the standard deviation to 0.0. Using the chi-square goodness-of-fit test, determine whether these numbers come from a gaussian distribution. (*Hint*: Use MATHCAD or MATLAB.)

18. Repeat Problem 17 using the Kolmogoroff-Smirnoff goodness-of-fit test. (*Hint*: Use MATHCAD or MATLAB.)

19. The data file *eqk* is data from an earthquake on the California coast. For the first 100 values of the data, plot the data and compute the mean and variance. Determine whether the data pass the run test. If the data pass the run test, determine if they are gaussian distributed.

20. For the same data of Problem 19, compute the mean and the variance, using a window of time. Choose a window that is 20 points long and does not overlap with adjacent windows. Plot the mean and the variance as a function of the window number. What deductions can you make about the data?

3.18 COMPUTER EXAMPLES

MATHCAD Computer Example 3.1
THE GAUSSIAN JOINT PROBABILITY DENSITY FUNCTION

This document will aid in your understanding of the joint probability density function. We will use the gaussian density function of two variables X and Y which will come from independent populations. For independent random variables X and Y we can state the joint probability density function as

$$f(x, y) = f(x)f(y)$$

where $f(x)$ is the marginal density function of X and $f(y)$ is the marginal density function of Y. We know the general form of the gaussian density function, so given that X and Y are independent, we can state the gaussian joint probability density function as

$$p(x, y) := \frac{1}{2 \cdot \pi \cdot \sigma_1 \cdot \sigma_2} \cdot \exp\left[\frac{-1}{2.0} \cdot \left[\frac{(x - xmean)^2}{(\sigma_1)^2} + \frac{(y - ymean)^2}{(\sigma_2)^2} \right] \right]$$

$\sigma_1 \equiv 1.0$ standard deviation of x

$xmean \equiv 1.0$ mean of x

$\sigma_2 \equiv 1.0$ standard deviation of y

$ymean \equiv 0.0$ mean of y

Now we will generate the surface plot which will be displayed below. Try changing the mean and standard deviation of x and y and observe the changes in the plot. Watch the ranges of +5 and –5 on x and y so that your plot will not be off the screen.

$$M_{i,j} := p\left(x_i, y_j\right)$$

$N \equiv 20$

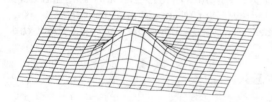

M

$i \equiv 0 \,..$

$j \equiv 0 \,..$

$x_i = -5 + .5 \cdot i$

$y_j = -5 + .5 \cdot$

MATHCAD Computer Example 3.2
CORRELATION BETWEEN TWO RANDOM VARIABLES

In this document we will demonstrate the calculation of the sample correlation coefficient r. This is an estimate of the population correlation coefficient rho based on sampled data. This coefficient will give us a measure of the strength of association or likeness between two random variables. We will calculate r for a few different pairs of data samples. The samples are located in the files corr1, corr2, corr3, corr4, and corr5, we can try any pair of these. Suppose we wish to find the correlation between two random variables X and Y and we have 30 sample points from each process. First, we read in the data:

$i := 0 .. 29$

We read in data files:

$Y_i := \textbf{READ}(\text{corr5})$

$X_i := \textbf{READ}(\text{corr4})$

First, let's look at the two data sets plotted side by side; the X values are the solid line, and the Y values have plus signs:

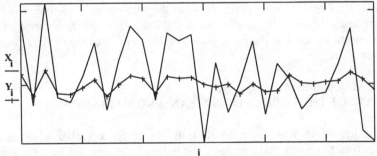

We will need the sample means; also for comparison the standard deviation of each sample is shown below.

$mx := \textbf{mean}(X)$

$mx = 1.2$

stdev(X) = 5.0

my := mean(Y)

my = 0.0

stdev(Y) = 0.9

Then the sample correlation coefficient is defined as follows:

$$j := 0..29 \qquad r := \frac{\displaystyle\sum_{j} \left(X_j - mx\right) \cdot \left(Y_j - my\right)}{\sqrt{\displaystyle\sum_{j} \left(X_j - mx\right)^2 \cdot \sum_{j} \left(Y_j - my\right)^2}}$$

Value of the sample correlation coefficient:

r = 0.662

Now r by definition will be between -1 and $+1$: $-1.0 \le r \le 1.0$. A value of 0 means that the two variables are uncorrelated, a value of 1.0 means that they are perfectly correlated, and a value of -1 means that they are negatively correlated. Values in between these indicate degrees of partial correlation either positive or negative. Note in this first example the waveforms are very similar in shape but not in amplitude. Their mean and standard deviation are quite different, yet they are perfectly correlated ($r = 1$). Try changing the files to be read in, using corr1 through corr5; observe the waveforms and their r values. Try comparing corr4 with corr5; you should see that the first parts of the waveforms are similar but the latter parts are different. This results in a positive partial correlation.

MATHCAD Computer Example 3.3
ESTIMATE OF THE POPULATION MEAN AND VARIANCE

In this example we will compute an estimate of the population mean and variance from a random data sample. We will start by generating 300 random numbers from the computer and saving these in an array.

N := 300   total number of samples generated

i := 1..   index of summation

x_i := rnd(5)   random numbers uniformly distributed between 0 and 5

We can observe what the data look like in the plot below, in which each dot represents one data point:

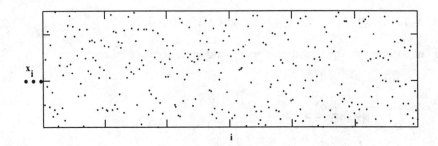

It might be easier to visualize the data by connecting the dots:

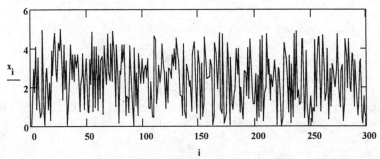

Now since we have all the data values in this given population we can compute the actual population mean and variance directly using all 300 points as follows:

$$\mathbf{xmean} := \frac{1}{N} \cdot \sum_{i} x_i$$

$\mathbf{xmean} = 2.3$

$$\mathbf{xvar} := \frac{1}{N-1} \cdot \sum_{i} \left(x_i - \mathbf{xmean} \right)$$

$\mathbf{xvar} = 2.05$

Now we will make an estimate of the actual population mean and variance. This is done using the same equations as above; however, now we will only use a portion of the original 300 points. This is more typically the case in the real world where we have only a limited sample of the population we wish to estimate.

NSAMP := 200   Number of samples from original 300

j := 1 .. **NSAM**   Index of summation used below

The estimate of the population mean and variance is as follows:

$$\text{estmean} := \frac{1}{\text{NSAMP}} \cdot \sum_j x$$

$$\text{estvar} := \frac{1}{\text{NSAMP}} \cdot \sum_j \left(x_j - \text{estmean} \right)$$

ESTIMATES: **estmean = 2.461**   **estvar = 1.946**

ACTUAL: **xmean = 2.385**   **xvar = 2.052**

Try increasing the value of NSAMP. What happens? You should find that the estimates get closer and closer to the actual population values as you increase NSAMP.

MATHCAD Computer Example 3.4
GENERATION OF GAUSSIAN DISTRIBUTED RANDOM VARIABLES

Use of the Central Limit Theorem

In this document we will show how to generate gaussian distributed random variables from a uniform distribution by using the central limit theorem. We can use this theorem to generate gaussian random variables from a sequence of uniform random variables as follows (see Hamming, 1962):

K := 12   *K* is number of uniform random variables to be added

i := 1 ..

k := 1 .. 75   number of gaussian random variables to be generated

$$X_{i,k} := \text{rnd}(1)$$

The X's are approximately uniformly distributed random variables distributed between 0 and 1. Note the use of the rnd function in MATHCAD. Each Y is now approximately gaussian distributed with a mean equal to 0 and a standard deviation equal to 1.

$$Y_k := \frac{\sum_i X_{i,k} - \frac{K}{2}}{\sqrt{\frac{K}{12}}}$$

As K approaches infinity, Y asymptotically approaches a true gaussian distribution. It is possible to generate a good approximation to the gaussian distribution using $K = 12$. In order to adjust for the desired mean and variance, we do as follows:

$m := 1 .. 75$

$\mu \equiv 0$ desired value of the mean

$\sigma \equiv 1$ desired value of the standard deviation

$YD_m := Y_m \cdot \sigma + \mu$ conversion to the desired mean and variance

Now we plot the 100 approximately gaussian distributed random variables just generated with the above-defined mean and standard deviation. Note that the actual mean and standard deviation are shown to the right of the plot for the numbers generated. To see another set of numbers, press the Calculate Document option in the Math window.

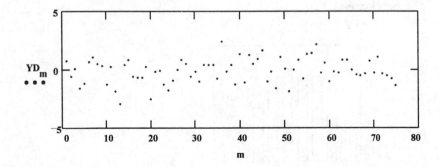

It will be easier to visualize the plot if we connect each of the points with a straight line:

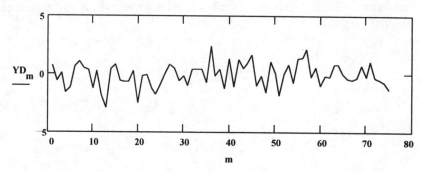

Now we construct a histogram of the points values generated. The intervals about to be defined are valid only for the case where the mean equals 0 and the standard deviation equals 1.

$interval_0 := -2.0$

$interval_1 := -1.6$

$interval_2 := -1.2$

$interval_3 := -0.8$

$interval_4 := -0.4$

$interval_5 := 0.0$

$interval_6 := 0.4$

$interval_7 := 0.8$

$interval_8 := 1.2$

$interval_9 := 1.6$

$interval_{10} := 2.0$

$j := 0..9$

$f := hist(\,interval,\, YD)$

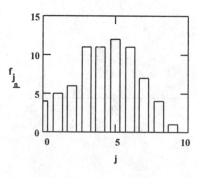

	0
0	4
1	5
2	6
3	11
4	11
5	12
6	11
7	7
8	4
9	1

$f =$

MATHCAD Computer Example 3.5
GENERATION OF GAUSSIAN RANDOM VARIABLES ... The Box-Muller Method

This document will help you to understand the gaussian probability density function and give you some insight into what gaussian distributed random variables are. In addition, you will learn a method for generating gaussian random deviates from uniform deviates.

We start by generating 100 random variables which are gaussian distributed with the following mean and standard deviation:

NRAND $\equiv$ 100   number of variables to generate

$\mu \equiv 0$ value of the mean

$\sigma \equiv 2$ value of the standard deviation, equal to the square root of the variance

Now we define a function to generate normal or gaussian deviates from uniform deviates using the Box-Muller method

$$\text{NORM}(\mu, \sigma, u, v) \equiv \mu + \sigma \cdot \sqrt{-2 \cdot \ln(u)} \cdot \cos(2 \cdot \pi \cdot v)$$

$k \equiv 0 .. \text{NRAND} - 1$ index for array to contain variables

$V_k \equiv \text{NORM}(\mu, \sigma, \text{rnd}(1), \text{rnd}(1))$ array to contain variables generated

We plot the numbers we just generated, one dot for each number generated. The actual mean and standard deviation for this set of numbers are shown to the right of the plot.

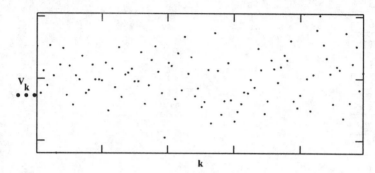

mean (V) = 0.01

stdev (V) = 2.02

Note how the dots are distributed in the plot. To generate another set of 100 points, put the cursor on rnd(1) and hit [F9]. Try changing NRAND to 1000 instead of 100; note again the distribution of points. The points are densest near the mean, zero in our example, and become less and less dense as we move away from the mean.

Now we change the range variable i back to 100. Next we create a histogram of the data points we just generated. The histogram will be displayed as a bar chart where the width of each bar represents a number interval and the height of the bar represents the number of times a variable within that interval was generated. The following definitions and equations will be used to generate the histogram:

NBIN ≡ 20   Number of bins or intervals in the histogram

W ≡ 3   plotting window on either side of m; plot out to 3 standard deviations on each side of mean (see plot below)

We establish the interval covered by each bin:

U ≡ max(V)

L ≡ min(V)

$$H \equiv \frac{U - L}{NBIN}$$

$I \equiv 0 .. NBI$ index to array containing the bins

$BINS_I \equiv L + I \cdot$

$BINS_{last(BINS)} \equiv U + 1$ array containing the bins

We create an array containing the number of deviates falling within the intervals specified in the BINS array, then we normalize it:

$$D \equiv \frac{hist(BINS, V)}{NRAND \cdot H}$$

$J \equiv 0 .. last(BINS) - 1$ index variable for plotting above array

$BINS \equiv BINS + 0.5 \cdot$ center bars on appropriate interval

And finally we can plot the histogram:

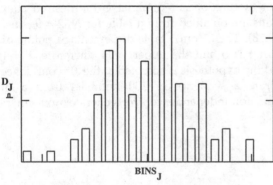

$\mu - W \cdot \sigma = -6$
$\mu + W \cdot \sigma = 6$

Note the heights of the bars. To see another set of 100 numbers, depress the Calculate Document option in the Math window. Try changing the range variable i in the top of the document to 1000, then look at the bar chart again.

MATHCAD Computer Example 3.6
RUN TEST . . . A Test for Independence of Data Samples

In this document we will generate and test a random sequence of numbers for independence by using the run test. Consider that we have a sequence of N

observations of a random variable and we wish to test these values for independence. We first classify each variable into one of two mutually exclusive classes which can be identified with a plus or a minus. This can be done for many distributions by calculating the mean (xmean) and classifying each variable as a plus (+) if it is above the mean value or a minus (−) if it is below it. Then the sequence of plus and minus observations is analyzed for runs. A run is a sequence of identical observations that is followed or preceded by a different observation. For example, consider the sequence of plus and minus observations below:

```
+ − + + − − + − + + + + − − + − − −
^ ^ ^   ^   ^ ^    ^        ^     ^   ^
1 2 3   4   5 6    7        8     9  10
```

There are 10 runs in this sequence of 18 observations. Next we hypothesize that the observations are independent. Then the acceptance region for this hypothesis is defined as

$$r_{9,1-\frac{\alpha}{2}} < r \le r_{9,\frac{\alpha}{2}}$$

Here r is the number of runs observed, alpha is the desired confidence level, and the outer regions are obtained from a table for $N/2 = 9$, since there are 18 observations ($N = 18$). Then from a table of percentage points of a run distribution, the lower limit is 5 and the upper is 14; therefore, $r = 10$ falls between these values, and the hypothesis is accepted at the .05 confidence level.

Now we will generate a sequence of 100 numbers from a uniform distribution and test them for independence. We generate uniform random variables between 0 and 2:

$i := 0 .. 99$

$uni_i := rnd(2)$

We calculate the median value:

$median(uni) = 1.013$

$k := 0 .. 99$

We assign a value of 1 if the number is greater than the median, 0 otherwise.

$count_k := if\left(uni_k > median(uni), 1, 0\right)$

$j := 0 .. 98$

$\mathbf{run_j} := \mathbf{if}\left(\mathbf{count_j} - \mathbf{count_{j+1}} = 0, 0.0, 1.0\right)$

$i := 0..98$

$$\sum_i \mathbf{run_i} = 48$$

The acceptance region for alpha = .05 is between 40 and 61. Calculate the document again. It should pass the test if the computer-generated rnd function has been well designed.

MATHCAD Computer Example 3.7
CHI-SQUARE GOODNESS-OF-FIT TEST ... Test for Normality

This document demonstrates the use and calculation of the chi-square goodness-of-fit test. We start by reading in a data file of 200 points which we wish to test to see if the data came from a normal distribution. We will later test other supplied data sets as well. We will conduct the test at the .05 level of significance, which requires 16 class intervals for 200 data points. We will use the equal-frequency procedure, which for 16 classes requires the probability associated with each interval to be $P = 1/16 = .0625$. We find the required interval limits from a table of areas under standardized normal density function. These limits are shown below in array Z.

$k := 0..199$ number of points to read

$m_k := \mathbf{READ(\,uni1\,)}$ read in data file

We compute the mean and the standard deviation of data:

$\mathbf{mean(\,m\,)} = 0.411$

$\mathbf{stdev(\,m\,)} = 3.39$

We compute the mean and standard deviation of the input data in order to scale the interval limits Z to fit the actual data. This is done below in the equation for X, and the values are displayed to the right of the Z values.

$i := 0..14$

X_i	$Z_i :=$
- 4.776	- 1.53
- 3.488	- 1.15
- 2.607	- 0.89
- 1.861	- 0.67
- 1.251	- 0.49
- 0.674	- 0.32
- 0.132	- 0.16
0.411	0
0.953	0.16
1.495	0.32
2.072	0.49
2.682	0.67
3.428	0.89
4.309	1.15
5.598	1.53

$X_i := stdev(m) \cdot Z_i + mean(m)$

Interval limits from the standard normal table scaled limits to fit actual data. Now we will count the number of occurrences in each interval above by forming a histogram with the above interval limits. We compute the minimum and maximum values of input:

$min(m) = {}^-5.$ data for histogram interval limits

$max(m) = 5.9$

$interval_0 := min(m) - 1.0$ lower limit of histogram

$interval_{16} := max(m) + 1.0$ upper limit of histogram

$j := 1 .. 15$

We assign remaining intervals based on

$interval_j := X_{j-}$ values calculated above for X

$p := 0 .. 15$

We compute histogram and display:

f := hist(interval, m)

f_p

17
18
7
13
14
9
15
4
8
8
12
17
8
19
24
7

$\Sigma f = 200$

Now we can compute the chi-square statistic given that the expected frequency for each interval is 12.5 = 200/16.

$i := 0 .. 15$

$$\text{chisquare} := \sum_i \left[\frac{(12.5 - f_i)^2}{12.5} \right] \qquad \text{formula for chi-square statistic}$$

$\text{chisquare} = 36.8$ \qquad value of chi-square statistic

Now the number of degrees of freedom is $n = K - 3 = 13$, where K is 16 from the number of intervals and there are three different restrictions in this case. One is due to the fact that the frequency in the last class interval is known once the others are computed, and one is for each of the estimates (mean and standard deviation) that had to be made to fit the data to a normal distribution. Given that, we can then look up the acceptance level in a table for the .05 level of significance. The acceptance level is chisquare < 22.36, so the first set of data passes the test for normality at the .05 level of significance.

Now go back to the top of the document and change the input file from normb to norm1. Then scroll down and observe the changes, Do these data pass the test? Also try the input files uni1 and exp1. You can view the input data in the plot below.

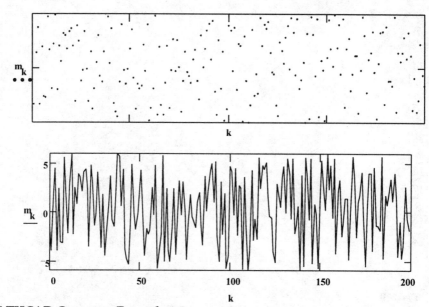

MATHCAD Computer Example 3.8
THE KOLMOGOROV-SMIRNOFF GOODNESS-OF-FIT-TEST . . . A Test for Normality

This document demonstrates the calculation and use of the Kolmogorov-Smirnoff goodness-of-fit-test. We will test several data samples for normality using this test. You should read the section in the text covering this test before proceeding with the document.

We read in data file to be tested

$i := 0 .. 199$

$m_i := READ(norm2)$

We compute mean and standard deviation:

$xmean := mean(m)$

$sd := stdev(m)$

We normalize the data to a mean of 0 and standard deviation of 1 because we will be comparing these data to a theoretical normal distribution of mean 0 and standard deviation 1.

We normalize the input data to a mean of 0 and standard deviation of 1:

$j := 0 .. 199$

$xmean = 0.176$

$sd = 2.735$

$$norm_j := \frac{m_j - xmea}{sd}$$

Next we sort the data into ascending order to form the cumulative distribution function of the data to be tested.

$x := sort(norm)$

$k := 0 .. 199$

Now the cumulative distribution function of the input data can easily be formed as follows:

$$Fo_k := \frac{k + 1}{200}$$

Next we form the theoretical cumulative distribution which we wish to test against, using the values of

x_k

as input to

$Fe_k := cnorm(x_k)$ cnorm(...) is cumulative normal distribution function

Now we can compute the maximum difference between the two distributions. We find the absolute value of the differences:

$$d_k := \left| Fo_k - Fe_k \right|$$

We find the maximum difference:

$\mathbf{D} := \mathbf{max(\ d\ )}$

$\mathbf{D\ =\ 0.03}$ value of K-S statistic

   Next we use the following formula for the .05 level of significance test for normality. If D is less than the following calculated value, then the data cannot be rejected at the level of significance 0.05. To pass the test

$$D\ <\ \frac{\mathbf{0.866}}{\sqrt{\mathbf{200}}} = \mathbf{0.061}$$

A plot of the two distributions is shown below.

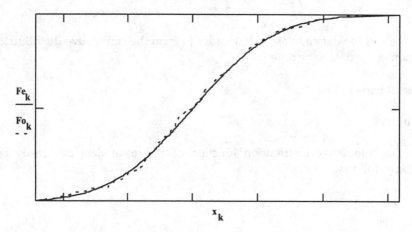

MATLAB Computer Example 3.1
THE GAUSSIAN JOINT PROBABILITY DENSITY FUNCTION . . . of Two Variables

This document will aid in your understanding of the joint probability density function. We will use the gaussian density function of two variables X and Y which will come from independent populations.

   For independent random variables X and Y we can state the joint probability density function as

$$f(x, y) = f(x)f(y)$$

where $f(x)$ is the marginal density function of X and $f(y)$ is the marginal density function of Y.

%===
% Input parameters

```
            meanx=0;
            sigmax=1;
            meany=0;
            sigmay=1;
%=============================================
mean_x=meanx;
sigma_x=sigmay;
mean_y=meany;
sigma_y=sigmay;
x=[];y=[];
inc_x=((mean_x+3*sigma_x)-(mean_x-3*sigma_x))/40;
inc_y=((mean_y+3*sigma_y)-(mean_y-3*sigma_y))/40;
t_x=(mean_x-3*sigma_x):inc_x:(mean_x+3*sigma_x);
t_y=(mean_y-3*sigma_y):inc_y:(mean_y+3*sigma_y);
[X,Y]=meshgrid(t_x,t_y);
s1=subplot(1,1,1);
z=normpdf(X,mean_x,sigma_x).*normpdf(Y,mean_y,sigma_y);
mesh(X,Y,z)
title('Joint Gaussian probability density function','Fontsize',[10]);grid
set(s1,'Fontsize',[10]);clear
```

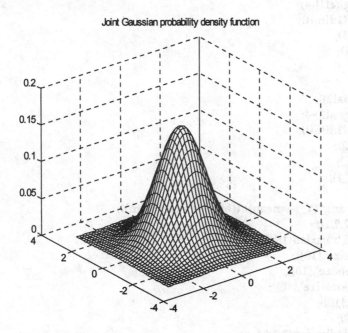

Try changing the mean and standard deviation of x and y and observe the changes in the plot. Watch the ranges of +5 and –5 on x and y so that your plot will not be off the screen.

MATLAB Computer Example 3.2
CORRELATION BETWEEN TWO RANDOM VARIABLES . . .
The Correlation Coefficient

In this document we will demonstrate the calculation of the sample correlation coefficient r. This is an estimate of the population correlation coefficient rho based on sampled data. This coefficient will give us a measure of the strength of association or likeness between two random variables. We will calculate r for a few different pairs of data samples. The samples are located in the files corr1, corr2, corr3, corr4, and corr5; you can try any pair of these. Suppose we wish to find the correlation between two random variables X and Y and we have 30 sample points from each process.

```
%===============================================
%Input parameters
            signal1='corr4.dat';
            signal2='corr5.dat';
%===============================================
eval(['load ',signal1]);
limit=length(signal1)-4;
signal1=signal1(1:limit);
X_i=eval(signal1);
M_x=length(X_i);
xmean=mean(X_i);
xstd=std(X_i);
eval(['load ',signal2]);
limit=length(signal2)-4;
signal2=signal2(1:limit);
Y_i=eval(signal2);
M_y=length(Y_i);
ymean=mean(Y_i);
ystd=std(Y_i);
r=sum((X_i-xmean).*(Y_i-ymean))/((M_x-1)*(xstd*ystd));
clg;s1=subplot(2,2,1);
plot(1:1:M_x,X_i,'r');axisn(1:1:M_x);
xlabel(' i ','Fontsize',[10]);
ylabel('X_i','Fontsize',[10]);
title('X signal','Fontsize',[10]);
set(s1,'Fontsize',[10]);
s2=subplot(2,2,2);
plot(1:1:M_x,Y_i,'r');axisn(1:1:M_x);
xlabel(' i ','Fontsize',[10]);
ylabel('Y_i','Fontsize',[10]);
title('Y signal','Fontsize',[10]);
set(s2,'Fontsize',[10]);
```

```
s3=subplot(2,2,3);
plot(1:1:M_x,X_i,'r',1:1:M_x,Y_i,'k'); axisn(1:1:M_x);
v=axis;
v(3)=min([min(X_i) min(Y_i)]);
v(4)=max([max(X_i) max(Y_i)]);
axis(v);
ylabel('X _i     Y_i ','Fontsize',[10]);
xlabel('i','Fontsize',[10]);
set(s3,'Fontsize',[10]);
s4=subplot(2,2,4);
axis('off');
text(0,1.0,['Xmean value: ',num2str(xmean)],'Fontsize',[10]);
text(0,0.8,['Xstandard deviation: ',num2str(xstd)],'Fontsize',[10]);
text(0,0.6,['Ymean value: ',num2str(ymean)],'Fontsize',[10]);
text(0,0.4,['Ystandard deviation: ',num2str(ystd)],'Fontsize',[10]);
text(0,0.2,['Correlation coefficient: ',num2str(r)],'Fontsize',[10]);
clear;
```

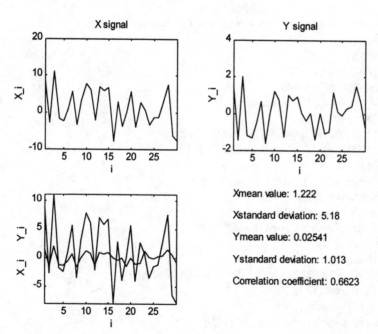

By definition r will be between -1 and $+1$: $-1.0 \leq r \leq 1.0$. A value of 0 means that the two variables are uncorrelated, a value of 1.0 means that they are perfectly correlated, and a value of -1 means that they are negatively correlated. Values in between these indicate degrees of partial correlation either positive or negative.

Note in this first example that the waveforms are very similar in shape but not in amplitude. Their mean and standard deviation are quite different, yet they are perfectly correlated ($r = 1$). Try changing the files to be read in, using corr1 through corr5, and observe the waveforms and their r values. Try comparing corr4 with corr5; the first parts of the waveforms are similar but the latter parts are different. This results in a positive partial correlation.

MATLAB Computer Example 3.3
ESTIMATE OF THE POPULATION MEAN AND VARIANCE

In this example we will compute an estimate of the population mean and variance from a random data sample. We start by generating 300 random numbers from the computer and saving these in an array.

```
%===========================================
% Input parameters
                    N=300;
                    NSAMP=100;
%===========================================
samples=N;
samples1=300;
x=5*rand(samples,1);
xmean=(1/samples)*sum(x);
xvar=sum((x-xmean).^2)/samples;
if NSAMP <= samples
 estmean=(1/NSAMP)*sum(x(1:NSAMP)    );
 estvar=sum((x(1:NSAMP)-estmean).^2)/NSAMP;
clg;s1=subplot(2,2,1);
plot((1:1:samples),x,'r');axisn(1:1:samples);
xlabel('i','Fontsize',[10]);
ylabel('x( i )','Fontsize',[10]);
title('Random numbers','Fontsize',[10]);
set(s1,'Fontsize',[10]);
s2=subplot(2,2,2);
plot((1:1:NSAMP),x(1:NSAMP),'.r');axisn(1:1:NSAMP);
xlabel('i','Fontsize',[10]);
ylabel('x( i )','Fontsize',[10]);
title('Random  numbers','Fontsize',[10]);
set(s2,'Fontsize',[10]);
s3=subplot(2,2,3);
axis('off')
text(0.3,0.8,['Initial length : ',num2str(samples)],'Fontsize',[10]);
text(0.3,0.5,['Mean :',num2str(xmean)],'Fontsize',[10]);
text(0.3,0.2,['Variance :',num2str(xvar)],'Fontsize',[10]);
s4=subplot(2,2,4);
axis('off')
```

```
text(0.3,0.8,['Final length : ',num2str(NSAMP)],'Fontsize',[10]);
text(0.3,0.5,['Mean :',num2str(estmean)],'Fontsize',[10]);
text(0.3,0.2,['Variance :',num2str(estvar)],'Fontsize',[10]);
end;
clear
```

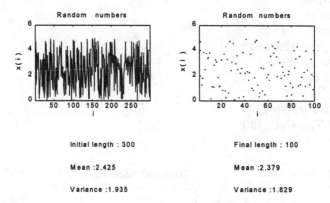

Initial length : 300   Final length : 100

Mean :2.425   Mean :2.379

Variance :1.935   Variance :1.829

MATLAB Computer Example 3.4
GENERATION OF GAUSSIAN DISTRIBUTED RANDOM VARIABLES . . .
Use of the Central Limit Theorem

In this example we will show how to generate gaussian distributed random
variables from a uniform distribution using an approximation to the central
limit theorem.

```
%=============================================
% Input parameters
k = 12;
mu = 2;
sigma = 4;
j = 100;
%=============================================
length_data=j;
X_ij=rand(k,length_data);
Y_j=sum(X_ij);
Y_j=(Y_j-k/2)/sqrt(k/12);
clg;s1=subplot(2,2,1);
plot(1:1:length_data,X_ij(1,:),'.r'); axisn(1:1:length_data)
xlabel('j','Fontsize',[10]);
ylabel('X_ij(1)','Fontsize',[10]);
title('Uniform random variable','Fontsize',[10]);
set(s1,'Fontsize',[10]);
s2=subplot(2,2,2);
plot(1:1:length_data,Y_j,'.r'); axisn(1:1:length_data)
```

```
xlabel('j','Fontsize',[10]);
ylabel('Y_j','Fontsize',[10]);
title('Gaussian distribution function','Fontsize',[10])
set(s2,'Fontsize',[10]);
s3=subplot(2,2,3);
hist(X_ij(1,:),10);
xlabel('kk','Fontsize',[10]);
ylabel('fx_kk','Fontsize',[10]);
set(s3,'Fontsize',[10]);
s4=subplot(2,2,4);
hist(Y_j,10);
xlabel('kk','Fontsize',[10]);
ylabel('fy_kk','Fontsize',[10]);
set(s4,'Fontsize',[10]);
```

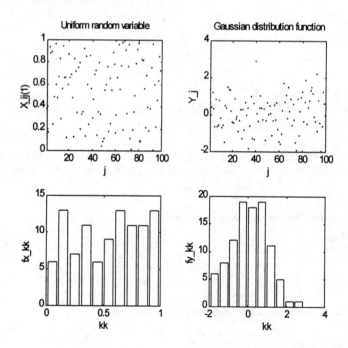

```
Z_j=Y_j*sigma + mu;
mu=mean(Z_j);
sigma=std(Z_j);
clg;s1=subplot(2,2,1);
plot(1:1:length_data,Z_j,'.r');axisn(1:1:length_data);
xlabel('j','Fontsize',[10]);
ylabel('Z_j','Fontsize',[10]);
title('Gaussian random variable','Fontsize',[10]);
```

```
set(s1,'Fontsize',[10])
s2=subplot(2,2,2);
plot(1:1:length_data,Z_j,'r');axisn(1:1:length_data);
xlabel('j','Fontsize',[10]);
ylabel('Z_j','Fontsize',[10]);
title('Gaussian random variable','Fontsize',[10]);
set(s2,'Fontsize',[10])
s3=subplot(2,2,3);
hist(Z_j,10);
xlabel('kk','Fontsize',[10]);
ylabel('fz_kk','Fontsize',[10]);
set(s3,'Fontsize',[10])
s4=subplot(2,2,4);
[intervals,bins]=hist(Z_j,10);
axis('off');
set(text(0,0.6,['Mean: ',num2str(mu)]),'Fontsize',[10]);
set(text(0,0.45,['std: ',num2str(sigma)]),'Fontsize',[10]);
set(text(0.55,1,'intervals : '),'Fontsize',[10]);
for i=1:10,
 set(text(0.7,0.1*(10-i),['
',num2str(intervals(i))]),'Fontsize',[10]),'HorizontalAlignment','right');
end
clear
```

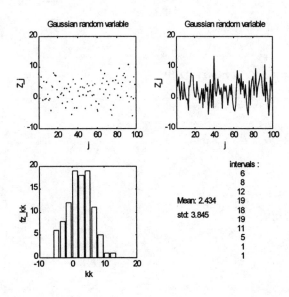

MATLAB Computer Example 3.5
GENERATION OF GAUSSIAN RANDOM VARIABLES . . . The Box-Muller Method

This document will help you to understand the gaussian probability density function and give you some insight into what gaussian distributed random variables are. In addition, you will learn a method for generating gaussian random deviates from uniform deviates.

```
%===============================================
% Input parameter
                        NRAND = 100;
                        mu = 0;
                        sigma = 2;
                        NBIN = 20;
%===============================================
samples = NRAND;
u=rand(samples,1);
v=rand(samples,1);
clg;s1=subplot(221);
plot(1:1:samples,u,'r'),xlabel('j','Fontsize',[10]);axisn(1:1:samples);
ylabel('u','Fontsize',[10]);
title('Uniform random variable u','Fontsize',[10]);
set(s1,'Fontsize',[10])
s2=subplot(222);
plot(1:1:samples,v,'r');axisn(1:1:samples);
xlabel('j','Fontsize',[10]);
ylabel('v','Fontsize',[10]);
title('Uniform random variable v','Fontsize',[10]);
set(s2,'Fontsize',[10])
s3=subplot(223);
hist(u,10);
xlabel('BINS_j','Fontsize',[10]);
ylabel('f_u','Fontsize',[10]);
set(s3,'Fontsize',[10]);
s4=subplot(224);
hist(v,10);
xlabel('BINS_j','Fontsize',[10]);
ylabel('f_v','Fontsize',[10]);
set(s4,'Fontsize',[10])
```

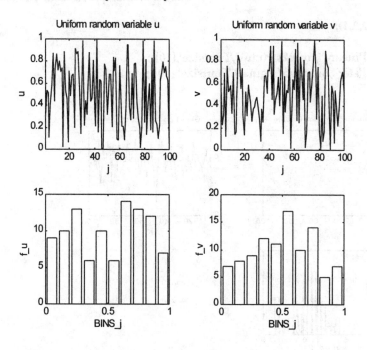

```
V_k=mu+sigma*(sqrt(-2*log(u)).*cos(2*pi*v));
mean_V=mean(V_k);
stdev_V=std(V_k);
[D, BINS]=hist(V_k,NBIN);
D=D/samples;
clg;s1=subplot(2,2,1);
plot(1:1:samples,V_k,'.r');axisn(1:1:samples);
xlabel('k','Fontsize',[10]);
ylabel('V_k','Fontsize',[10]);
title('Gaussian random variable','Fontsize',[10])
set(s1,'Fontsize',[10]);
s2=subplot(2,2,2);
plot(1:1:samples,V_k,'r');axisn(1:1:samples);
xlabel('k','Fontsize',[10]);
ylabel('V_k','Fontsize',[10]);
title('Gaussian random variable','Fontsize',[10])
set(s2,'Fontsize',[10])
s3=subplot(2,2,3);
bar(BINS,D);
xlabel('BINS_j','Fontsize',[10]);
ylabel('D_j','Fontsize',[10]);
set(s3,'Fontsize',[10]);
```

```
s4=subplot(2,2,4);
axis('off')
text(0.15,0.6,['mean: ',num2str(mu)],'Fontsize',[10]);
text(0.15,0.4,['std: ',num2str(sigma)],'Fontsize',[10]);
clear
```

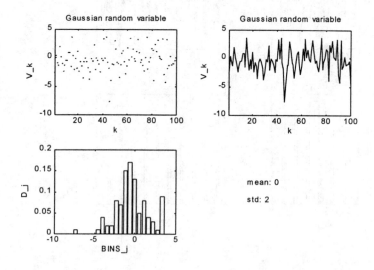

Note the heights of the bars. To see another set of 100 numbers, recalculate the worksheet. Try changing the range variable i in the top of the document to 1000, then look at the bar chart again.

MATLAB Computer Example 3.6
RUN TEST . . . A Test for Independence of Data Samples

In this document we will generate and test a random sequence of numbers for independence by using the run test. If you have not read about the run test yet, you should do so before continuing with this document. Consider that we have a sequence of N observations of a random variable and we wish to test these values for independence. We first classify each variable into one of two mutually exclusive classes which can be identified with a plus or a minus. This can be done for many distributions by calculating the mean (xmean) and classifying each variable as a plus (+) if it is above the mean value or a minus (–) if it is below it. Then the sequence of plus and minus observations are analysed for runs. A run is a sequence of identical observations that is followed or preceded by a different observation.

For example, consider the sequence of plus and minus observations below:

```
+ - + + - - + - + + + + - - + - - -
^ ^ ^   ^   ^ ^   ^       ^   ^   ^
1 2 3   4   5 6   7       8   9   10
```

There are 10 runs in this sequence of 18 observations. Next we hypothesize that the observations are independent. Then the acceptance region for this hypothesis is defined as

$$r_{9,1-\frac{\alpha}{2}} < r \leq r_{9,\frac{\alpha}{2}}$$

Here r is the number of runs observed, alpha is the desired confidence level, and the outer regions are obtained from a table for $N/2 = 9$, since there are 18 observations ($N = 18$). Then from a table of percentage points of a run distribution, the lower limit is 5 and the upper is 14; therefore, $r = 10$ falls between these values, and the hypothesis is accepted at the .05 confidence level.

```
%===============================================
% Input parameter
                        i=100;
%===============================================
samples=i;
uni_i=2*rand(1,samples);
umed=median(uni_i);
for i=1:1:samples
 if (uni_i(i)>umed)
  count_k(i)=1;
 else count_k(i)=0;
 end;
end;
for j=1:1:samples-1
 if ((count_k(j)-count_k(j+1))==0)
  run_j(j)=0;
 else run_j(j)=1;
 end;
end;
rt=sum(run_j);
mean_uni=mean(uni_i);
stdev=std(uni_i);
if (samples==100)
 if (40<rt<61)
  str=str2mat(...
    'The acceptance region for a confidence level of 0.05, with data length of
100 points, is between 40 and',...
    '61. Because our result is in this range, the data:',...
    ' ',...
    '              PASS  the Run Test.',...
    ' ');
  disp(str)
```

```
        else
        str=str2mat(...
            ' ',...
        'The acceptance region for a confidence level of  0.05, with data length of
100 points, is between 40 and',...
        '61. Because our result is out of this range, the data:',...
            ' ',...
        '              DO NOT PASS  the Run Test.',...
        ' ');
        disp(str);
        end;
        elseif(samples~=100)
        str=str2mat(...
            ' ',...
            ' ',...
        'For this number of samples you need to look for the right range in tables
in order to check',...
        'independence. !!!');
        disp(str);
        end;
        clg;s1=subplot(2,2,1);
        plot(1:1:samples,uni_i,'.r');axisn(1:1:samples);
        xlabel('i','Fontsize',[10]);
        ylabel('uni_i','Fontsize',[10]);
        title('Random variable','Fontsize',[10])
        set(s1,'Fontsize',[10])
        s2=subplot(2,2,2);
        plot(1:1:samples,uni_i,'r');axisn(1:1:samples);
        xlabel('i','Fontsize',[10]);
        ylabel('uni_i','Fontsize',[10]);
        title('Random variable','Fontsize',[10])
        set(s2,'Fontsize',[10])
        s3=subplot(2,2,3);
        hist(uni_i,10);
        xlabel('BINS','Fontsize',[10]);
        ylabel('f_uni','Fontsize',[10]);
        set(s3,'Fontsize',[10])
        s4=subplot(2,2,4);
        axis('off')
        text(0.15,0.2,['mean: ',num2str(mean_uni)],'Fontsize',[10]);
        text(0.15,0.4,['std: ',num2str(stdev)],'Fontsize',[10]);
        text(0.15,0.6,['Run test : ',num2str(rt)],'Fontsize',[10]);
        clear
```

The acceptance region for a confidence level of 0.05, with data length of 100 points, is between 40 and 61. Because our result is in this range, the data:

PASS the Run Test.

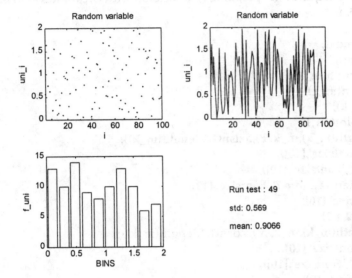

MATLAB Computer Example 3.7
CHI-SQUARE GOODNESS-OF-FIT-TEST . . . Test for Normality

This document demonstrates the use and calculation of the chi-square goodness-of-fit test. We start by reading in a data file of 200 points which we wish to test to see if the data came from a normal distribution. We will later test other supplied data sets as well. We will conduct the test at the .05 level of significance, which requires 16 class intervals for 200 data points. We will use the equal-frequency procedure, which for 16 classes requires the probability associated with each interval to be $P = 1/16 = .0625$. We find the required interval limits from a table of areas under the standardized normal density function. These limits are shown below in array Z.

```
%==========================================
% Input parameter
                i=100;

%==========================================
str=str2mat(...
    'We start by reading in data file of 200 points which we wish to test to see
if the data came from a',...
    'normal probability distribution or not. We will conduct the test at the
0.05 level of significance which',...
```

'requires 16 class intervals for 200 data points. We will use the equal
frequency procedure which for',...
    '16 classes requires the probability associated with each interval to be P =
1/16 =0.0625.');
disp(str);
load normbr.dat;
m_k=normbr;
samples=length(m_k);
xmean=mean(m_k);
xstd=std(m_k);
clg;s1=subplot(2,1,1);
plot(1:1:length(m_k),m_k,'r');axisn(1:1:length(m_k));
xlabel('k','Fontsize',[10]);
ylabel('m_k','Fontsize',[10]),
title('Gaussian variable ?','Fontsize',[10]);
set(s1,'Fontsize',[10])
s2=subplot(2,1,2);
plot(1:1:length(m_k),m_k,'.r');axisn(1:1:length(m_k));
xlabel('k','Fontsize',[10]);
ylabel('m_k','Fontsize',[10]);
set(s2,'Fontsize',[10])

We start by reading in data file of 200 points which we wish to test to see if
the data came from a normal probability distribution or not. We will conduct
the test at the 0.05 level of significance which requires 16 class intervals for
200 data points. We will use the equal frequency procedure which for 16
classes requires the probability associated with each interval to be P = 1/16
=0.0625.

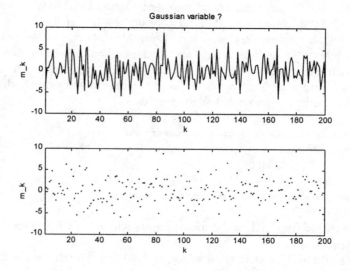

```
str=str2mat(...
    'We find the required interval limits from a table of areas under
standardized normal density function.',...
    'Afterward, the intervals are scaled to fit the actual data. In the next
figure, Z represents the original',...
    'intervals, X represents the mapped intervals, and f the number of points
in each interval.');
disp(str);
Z(1)=-1.53; Z(2)=-1.15; Z(3)=-0.89; Z(4)=-0.67; Z(5)=-0.49;
Z(6)=-0.32; Z(7)=-0.16; Z(8)=0.0; Z(9)=0.16; Z(10)=0.32;
Z(11)=0.49; Z(12)=0.67; Z(13)=0.89; Z(14)=1.15; Z(15)=1.53;
X=[];interval=[];
for i=1:15
X(i)=xstd*Z(i)+xmean;
end
interval(1)=min(m_k);
interval(17)=max(m_k);
for i=1:15
interval(i+1)=X(i);
end;
[f_p,bins]=histo(m_k,interval');
clg;s1=subplot(2,1,1);
histo(m_k,interval');
xlabel('p','Fontsize',[10]);
ylabel('f_p','Fontsize',[10])
set(s1,'Fontsize',[10])
s2=subplot(2,1,2);
axis('off')
text(0.0,1,'Z : ','Fontsize',[10]);
text(0.2,1,'X : ','Fontsize',[10]);
text(0.4,1,'Z : ','Fontsize',[10]);
text(0.6,1,'X : ','Fontsize',[10]);
text(0.85,1,'f_p :','Fontsize',[10]);
for j=0:7,
set(text(0.05,0.12*(7-j),['
',num2str(Z(j+1))]),'HorizontalAlignment','right','Fontsize',[10]);
set(text(0.25,0.12*(7-j),['
',num2str(X(j+1))]),'HorizontalAlignment','right','Fontsize',[10]);
set(text(0.85,0.12*(7-j),['
',num2str(f_p(j+1))]),'HorizontalAlignment','right','Fontsize',[10]);
end
for j=0:6,
set(text(0.45,0.12*(7-j),['
',num2str(Z(j+9))]),'HorizontalAlignment','right','Fontsize',[10]);
set(text(0.65,0.12*(7-j),['
',num2str(X(j+9))]),'HorizontalAlignment','right','Fontsize',[10]);
```

```
set(text(0.95,0.12*(7-j),['
',num2str(f_p(j+9))]),'HorizontalAlignment','right','Fontsize',[10]);
end
set(text(0.95,0,['
',num2str(f_p(16))]),'HorizontalAlignment','right','Fontsize',[10]);
```

We find the required interval limits from a table of areas under standardized normal density function. Afterwards, the intervals are scaled to fit the actual data. In the next figure, Z represents the original intervals, X represents the mapped intervals, and f the number of points in each interval.

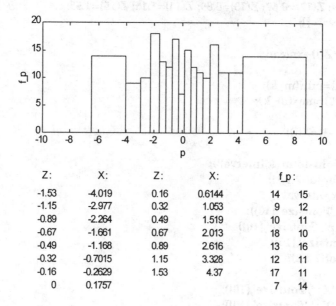

Z:	X:	Z:	X:	f_p:	
-1.53	-4.019	0.16	0.6144	14	15
-1.15	-2.977	0.32	1.053	9	12
-0.89	-2.264	0.49	1.519	10	11
-0.67	-1.661	0.67	2.013	18	10
-0.49	-1.168	0.89	2.616	13	16
-0.32	-0.7015	1.15	3.328	12	11
-0.16	-0.2629	1.53	4.37	17	11
0	0.1757			7	14

```
str=str2mat(...
   'Now we compute the chi-square statistic test given that the expected
frequency for each',...
   'interval is 200/16 = 12.5');
disp(str);
asum=sum(f_p);
chisquare=0;
for i=1:16
chisquare=chisquare+((12.5-f_p(i))^2)/12.5;
end
s1=subplot(2,2,1);
if chisquare < 22.36
plot(1:1:samples,m_k,'r');axisn(1:1:samples);
xlabel('k','Fontsize',[10]);
```

```
ylabel('m_k','Fontsize',[10]);
title('gaussian variable.','Fontsize',[10]);
else
plot(1:1:samples,m_k,'r');axisn(1:1:samples);
xlabel('k','Fontsize',[10]);
ylabel('m_k','Fontsize',[10]);
title('nongaussian variable.','Fontsize',[10]);
end
set(s1,'Fontsize',[10])
s2=subplot(2,2,2);
plot(1:1:samples,m_k,'.r');axisn(1:1:samples);
xlabel('k','Fontsize',[10]);
ylabel('m_k','Fontsize',[10]);
set(s2, 'Fontsize',[10])
s3=subplot(2,2,3);
hist(m_k,20);
title('Histogram of the signal','Fontsize',[10]);
set(s3, 'Fontsize',[10])
s4=subplot(2,2,4);
axis('off')
text(0,0.2,['sum_histograma=  ',num2str(asum)],'Fontsize',[10]);
text(0,0,['The sum check !!!'],'Fontsize',[10]);
text(0,0.4,['Mean value: ',num2str(xmean)],'Fontsize',[10]);
text(0,0.6,['standard deviation: ',num2str(xstd)],'Fontsize',[10]);
text(0,0.8,['chisquare : ',num2str(chisquare)],'Fontsize',[10]);
if chisquare < 22.36
str=str2mat(...
    ' ',...
    'The acceptance level is chisquare < 22.36 so the data:',...
    ' ',...
    '         PASS  the test for normality',....
    ' ',...
    'at the 0.005 level of significance');
disp(str);
else
str=str2mat(...
    ' ',...
    'The acceptance level is chisquare < 22.36 so the data:',...
    ' ',...
    '         DO NOT PASS  the test for normality',....
    ' ',...
    'at the 0.005 level of significance');
disp(str);
end;
clear
```

Now we compute the Chi-Square statistic test given that the expected frequency for each interval is 200/16 = 12.5

The acceptance level is chisquare < 22.36 so the data:

PASS the test for normality

at the 0.005 level of significance

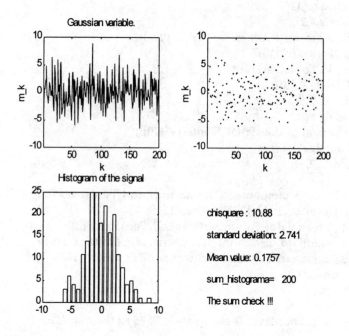

chisquare : 10.88

standard deviation: 2.741

Mean value: 0.1757

sum_histograma= 200

The sum check !!!

MATLAB Computer Example 3.8
THE KOLMOGOROV-SMIRNOFF GOODNESS-OF-FIT TEST . . . A Test for Normality

This document demonstrates the calculation and use of the Kolmogorov-Smirnoff goodness-of-fit test. We will test several data samples for normality using this test.

```
%==========================================
% Input parameter
                    load exp1.dat;

%==========================================
```

```
str=str2mat(...
    'We normalize the data to a mean of zero and standard deviation of one
because we will be',...
    'comparing these data to theoretical normal distribution of mean zero
and deviation of one.',...
    'We form the cumulative distribution function of the input data and the
theoretical cumulative',...
    'distribution function.');
disp(str);
m_i=exp1;
samples=length(m_i);
xmean=mean(m_i);
sd=std(m_i);
k=1:1:samples;
s1=subplot(2,1,1);
plot(1:1:samples,m_i,'r');axisn(1:1:samples);
xlabel('i','Fontsize',[10]);
ylabel('m_i','Fontsize',[10]);
title('Gaussian variable ?','Fontsize',[10])
set(s1,'Fontsize',[10])
s2=subplot(2,1,2);
plot(1:1:samples,m_i,'.r');axisn(1:1:samples)
xlabel('i','Fontsize',[10]);
ylabel('m_i','Fontsize',[10]);
set(s2,'Fontsize',[10]);
```

We normalize the data to a mean of zero and standard deviation of one because we will be comparing these data to theoretical normal distribution of mean zero and deviation of one. We form the cumulative distribution function of the input data and the theoretical cumulative distribution function.

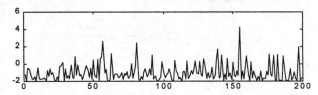

```
norm_j=(m_i-xmean)/sd;
[freq_re,bins]=hist(norm_j,length(norm_j));
```

```
Fo_k=cumsum(freq_re/length(norm_j));
maximun=max(norm_j);
minimun=min(norm_j);
x=(minimun:(maximun-minimun)/(length(norm_j)-1):maximun);
Fe_k=normcdf(x,0,1);
for k=1:1:samples;
 d_k(k)=abs(Fo_k(k)-Fe_k(k));
end
D=max(d_k);
str=str2mat(...
    'Now we can compute the maximum diference between the two
distributions.',...
    ' ',...
    '>>d(k) = max(abs(Fo_k-Fe_k));',...
    ' ',...
    ['Maximum difference =   ',num2str(D)]);
disp(str);
clg;s1=subplot(1,1,1);
plot(bins,Fo_k,'r',bins,Fe_k,'g');
xlabel('x_k','Fontsize',[10]);
ylabel('Fe_k: green        Fo_k: red','Fontsize',[10]);
title('Kolmogoroff-Smirnoff cdf test-Normal case','Fontsize',[10]);
set(s1,'Fontsize',[10]);
```

Now we can compute the maximum difference between the two
distributions.

>>d(k) = max(abs(Fo_k-Fe_k));

Maximum difference = 0.1641

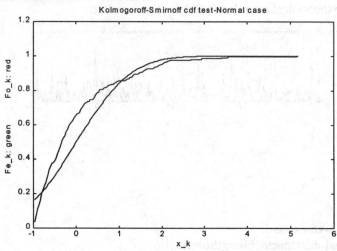

```
str=str2mat(...
    'We use the formula for the 0.05 level of significance test for
normality. If D is less than the',...
    'next calculated value then the data has to be accepted as normal
distributed data.',...
    ' ',...
    '>> 0.866/sqrt(length(data))',...
    'D < 0.061235    ....To pass test.');
disp(str);
if D < 0.061
 str=str2mat(...
    ' ',...
    '   The data pass the Kolmogorov-Smirnoff test for normality.');
 disp(str);
else;
 str=str2mat(...
    ' ',...
    '   The data do not pass the Kolmogorov-Smirnoff test for normality.');
 disp(str);
end
clg;s1=subplot(2,2,1);
if D< 0.061
 plot(1:1:samples,m_i,'r');axisn(1:1:samples);
 xlabel('i','Fontsize',[10]);
 ylabel('m_i','Fontsize',[10]);
 title('gaussian variable.','Fontsize',[10])
else
 plot(1:1:samples,m_i,'r');axisn(1:1:samples);
 xlabel('i','Fontsize',[10]);
 ylabel('m_i','Fontsize',[10]);
 title('No Gaussian variable.','Fontsize',[10]);
end
set(s1,'Fontsize',[10])
s2=subplot(2,2,2);
plot(1:1:samples,m_i,'.r');axisn(1:1:samples);
xlabel('i','Fontsize',[10]);
ylabel('m_i','Fontsize',[10]);
set(s2,'Fontsize',[10]);
s3=subplot(2,2,3);
hist(m_i,20);
xlabel('kk','Fontsize',[10]);
ylabel('f_m','Fontsize',[10]);
set(s3,'Fontsize',[10]);
s4=subplot(2,2,4);
axis('off')
```

text(0,0.6,['Mean value: ',num2str(xmean)],'Fontsize',[10]);
text(0,0.4,['standard deviation: ',num2str(sd)],'Fontsize',[10]);
clear

We use the formula for the 0.05 level of significance test for normality. If D is less than the next calculated value then the data have to be accepted as normal distributed data.

>> 0.866/sqrt(length(data))
D < 0.061235   ....To pass test.

The data do not pass the Kolmogorov-Smirnoff test for normality.

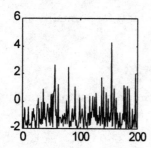

Random Processes 4

4.0 RANDOM PROCESSES: BASIC DEFINITIONS

In many cases, the random data that are collected from a physical phenomenon are a function of time. For example, in the analysis of brain waves or electroencephalograms, the fluctuations of voltage with respect to time represent the random discharges of thousands or perhaps millions of brain cells. Consider, e.g., Figure 4.1. These data were collected in a brain wave experiment where electrodes were positioned on the scalp of a subject, the voltage was differentially amplified, and the resulting information was digitized as a function of time and was stored in a computer. An arbitrary dc value has been added to each set of data to more easily tell them apart. All the three sets of data result from the same experiment, performed on the same subject, and the data were recorded using the same electrode.

In this example, one cannot predict the time course of the fluctuations of voltage. The subject was retested within seconds, which yielded each of the three different data files. Although there are similarities and trends between the different traces, there are enough differences to tell them apart. The experiments performed were identical, the system (i.e., the subject) was the same, and no signal processing was done to the data.

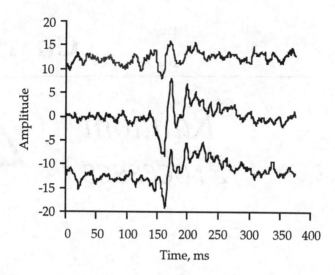

Figure 4.1 Brain wave data obtained from a subject. The same experiment was performed three times, each time yielding different data.

We are dealing here with a *random process*. The concept of random variable involved mapping of the elements of a sample space onto a set of real numbers. A random process is the mapping of the elements of the sample space to a space of continuous functions of time. Each of the data sets pictured is a different realization of the random process. Now consider Figure 4.2. This figure represents some of the characteristics of a random process. Each realization is an infinitely long time sequence of random variables. Each realization of the random process is called a *sample function*, and there are an infinite number of sample functions forming a random process. If we pick an arbitrary time, say, t_1, we can define a *random variable* associated with time t_1. We call this random variable $x(t_1)$, or x_1 for convenience. Likewise at arbitrary time t_2 we define the random variable x_2. Therefore a random process is a multidimensional function of sample space and time.

Of interest in a random process is the specification of the joint behavior of the random variables within the sample functions of the random process. Consequently, it will be of interest to determine not only the single-variable statistics but also joint-variable statistics. At each time instant, say, t_1, there is a probability density function associated with the random variable x_1. We call this probability density function $f(x_1, t_1)$, or simply $f(x_1)$. At another time instant, say, t_2, the random variable associated with that time instant x_2 will have $f(x_2, t_2)$, or simply $f(x_2)$, as its probability density function.

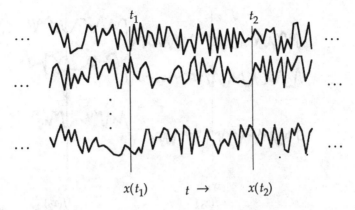

Figure 4.2 Sample functions of a random process. At times t_1 and t_2, the random variables x_1 and x_2 are defined.

Let the probability density functions at times t_1 and t_2 be described by the following equations:

$$f_X(x_1) = \frac{1}{2\sqrt{2\pi}} \exp\left[\frac{-(x_1)^2}{8}\right] \tag{4.1}$$

and

$$f_X(x_2) = \frac{1}{2\sqrt{2\pi}} \exp\left[\frac{-(x_2 - 1.0)^2}{8}\right] \tag{4.2}$$

These are the well-known gaussian probability density functions. In the first case, the random variable x_1 has an expected value or mean value equal to 0 and a standard deviation equal to 2. In the second case, the random variable x_2 has an expected value or mean value equal to 1 and a standard deviation equal to 2. Figure 4.3 illustrates a random process possessing these properties.

Consider now the following probability density functions:

$$f_X(x_1) = \frac{1}{2\sqrt{2\pi}} \exp\left[\frac{-(x_1)^2}{8}\right] \tag{4.3}$$

and

$$f_X(x_2) = \frac{1}{3\sqrt{2\pi}} \exp\left[\frac{-(x_2)^2}{18}\right] \tag{4.4}$$

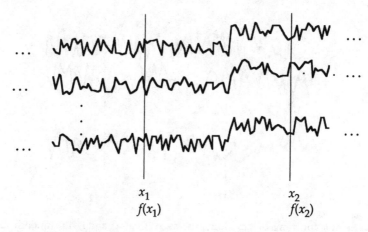

$$x_1 \qquad\qquad\qquad\qquad x_2$$
$$f(x_1) \qquad\qquad\qquad\qquad f(x_2)$$

Figure 4.3 At times t_1 and t_2, the random variables x_1 and x_2 have probability density functions given by Equations (4.1) and (4.2), respectively. Observe that there is a dc shift in the random process, causing the statistics of the random variables to differ from each other.

In this particular case, the statistics of the two random variables differ in their standard deviations. In the case of x_1, the standard deviation is 2, and in the case of x_2, the standard deviation is 3. Figure 4.4 illustrates a random process with these properties.

The joint statistical behavior of the random variables x_1 and x_2 can be specified by considering the joint probability function $f(x_1, x_2)$. It is also possible to consider joint probability density functions of the type $f(x_1, x_2, x_3, \ldots)$ which specify joint characteristics of the random variables $x_1, x_2, x_3, \ldots$.

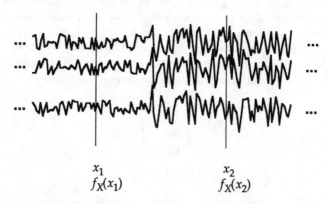

$$x_1 \qquad\qquad\qquad\qquad x_2$$
$$f_X(x_1) \qquad\qquad\qquad\qquad f_X(x_2)$$

Figure 4.4 At times t_1 and t_2, the random variables x_1 and x_2 have probability density functions given by Equations (4.3) and (4.4), respectively. As may be observed, there is a standard deviation change in the random process, causing the statistics of the random variables to differ from each other.

4.1 BASIC PROPERTIES OF RANDOM PROCESSES

There is a set of basic definitions associated with random processes. These definitions help us understand the basic characteristics of the various processes. Note that when we are dealing with random data sampled from a random process, our knowledge of the characteristics of the process is often incomplete. The probability density functions and joint probability density functions are often not known. Consequently, as researchers, we are frequently required to estimate or guess at the type of process we are dealing with.

STATIONARY RANDOM PROCESSES

A random process for which the nth-order joint probability density functions do not depend on the choice of time origin is called a *stationary random process*. This can be expressed as

$$f_X(x_1, \ldots, x_n; t_1, \ldots, t_n) = f_X(x_1, \ldots, x_n; t_1 + t \cdots t_n + t) \quad \text{for all } t, t_1, \ldots, t_n$$

This definition of stationariness implies that *all* the moments are equal and do not depend on the choice of time origin. Likewise, all joint moments such as $E[x_1^n x_2^n] = E[x_{t+1}^n x_{t+2}^n]$ are equal and do not depend on the choice of t.

WIDE-SENSE STATIONARY RANDOM PROCESSES

The requirements placed on a stationary random process are very restrictive. Indeed, it is difficult to prove, except in limited cases, whether a particular random process is stationary. For random processes with only limited observation times available, proof of stationariness is virtually impossible. We must, then, resort to a process definition that is somewhat less limiting than stationariness. A random process is said to be a *wide-sense stationary* process if the following conditions are met:

1. The *expected value* of the random process must be *independent* of the time origin. This condition is given by the following equation:

$$E[x_1] = E[x_{t+1}] \tag{4.5}$$

The random process illustrated by Figure 4.3 does not meet this condition. Note that when $t + 1 = t_2$, the expected value of the random process has increased from 0.0 to 1.0. This random process does not meet the first condition and is, therefore, not wide-sense stationary.

2. The correlation and the covariance between any two of the random variables must depend only on the *time difference* between the two random variables. This condition implies that $E[x_1 x_2]$ is a function only of the time separation between the two random variables, that is, $t_2 - t_1$. Based on this condition, we could write the following expression:

$$E[x_1 x_2] = E[x_3 x_4] \quad \text{as long as} \quad t_2 - t_1 = t_4 - t_3$$

In summary, a random process is wide-sense stationary if

$$E[x_t] = \text{constant}$$

and

$$E\left[x_{t_1} x_{t_2}\right] = R_{xx}(t_2 - t_1)$$

Example 4.1 A random process has sample functions of the form

$$X(t) = A \sin \omega t$$

where A is a random variable which is uniformly distributed between -1 and 1. Is this process wide-sense stationary? Prove your answer.

First we need to find the probability density function of the random variable A. Since A is uniformly distributed between -1 and 1, $f(A) = .5$ for $-1 \le A \le 1$.

Next, we find $E[x_1]$:

$$E[x_1] = E[x(t_1)] = \int_{-1}^{1} A \sin \omega t_1 \cdot f(A) \, dA = .5 \sin \omega t_1 \cdot \left. \frac{A^2}{2} \right|_{-1}^{1} = 0$$

Note that the integral is with respect to the random variable. Now we find the value of the correlation:

$$E[x_1 x_2] = E[x(t_1)x(t_2)] = E[A \sin \omega t_1 \cdot A \sin \omega t_2] = \sin \omega t_1 \cdot \sin \omega t_2 \cdot E[A^2]$$

which equals

$$E[x_1 x_2] = \frac{1}{3} \cdot \frac{1}{2} [\cos \omega(t_1 - t_2) - \cos \omega(t_1 + t_2)]$$

The random process of this example does not meet the criteria for wide-sense stationariness. The first moment equals zero, but the correlation value is a function of both the sum and the difference of the time between the two random variables. Figure 4.5 shows several sample functions of this random process.

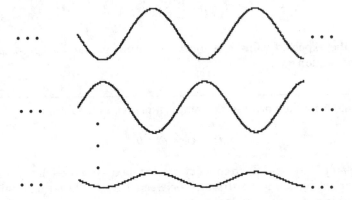

Figure 4.5 Sample functions of the random process $X(t)=A\sin\omega t$. The random variabl is A, and the frequency is a constant. Whitin a sample function, once the amplitude A has been picked from $f(A)$, the rest of the sample function is fixed.

Example 4.2 A random process has sample functions of the form

$$x(t) = A \sin(\omega t + \theta)$$

where A and θ are statistically independent random variables and the frequency is constant. The random variable A is uniformly distributed from 0 to 10, and the random variable θ is uniformly distributed from 0 to π. Is this process wide-sense stationary?

First, we find $E[x(t)]$:

$$E[x(t)] = E[A]E[\sin(\omega t + \theta)]$$

Note that the two random variables are independent. We now find the expected value of each of the two factors:

$$E[A] = \int_0^{10} Af(A) \, dA = \int_0^{10} A\left(\frac{1}{10}\right) dA = 5$$

and

$$E[\sin(\omega t + \theta)] = \int_0^\pi \sin(\omega t + \theta) f(\theta)\, d\theta = \frac{2}{\pi} \cos(\omega t)$$

Consequently,

$$E[x(t)] = \frac{10}{\pi} \cos(\omega t)$$

Since the expected value is a function of time, the random process is not wide-sense stationary.

Example 4.3 A random process has sample functions of the form

$$X(t) = A \cos(\omega t + \theta)$$

where the amplitude and the frequency are constants and the phase angle is a random variable uniformly distributed between 0 and 2π. Determine whether this process is wide-sense stationary.

First, we determine the expected value of the process:

$$E[X(t)] = E[A \cos(\omega t + \theta)] = A \cdot E[\cos(\omega t + \theta)]$$

$$= A \int_0^{2\pi} \cos(\omega t + \theta) f(\theta)\, d\theta = A \int_0^{2\pi} \cos(\omega t + \theta) \cdot \frac{1}{2\pi}\, d\theta$$

$$= \frac{A}{2\pi} \cdot \int_0^{2\pi} \cos(\omega t + \theta)\, d\theta = 0$$

Next, we determine the value of the correlation:

$$E[X(t_1)X(t_2)] = E[A \cos(\omega t_1 + \theta) \cdot A \cos(\omega t_2 + \theta)]$$

$$= A^2 \left\{ E\left[\frac{1}{2} \cos(\omega t_1 + \theta + \omega t_2 + \theta) \right] + E\left[\frac{1}{2} \cos \omega(t_1 - t_2) \right] \right\}$$

$$= \frac{A^2}{2} \cos \omega(t_1 - t_2)$$

The expected value of the random process is constant (zero), and the correlation depends only on the time difference. Therefore, this random process is wide-sense stationary.

DETERMINISTIC RANDOM PROCESSES

Occasionally, each sample function of a random process is of a deterministic nature. For example, consider the sample functions of the random process $X(t) = A \sin \omega t$ where A is a random variable uniformly distributed from -1 to 1. This is the same random process used in Example 4.1. Note that the frequency is a constant. Figure 4.5 shows several sample functions from this random process. The randomness in $x(t) = A \sin \omega t$ is determined by A, whereas the time behavior is determined by $\sin \omega t$. Once this process is observed for some time t, then all the future values can be predicted exactly. A random process is called a *deterministic random process* if all the future values can be predicted from past observations.

ERGODIC RANDOM PROCESSES

In Chapter 3, the concepts of estimate of the population mean and variance were explained. All the expectations defined for the random process so far are in terms of ensemble averages or integrals over the probability density functions. However, the nth-order joint probability density functions are seldom known. In addition, the observations of random processes are sample functions in time, and they may be the only data we have to work with. Therefore, there is a need to infer the statistical properties of the random processes just from observations over time. In Chapter 3 we were able to estimate the mean of a random variable based on samples from the population. We used the following:

$$\hat{\mu} = \frac{1}{n} \sum_{i=1}^{n} x_i \tag{4.6}$$

which is an ensemble average. In a similar way we can define *time average* of a random process over an interval $[-T, T]$ as

$$\frac{1}{2T} \int_{-T}^{T} x(t) \, dt$$

For many random processes, as T goes to infinity, this time average will approach the ensemble mean. Such random processes are called *ergodic* (in the mean) random processes. The time average of a random process is defined as an average over an infinite interval

$$<X(t)> = \lim_{T \to \infty} \frac{1}{2T} \int_{-T}^{T} x(t) \, dt \tag{4.7}$$

where < > designates the time average of the function in between the symbols. In a similar manner, we could define the time-average second moment as

$$< X^2(t) > = \lim_{T \to \infty} \frac{1}{2T} \int_{-T}^{T} x^2(t)\, dt \tag{4.8}$$

It is obvious from these definitions that the random process must be wide-sense stationary. Therefore, an ergodic random process is also a wide-sense stationary random process. The converse is not necessarily true; i.e., not all wide-sense stationary processes are ergodic.

Note that when dealing with sample functions of a random process, we are often faced with incomplete information about the statistics of the process. Often the probability density function of the process is not known, and any parameters of the process must be estimated. (This estimation process was covered in Chapter 2.) Suppose now that we have obtained a limited time duration record from a random process. Based on the data we now have, we would like to estimate some of the statistical parameters of the process. Based on the information covered in Chapter 3, we could estimate the mean value of the process by computing the following:

$$\hat{\mu} = \frac{1}{n} \sum_{i=1}^{n} x_i \tag{4.9}$$

The assumption has been made that the random process is ergodic and, of course, wide-sense stationary. If this assumption were not made, we would not be in a position to estimate the mean value or any other statistical parameters of the random process.

4.2 AUTOCORRELATION FUNCTION

The correlation function is one of the statistical parameters of random processes that will give us unique insights into the properties and behavior of a random process.

THE CORRELATION COEFFICIENT AND THE AUTOCORRELATION FUNCTION

In Equation (3.16) the *correlation* between two jointly distributed random variables was defined as

$$E[XY] = \int\!\!\int_{-\infty}^{\infty} xyf(x,y)\,dx\,dy \tag{4.10}$$

Note that if the random variables are statistically independent, then

$$E[XY] = \int\!\!\int_{-\infty}^{\infty} xy\,f(x)f(y)\,dx\,dy = \mu_x\mu_y \tag{4.11}$$

This result implies that the random variables are uncorrelated. However, if the value of the correlation equals zero, then it is said that they are orthogonal.

Consider now a random process such the one shown in Figure 4.6. Assume that the random process is wide-sense stationary. The random variable at time t_1 is $X(t_1)$, and at time t_2 the random variable is $X(t_2)$. The correlation between these two random variables is then defined as

$$E[X(t_1)X(t_2)] = \int\!\!\int_{-\infty}^{\infty} x(t_1)x(t_2)f[x(t_1)x(t_2)]\,d[x(t_1)]\,d[x(t_2)] \tag{4.12}$$

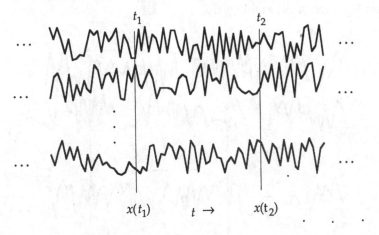

Figure 4.6 Sample functions of a random process. At times t_1 and t_2, the random variables $X(t_1)$ and $X(t_2)$ are defined.

Letting $X(t_1) = X_1$, we can simplify the above equation as follows:

$$E[X_1 X_2] = \int\limits_{-\infty}^{\infty}\!\!\!\int x_1 x_2 f(x_1, x_2)\, d(x_1)\, d(x_2) \tag{4.13}$$

Consider now that t_1 and t_2 may be totally arbitrary. We can see that Equation (4.13) describes, in the most general terms, the *autocorrelation* of the random process. This definition is valid whether the process is wide-sense stationary or not. More formally, we define the autocorrelation of a random process by the following equation:

$$R_X(t_1 t_2) = E[X(t_1) X(t_2)] = \int\limits_{-\infty}^{\infty}\!\!\!\int x_1 x_2 f(x_1, x_2)\, d(x_1)\, d(x_2) \tag{4.14}$$

In the most general case R_X is a function of t_1 and t_2. If the random process is at least wide-sense stationary, then the result given by Equation (4.14) depends only on the time difference $t_2 - t_1$. We call this time difference τ. Figure 4.7 shows the random process with the time difference denoted by the variable τ_1.

Equation (4.14) may be simplified for a wide-sense stationary random process since $R(t_1 t_2)$ is only a function of the time difference τ. Consequently, for a given τ, $R(\tau)$ has the same value throughout the random process. We express Equation (4.14) as follows:

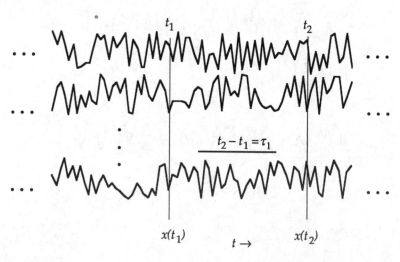

Figure 4.7 Sample functions of a random process. The time difference between t_1 and t_2 is denoted by the variable τ_1.

$$R(\tau) = E[X(t)X(t + \tau)] \tag{4.15}$$

For a given time difference τ_1, as depicted in Figure 4.7, there is a unique value for $R(\tau)$, namely, $R(\tau_1)$. Equation (4.16) expresses this relationship:

$$R(\tau_1) = E[X(t)X(t + \tau_1)] \tag{4.16}$$

Note that there is no subscript on t. This emphasizes the fact that the value of $R(\tau)$ depends exclusively on the value of τ. As mentioned earlier, this is true for all wide-sense stationary random processes.

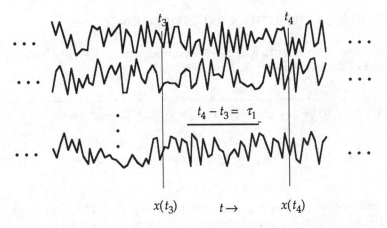

Figure 4.8 Sample functions of a random process. Since the difference $t_4 - t_3$ is still τ_1, the value of $R(\tau)$, namely, $R(\tau_1)$, remains constant.

Figure 4.8 shows the random process with a different start time t_3. Since the difference $t_4 - t_3$ is still τ_1, the value of $R(\tau)$, namely, $R(\tau_1)$, remains constant. Figure 4.9 shows the same start time as t_3 and a different end time t_5. The difference $t_5 - t_3$ has changed and is now called τ_2, and the value of $R(\tau)$, namely, $R(\tau_2)$, will be different from $R(\tau_1)$.

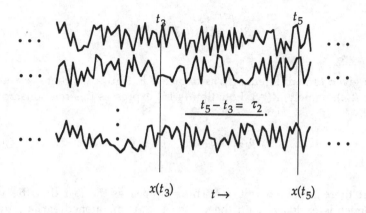

Figure 4.9 Sample functions of a random process. The difference $t_5 - t_3$ is now τ_2; the value of $R(\tau)$, namely, $R(\tau_2)$, has changed.

COMPUTATION OF THE AUTOCORRELATION FUNCTION

We begin this section with a simple example that will illustrate the computation of the autocorrelation function (ACF).

Example 4.4 Let $X(t)$ be a random process having the following form:

$$X(t) = A\cos(\omega_0 t + \theta) \tag{4.17}$$

where A and ω_0 are constants, and θ is a random variable. The random variable θ has a probability density function of the following type:

$$f(\theta) = \begin{cases} B & 0 \le \theta \le 2\pi \\ 0 & \text{elsewhere} \end{cases} \tag{4.18}$$

The value of B is equal to $1/(2\pi)$. Note that the only random variable in Equation (4.17) is θ. Figure 4.10 illustrates sample functions from this random process. Note that the amplitude and the fundamental frequency are constants and do not change from sample function to sample function. Only the phase angle changes from one sample function to the next.

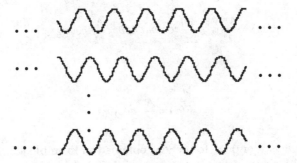

Figure 4.10 Sample functions of the random process having the form of Equation (4.17).

Now the autocorrelation function of the random process can be computed as follows. Using Equation (4.15), we obtain

$$R(\tau) = E[X(t)X(t+\tau)] = \int_0^{2\pi} X(t)X(t+\tau)f(\theta)\, d\theta \qquad (4.19)$$

We now substitute for the values of $X(t)$ and $f(\theta)$:

$$R(\tau) = \int_0^{2\pi} [A\cos(\omega_0 t + \theta)][A\cos(\omega_0 t + \theta + \omega_0\tau)]\frac{1}{2\pi}\, d\theta \qquad (4.20)$$

which yields

$$R(\tau) = \frac{A^2}{2}\int_0^{2\pi}\frac{1}{2\pi}[\cos(2\omega_0 t + 2\theta + \omega_0\tau) + \cos\omega_0\tau]\, d\theta \qquad (4.21)$$

where we have used the identity

$$\cos A \cdot \cos B = \frac{1}{2}[\cos(A+B) + \cos(A-B)] \qquad (4.22)$$

Equation (4.21) is now solved as follows:

$$R(\tau) = \frac{A^2}{2}\int_0^{2\pi}\frac{1}{2\pi}[\cos(2\omega_0 t + 2\theta + \omega_0\tau)]\, d\theta + \frac{A^2}{2}\int_0^{2\pi}\frac{1}{2\pi}[\cos(\omega_0\tau)]\, d\theta \qquad (4.23)$$

The first integral is equal to zero since we are integrating a cosine over an entire period, i.e., from 0 to 2π. Solution of the second integral yields the following result:

$$R(\tau) = \frac{A^2}{2} \int_0^{2\pi} \frac{1}{2\pi} \left(\cos \omega_0 \tau \right) d\theta = \frac{A^2}{2} \cos \omega_0 \tau \int_0^{2\pi} \frac{d\theta}{2\pi} \qquad (4.24)$$

and finally,

$$R(\tau) = \frac{A^2}{2} \cos \omega_0 \tau \qquad (4.25)$$

This result is identical to the one obtained in Example 4.3. It is cast in a different light since, in this case, we are interested in obtaining the value of the autocorrelation function of the random process.

Let us assume now that the random process is corrupted with an additive zero-mean independent random noise $N(t)$. We may assume that the resulting random process is modeled as follows:

$$Y(t) = X(t) + N(t) \qquad (4.26)$$

where

$$X(t) = A \cos(\omega_0 t + \theta) \qquad (4.27)$$

when A and ω_0 are constants and a random variable θ is uniformly distributed between 0 and 2π. A sample function of such a random process for an arbitrary signal-to-noise ratio is shown in Figure 4.11.

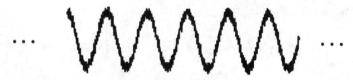

Figure 4.11 A possible sample function from the random process of Equation (4.26).

Now we compute the autocorrelation function of the process

$$R(\tau) = E[Y(t)Y(t+\tau)] = E\big[[X(t)+N(t)][X(t+\tau)+N(t+\tau)]\big] \qquad (4.28)$$

which yields

$$R(\tau) = E[X(t)X(t+\tau) + X(t)N(t+\tau) + N(t)X(t+\tau) + N(t)N(t+\tau)] \qquad (4.29)$$

Since the random noise $N(t)$ and the random process $X(t)$ are independent, the two middle terms of Equation (4.29) are equal to zero. Equation (4.29) yields the following result:

$$R(\tau) = E[X(t)X(t+\tau) + N(t)N(t+\tau)] = R_X(\tau) + R_N(\tau) \qquad (4.30)$$

We have already computed the autocorrelation function of $X(t)$; therefore, the autocorrelation function of $Y(t)$ is the following:

$$R(\tau) = \frac{A^2}{2}\cos\omega_0\tau + R_N(\tau) \qquad (4.31)$$

These two examples illustrate one of the most important properties of the autocorrelation function. The autocorrelation function gives us insight into the degree of self-similarity of the random process as a function of time. Notice that the random process studied in Example 4.3 was described by Equation (4.17). Physically, this is a random process composed of a periodic waveform and a random phase angle. The periodicity of the waveform is fixed as ω_0. The autocorrelation function of the process, given by Equation (4.25), is also a periodic waveform of the same fixed periodicity ω_0.

PROPERTIES OF THE AUTOCORRELATION FUNCTION

The autocorrelation function has a number of properties that relate to the physical properties of the random process.

Let us consider again the defining equation for the autocorrelation function:

$$R(\tau) = E[X(t)X(t+\tau)] \qquad (4.32)$$

Notice that the autocorrelation function $R(\tau)$ is a function of the variable τ. Consider now the case when $\tau = 0$:

$$R(\tau)\big|_{\tau=0} = E[X(t)X(t)] = E\left[X^2(t)\right] \qquad (4.33)$$

This equals the mean square value of the random process. Note that in order to compute Equation (4.33), we must have complete knowledge of the probability density function of the random process. Referring to Figure 4.6, we see that this is equivalent to computing the following equation:

$$R(\tau)|_{\tau=0} = E[X(t_1)X(t_1)] = \int\int_{-\infty}^{\infty} x^2(t_1)f[x(t_1)]\,d[x(t_1)] \qquad (4.34)$$

For a wide-sense stationary random process, Equation (4.34) can be simplified as

$$R(\tau)|_{\tau=0} = E[X(t)X(t)] = \int\int_{-\infty}^{\infty} x^2(t)f_X(x)\,dx \qquad (4.35)$$

Consequently, the *first property* of the autocorrelation function may be expressed by the following equation:

$$R(\tau)|_{\tau=0} = E[X^2(t)] = \text{mean square value of random process} \qquad (4.36)$$

Consider Example 4.3. The random process studied in this example had the form

$$X(t) = A\cos(\omega_0 t + \theta) \qquad (4.37)$$

where A and ω_0 are constants and θ is a random variable such that

$$f(\theta) = \begin{cases} \dfrac{1}{2\pi} & 0 \le \theta \le 2\pi \\ 0 & \text{elsewhere} \end{cases} \qquad (4.38)$$

The autocorrelation function of this random process, given by Equation (4.25), stated by

$$R(\tau) = \frac{A^2}{2}\cos\omega_0\tau \qquad (4.39)$$

Based on the first property of the autocorrelation function, we can immediately say that the mean square value of the random process equals

$$R(\tau)|_{\tau=0} = \frac{A^2}{2} \qquad (4.40)$$

The *second property* of the autocorrelation function is illustrated in Example 4.4. Referring to Equations (4.37) and (4.39), we observe that the autocorrelation function of a random process that has a periodic component also has a periodic component at the same frequency. This very important property allows us to measure the periodic components of a random process.

Consider now a random process which is at least wide-sense stationary. For such a process, the autocorrelation function may be written as follows:

$$R(\tau) = E[X(t)X(t + \tau)] = E[X(t + \tau)X(t)] \tag{4.41}$$

Based on this property of random processes which are at least wide-sense stationary, we are able to express the *third property* of the autocorrelation function as follows:

$$R(\tau) = R(-\tau) \tag{4.42}$$

This property implies that the autocorrelation function is an even function of the variable τ. This property has a number of important implications and uses. Later in this chapter, when we consider the actual computation of the autocorrelation function of a time-limited system, we will see that since the autocorrelation function is an even function, we need only compute the function for either positive values or negative values of the variable τ. Another important implication arises from the fact that the Fourier transform of an even function of time is a real function of frequency. In Chapter 5 we will use this property to gain insight into the *frequency domain* properties of a random process.

Consider now a random process composed of a random component and a constant value. One could model such a process as

$$X(t) = A + B(t) \tag{4.43}$$

where $B(t)$ is a wide-sense stationary random process such that

$$E[B(t)] = 0 \tag{4.44}$$

The autocorrelation function of $X(t)$ may be computed as follows:

$$R(\tau) = E\big[[A + B(t)][A + B(t + \tau)]\big] \tag{4.45}$$

$$R(\tau) = E\Big[A^2 + AB(t + \tau) + AB(t) + B(t)B(t + \tau)\Big] \tag{4.46}$$

which equals

$$R(\tau) = A^2 + AE[B(t + \tau)] + AE[B(t)] + E[B(t)B(t + \tau)] \tag{4.47}$$

Since $E[B(t)] = 0$, Equation (4.47) is simplified as follows:

$$R(\tau) = A^2 + R_B(\tau) \tag{4.48}$$

This result implies the following: If a random process which is wide-sense stationary is composed of a constant component and a random component such that its expected value equals zero, then the autocorrelation function of the random process equals the sum of the autocorrelation function of the random component and the square of the constant value. This result is the *fourth property* of the autocorrelation function.

Example 4.5 Consider the wide-sense stationary random process of Example 4.4 and add a constant parameter to the same:

$$X(t) = A\cos(\omega_0 t + \theta) + B \tag{4.49}$$

Let us assume $\omega_0 = 2$, $A = 2$, and $B = 1$. Then

$$X(t) = 2\cos(4\pi t + \theta) + 1 \tag{4.50}$$

The autocorrelation of $X(t)$ equals

$$R(\tau) = 1 + 2\cos 4\pi\tau \tag{4.51}$$

Figure 4.12 shows the autocorrelation function of the random process. Note that the largest value of the autocorrelation function is present at $\tau = 0$.

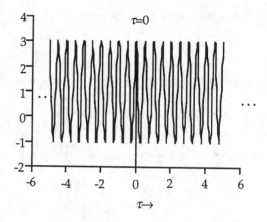

Figure 4.12 Autocorrelation function of the random process $X(t) = 2\cos(4\pi t + \theta) + 1$.

An interesting observation is the following: Suppose we know that the autocorrelation function of a wide-sense stationary random process is of the form:

$$R(\tau) = A^2 + R_B(\tau) \tag{4.52}$$

where $R_B(\tau)$ is periodic. From this information it is possible to give a qualified answer to the question, What is the dc component present in the random process? The autocorrelation functions of the following two wide-sense stationary random processes are identical:

$$\begin{cases} Y(t) = A + X(t) \\ Y(t) = -A + X(t) \end{cases} \tag{4.53}$$

And the autocorrelation function for either of these processes is

$$R_Y(\tau) = A^2 + R_X(\tau) \tag{4.54}$$

Therefore, the dc component of the random process is $\pm A$. Without some type of prior knowledge it would be possible to determine the value of the dc component, but *not* the sign.

Example 4.6 Find the mean value and the mean square value of the wide-sense stationary process with the following autocorrelation function (a plot is shown in Figure 4.13):

$$R(\tau) = 8.0 + \exp(-2|\tau|) + \cos 4\pi\tau \tag{4.55}$$

As can be observed from Figure 4.13, the autocorrelation function is made up of the sum of three distinct processes: the constant value of 8, an even exponential function, and a periodic waveform. The exponential waveform dies out rather quickly, so that after a few values of τ it does not affect the form of $R(\tau)$. At large values of the variable τ, only the constant value and the periodic waveform are significant. If Equation (4.55) were not available and the only information available about the random process were given by Figure 4.13, the computation of the dc value of the process would have to take into account the confounding effect of the periodic function. This implies that the effect of waveforms which are decaying, or are periodic in nature, must be subtracted when the constant value of an autocorrelation function is computed.

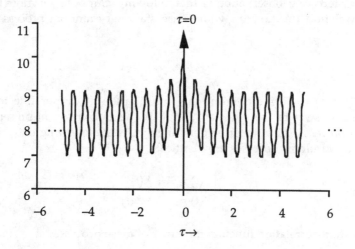

Figure 4.13 Plot of the autocorrelation function given by Equation (4.55).

TIME-AVERAGE AUTOCORRELATION FUNCTION

In Equation (4.7) we defined the time average of a random variable. Similarly, we can define the *time-average autocorrelation function* by the following equation (note the different notation):

$$R(\tau) = \lim_{T \to \infty} \frac{1}{2T} \int_{-T}^{T} x(t)x(t+\tau)\, dt \qquad (4.56)$$

If the above definition equals the probabilistic definition of Equation (4.14), then the random process is ergodic in autocorrelation function:

$$R(\tau) = R(\tau) \qquad (4.57)$$

Often the term *ergodicity* is used loosely to indicate ergodic in the mean and autocorrelation.

When dealing with engineering systems, we only have access to time-limited random data. We are often called upon to make estimates of the statistical properties of random processes when the only information available is a time-limited record of data. In such cases we will use the definition given by Equation (4.56) and *assume* that the process is ergodic. Later on in this chapter these concepts and the estimation methods will be explained in greater detail.

4.3 THE CROSS-CORRELATION FUNCTION

The cross-correlation function (CCF) describes the correlation between two random variables from different random processes. Figure 4.14 illustrates two different random processes with random variables $x(t_1)$ and $y(t_2)$.

The cross-correlation between these two random variables is defined as

$$E[X(t_1)Y(t_2)] = \int\limits_{-\infty}^{\infty}\!\!\int x(t_1)y(t_2)f[x(t_1)y(t_2)]\,d[x(t_1)]\,d[y(t_2)] \qquad (4.58)$$

Similarly to Equation (4.13), the present equation may be simplified as follows:

$$E[X_1Y_2] = \int\limits_{-\infty}^{\infty}\!\!\int x_1y_2f(x_1y_2)\,d(x_1)\,d(y_2) \qquad (4.59)$$

The formal definition of the cross-correlation function of two random processes is then given by the following equation:

$$R_{XY}(t_1,t_2) = E[X(t_1)Y(t_2)] = \int\limits_{-\infty}^{\infty}\!\!\int x_1y_2f(x_1y_2)\,d(x_1)\,d(y_2) \qquad (4.60)$$

The subscripts X and Y are added to R to indicate that this is a cross-correlation function between two different random processes. If the two processes are jointly wide-sense stationary, then the cross-correlation function $R(t_1, t_2)$ is dependent only on the time difference $\tau = t_2 - t_1$. Equation (4.60) may then be expressed as follows:

$$R_{XY}(\tau) = E[X_1Y_2] = \int\limits_{-\infty}^{\infty}\!\!\int x_1y_2f(x_1y_2)\,d(x_1)\,d(y_2) \qquad (4.61)$$

Just as with the autocorrelation function, the *time-average cross-correlation function* may be defined as follows:

$$R_{XY}(\tau) = \lim_{T\to\infty} \frac{1}{2T}\int_{-T}^{T} x(t)y(t+\tau)\,dt \qquad (4.62)$$

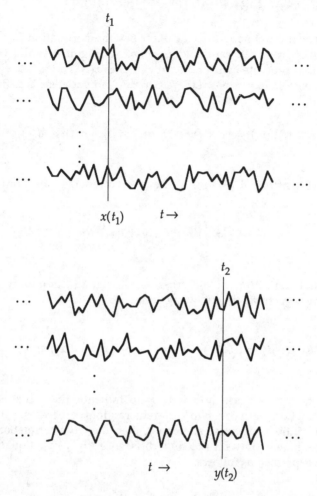

Figure 4.14 Sample functions of two random processes $X(t)$ and $Y(t)$. At times t_1 and t_2, the random variables $x(t_1)$ and $y(t_2)$ are defined.

PROPERTIES OF THE CROSS-CORRELATION FUNCTION

If the cross-correlation function between two random processes equals zero, then it is said that the processes are *orthogonal* to each other. If the two processes are independent, then the cross-correlation function equals a constant. This may be proved as follows:

$$R_{XY}(t_1 t_2) = \int\int_{-\infty}^{\infty} x_1 y_2 f(x_1) f(y_2) \, d(x_1) \, d(y_2) = E[x_1] E[y_2] \tag{4.63}$$

For processes that are jointly wide-sense stationary, the cross-correlation function has the following properties:

Property 1: The autocorrelation function is an even function of time. The cross-correlation function is not. This is a direct consequence of the defining equation for the cross-correlation function. This property is defined by the following equation:

$$R_{XY}(-\tau) = R_{YX}(\tau) \tag{4.64}$$

Property 2: In the case of the autocorrelation function, the value at zero was equal to the mean square value of the process. The value at zero of the cross-correlation function does not have any special significance.

Finally, there are two relationships between the mean square value of the processes and the value of the cross-correlation function. These relationships are defined by the following equations:

$$|R_{XY}(\tau)| \le [R_X(0) R_Y(0)]^{1/2} \tag{4.65}$$

$$|R_{XY}(\tau)| \le \frac{1}{2} |R_X(0) + R_Y(0)| \tag{4.66}$$

These relationships place limits on the maximum absolute value of the cross-correlation function.

Example 4.7 Two jointly wide-sense stationary random processes have sample functions of the form

$$X(t) = A \cos(\omega_0 t + \theta) \tag{4.67}$$

and

$$Y(t) = B \cos(\omega_0 t + \theta + \phi) \tag{4.68}$$

where θ is a random variable uniformly distributed between 0 and 2π and A, B, and ϕ are constants.

(a) Find the cross-correlation function $R_{XY}(\tau)$.

The cross-correlation function is computed as follows:

$$R_{XY}(\tau) = E[A\cos(\omega_0 t + \theta) \cdot B\cos\omega_0(t + \tau) + \theta + \phi] \qquad (4.69)$$

which yields

$$R_{XY}(\tau) = \frac{AB}{2} E[\cos(\omega_0\tau + \phi) + \cos(\omega_0 t + \omega_0\tau + \phi + 2\theta)] \qquad (4.70)$$

The second expectation in Equation (4.69) yields zero when it is integrated using θ as a random variable. The equation then yields

$$R_{XY}(\tau) = \frac{AB}{2}\cos(\omega_0\tau + \phi) \qquad (4.71)$$

(b) For what values of ϕ are $X(t)$ and $Y(t)$ orthogonal?

$X(t)$ and $Y(t)$ are orthogonal if $R_{XY}(\tau) = 0$

For values of the argument which are odd multiples of $\pi/2$, the cross-correlation function equals zero. Therefore

$$\omega_0\tau + \phi = \frac{n\pi}{2} \quad n \text{ odd}$$

which yields

$$\phi = \frac{n\pi}{2} - \omega_0\tau \quad n \text{ odd}$$

4.4 ESTIMATE OF AUTOCORRELATION FUNCTION OF SAMPLE RECORDS OF LIMITED DURATION

The process of estimating the autocorrelation function of a random process involves several assumptions and steps. An assumption that must be made is that the random process is at least wide-sense stationary. If it is not, use that the random process is at least wide-sense stationary. If it is not, use of time averages will not yield substantive results. Let us assume that we have a number of observations of a random process $X = \{x_1, x_2, x_3, \ldots, x_n\}$, where $x_1, \ldots, x_n$ are n samples of the random process observed at equispaced times $t_1, \ldots, t_n$ such that $t_i = t_1 + (i + 1)T_s$, where T_s is a fixed time step. This process of observing a random process at fixed time intervals is called *sampling*. During sampling, a continuous time signal is converted to a corresponding sequence of samples which are (typically) uniformly spaced in time. Let us assume that this random process is wide-sense stationary. Figure 4.15 shows a truncated 1-s data segment from a random process.

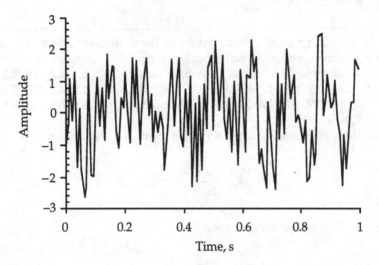

Figure 4.15 Truncated 1-s segment from a random process.

In order to estimate the autocorrelation function of the random process based on the limited observations available, we define the concept of discrete time autocorrelation

$$\hat{R}(m) = \frac{1}{n}\sum_{k=1}^{n-m}(x_k - \hat{\mu}_x)(x_{k+m} - \hat{\mu}_x) \quad 0 \le m \le n-1 \tag{4.72}$$

where

$$\hat{\mu}_x = \frac{1}{n}\sum_{i=1}^{n}x_i \tag{4.73}$$

The definition used in Equation (4.72) is that of the estimate of the *autocovariance* of the discrete time random process. And $\hat{R}(m)$ given by Equation (4.72) is actually the autocovariance function evaluated at lag mT_s. Therefore, the autocovariance of the discrete time process are samples of the continuous time process at sample interval T_s. The estimate of the *autocorrelation* function of the random process is then defined as follows:

$$\hat{\rho}(m) = \frac{\hat{R}(m)}{\hat{R}(0)} \tag{4.74}$$

It is important that we understand the type of operations involved in computing Equation (4.72).

Example 4.8 Fifty observations taken at fixed intervals from a random process are to be used to estimate the autocorrelation function of a random process. The observations are the following: {2.8, 7.8, 5.2, 6.6, 6.5, 5.4, 9.2, 4.3, 7.8, 5.1, 9.1, 9.4, 5.7, 4.9, 4.3, 6.3, 9.8, 9.8, 9.5, 9.0, 5.8, 7.1, 2.5, 4.8, 8.8, 3.3, 2.8, 3.2, 2.9, 9.3, 3.8, 3.9, 7.3, 2.8, 2.9, 9.3, 6.3, 8.5, 5.2, 4.8, 8.2, 4.8, 5.1, 8.8, 6.1, 5.7, 5.3, 8.2, 4.7, 8.5}.

First, let us compute the estimate of the mean value of the process:

$$\hat{\mu}_x = \frac{1}{50} \sum_{i=1}^{50} x_i = 6.2 \tag{4.75}$$

Equation (4.72) is now evaluated for $m = 0$:

$$\hat{R}(0) = \frac{1}{50} \sum_{k=1}^{50} (x_k - 6.2)^2 \tag{4.76}$$

Evaluation of Equation (4.76) yields

$$\hat{R}(0) = \frac{1}{50}\Big[(2.8 - 6.2)^2 + (7.8 - 6.2)^2 + \cdots + (8.5 - 6.2)^2\Big] = 5.1 \tag{4.77}$$

Equation (4.72) is now evaluated for $m = 1$:

$$\hat{R}(1) = \frac{1}{50} \sum_{k=1}^{49} (x_k - 6.2)(x_{k+m} - 6.2) \tag{4.78}$$

which yields

$$\hat{R}(1) = \frac{1}{50}\Big[(x_1 - 6.2)(x_2 - 6.2) + (x_2 - 6.2)(x_3 - 6.2)$$
$$+ \cdots + (x_{49} - 6.2)(x_{50} - 6.2)\Big] \tag{4.79}$$

which equals

$$R(1) = \frac{1}{50}[(2.8-6.2)(7.8-6.2)+(7.8-6.2)(5.2-6.2)$$
$$+ \cdots +(4.7-6.2)(8.5-6.2)] = 0.2 \qquad (4.80)$$

Note that the summation now goes from $k = 1$ to $k = 49$, that is, the last term of the summation is $(x_{49} - 6.2)(x_{50} - 6.2)$. Also note that the denominator in front of the summation sign always remains at $n = 50$ the total number of points. For completeness, the next few values of the autocovariance function are

$$\hat{R}(2) = .3$$
$$\hat{R}(3) = .4 \qquad (4.81)$$
$$\hat{R}(4) = -.9$$

From these values it is easy to compute the values of the autocorrelation function:

$$\rho(0) = \frac{\hat{R}(0)}{\hat{R}(0)} = 1.0$$

$$\rho(1) = \frac{\hat{R}(1)}{\hat{R}(0)} = .1$$

$$\rho(2) = \frac{\hat{R}(2)}{\hat{R}(0)} = .1 \qquad (4.82)$$

$$\rho(3) = \frac{\hat{R}(3)}{\hat{R}(0)} = .1$$

$$\rho(4) = \frac{\hat{R}(4)}{\hat{R}(0)} = -.4$$

It is important to emphasize that the measurements in Example 4.8 are all *estimates* of the autocovariance and autocorrelation function of the random process. The choice of n^{-1} as the term in front of the summation sign guarantees that properties of the estimates are consistent with the statistical properties of the random process.

One important aspect of the estimation process is the relationship between the number of observations n and the largest value of lags m one should consider taking. At one extreme, when m is close to n, the number of pairs used in the estimation process is very small. One could argue that the estimate in such a case will be unreliable because of the small number of pairs involved. Useful estimates of the autocorrelation (and cross-correlation) can be made as long as m is roughly 25 percent of n.

The number of computations required to compute the autocorrelation function depends directly on the number of observations. The actual time is primarily a function of the number of multiplications and the number of summations. Typical cycle times in a microprocessor or larger computer are mainly affected by multiplications and not as much by summations. In the case of an autocorrelation function the number of multiplications involved is the following:

$$\text{Number of multiplications } N = n + (n-1) + \cdots + (n-m) \qquad (4.83)$$

where n is the total number of points. In terms of a series, this equals

$$N = \sum_{k=0}^{m} (n-k) \qquad (4.84)$$

which equals

$$N = (m+1)n^2 + \sum_{k=0}^{m} k^2 - 2n \sum_{k=0}^{m} k \qquad (4.85)$$

This equation may be further evaluated as follows:

$$N = (m+1)n^2 + 1 + \frac{m(m+1)(2m+1)}{6} - 2n \cdot \frac{1}{2} [n(n+1)] \qquad (4.86)$$

For the given example, $n = 50$, $m = 10$, and the total number of multiplications equals 22,385. These are the values for computation of the autocorrelation function for positive values of τ. Since the autocorrelation function is an even function of time, the negative values of τ need not be computed.

4.5 CHAPTER SUMMARY

A random process for which the nth-order joint probability density functions do not depend on the choice of time origin is called a *stationary random process.* A random process is said to be a *wide-sense stationary* process if the following conditions are met:

1. The *expected value* of the random process must be *independent* of the time origin.
2. The correlation and the covariance between any two of the random variables must depend only on the *time difference* between the two random variables.

   A random process is called a *deterministic random process* if all the future values can be predicted from the past observations.

   For many random processes their time average as T goes to infinity will approach the ensemble mean. Such random processes are called *ergodic* (in the mean) random processes.

   The following equation describes, in the most general terms, the *autocorrelation* of a random process.

$$E[X_1X_2] = \int\int_{-\infty}^{\infty} x_1x_2 f(x_1, x_2)\, d(x_1)\, d(x_2)$$

For a wide-sense stationary random process the autocorrelation function may be expressed as follows:

$$R(\tau) = E[X(t)X(t + \tau)]$$

   The *autocorrelation function* has a number of properties that relate to the physical properties of the random process. The *first property* of the autocorrelation function may be expressed by the following equation:

$$R(\tau)\big|_{\tau=0} = E[X^2(t)] = \text{mean square value of random process}$$

   The *second property* of the autocorrelation function is the following: The autocorrelation function of a random process that has a periodic component also has a periodic component at the same frequency. This very important property allows us to measure the periodic components of a random process.

   The *third property* of the autocorrelation function is that the autocorrelation function is an even function of the variable τ. The *time average autocorrelation function* may be defined by the following equation:

$$R(\tau) = \lim_{T\to\infty} \frac{1}{2T} \int_{-T}^{T} x(t)x(t + \tau)\, dt$$

The *cross-correlation function* between two random variables is defined as

$$E[X(t_1)Y(t_2)] = \int\int_{-\infty}^{\infty} x(t_1)y(t_2)f[x(t_1)y(t_2)]\, d[x(t_1)]\, d[y(t_2)]$$

The formal definition of the cross-correlation function of two random processes is then given by the following equation:

$$R_{XY}(t_1,t_2) = E[X(t_1)Y(t_2)] = \int\int_{-\infty}^{\infty} x_1 y_2 f(x_1 y_2)\, d(x_1)\, d(y_2)$$

The *time average cross-correlation function* may be defined as follows:

$$R_{XY}(\tau) = \lim_{T \to \infty} \frac{1}{2T} \int_{-T}^{T} x(t)y(t+\tau)\, dt$$

Properties of the cross-correlation function: If the crosscorrelation function between two random processes equals zero, then it is said that the processes are *orthogonal* to each other. If the two processes are independent, then the cross-correlation function equals a constant.

For processes that are jointly wide-sense stationary, the cross-correlation function has the following properties:

1. The autocorrelation function is an even function of time. The cross-correlation function is not.
2. In the case of the autocorrelation function, the value at zero was equal to the mean square value of the process. The value at zero of the cross-correlation function does not have any special significance.

In order to estimate the autocorrelation function of the random process based on the limited observations available, we define the concept of discrete time autocorrelation.

$$\hat{R}(m) = \frac{1}{n} \sum_{k=1}^{n-m} (x_k - \hat{\mu}_x)(x_{k+m} - \hat{\mu}_x) \quad 0 \le m \le n-1$$

4.6 PROBLEMS

1. A random process has sample functions of the form

$$X(t) = At$$

where A is a random variable uniformly distributed between -3 and 3.

(a) Is the process discrete or continuous?

(b) Is the process deterministic or nondeterministic?

(c) Find $E[X(t)]$.

(d) Find the probability density function $f_{X_1}(x_1)$ where $X_1 = X(5)$.

2. A random process has sample functions of the form

$$X(t) = Y + Z$$

where Y and Z are statistically independent gaussian random variables with the following parameters:

$$\mu_Y = 0 \quad \sigma_Y^2 = 9$$
$$\mu_Z = 5 \quad \sigma_Z^2 = 16$$

(a) Is the process discrete or continuous?

(b) Is the process deterministic or nondeterministic?

(c) Find $E[X(t)]$.

(d) Find $E[X^2(t)]$.

3. A random process has sample functions of the form

$$X(t) = A \tan(\omega t + \theta)$$

where θ is a random variable uniformly distributed between $-\pi$ and π.

(a) Is the process discrete or continuous?

(b) Is the process deterministic or nondeterministic?

(c) Find $E[X(t)]$.

4. A random process has sample functions of the form

$$X(t) = Ae^{-|t|}$$

where A is a random variable which is uniformly distributed between 0 and 1.

(a) Is this process stationary in the wide sense? Prove your conclusion.

(b) Find the mean square value of the process.

5. A random process has sample functions of the form

$$X(t) = A \sin \omega t$$

where A is a random variable which is uniformly distributed between -1 and 1.

(a) Is this process stationary in the wide sense? Prove your conclusion.

(b) Find the mean square value of the process.

6. A sample function from an ergodic random process is sampled at 10 widely separated times with the following results:

$$\{6, \ 4, \ 8, \ 1, \ 1, \ 5, \ 4, \ 6, \ 5, \ 7\}$$

Estimate the mean and the variance of this process.

7. Consider a random process in which a typical sample function is given by

$$X(t) = A \sin(\omega t + \theta)$$

where A and θ are statistically independent random variables and ω is a constant. The random variable A is uniformly distributed from 0 to 10, and the random variable θ is uniformly distributed from 0 to π. What is the autocorrelation function of the process?

8. Suppose now that the sample functions are given by

$$X(t) = A \cos(\omega t + \theta)$$

where A and θ are statistically independent random variables, θ is uniformly distributed from 0 to 2π, and A is gaussian distributed with mean 0 and variance 4.

(a) Is the random process stationary? Explain.

(b) Is the random process ergodic? Explain.

(c) What is the autocorrelation function of the random process?

9. A noise voltage $V(t)$ has an autocorrelation function given by

$$R_V(\tau) = 4\delta(\tau) + 2e^{|-b\tau|}$$

The signal travels to a receiver by two different paths, as shown below.

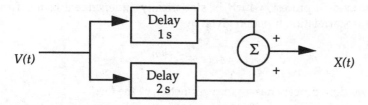

Find the autocorrelation function of $X(t)$, denoted by $R_X(t)$.

10. A noise voltage $n(t)$ has an autocorrelation function $R_n(\tau)$. It travels to a receiver by two different paths, in one of which there is a time delay of T s, as shown below. At the terminals of the receiver appears the voltage.

$$e(t) = \frac{1}{2}[n(t) + n(t - T)]$$

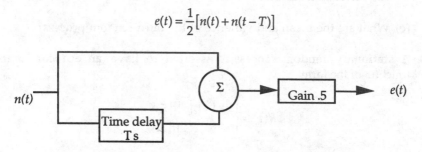

Find the autocorrelation of $e(t)$, denoted by $R_e(t)$, in terms of $R_n(t)$.

11. An ergodic random process has an autocorrelation function of the form

$$R_X(\tau) = 10e^{-|\tau|} - Ae^{-2|\tau|}$$

where A is a constant.

(a) Find the mean value of the process.

(b) Find the variance of the process.

(c) Find the largest value of A for which this expression can be a valid autocorrelation function.

12. A random process has an autocorrelation function given by

$$R_X(\tau) = e^{-2|\tau|}\cos\omega\tau + 25$$

A second random process, which is statistically independent of the first, has an autocorrelation function given by

$$R_Y(\tau) = 4\delta(\tau) + 2e^{-|b\tau|}$$

A third random process has a sample function of the form

$$Z(t) = 2X(t)Y(t)$$

(a) What are the mean and variance of the first random process? Sketch the correlation function.

(b) What are the mean and the variance of the second process? Sketch the correlation function.

(c) What are the mean and variance of the third random process?

13. A stationary random process is asserted to have an autocorrelation function of the form

$$R(\tau) = \begin{cases} 10(1-|\tau|) & |\tau| \le K \\ 0 & |\tau| > K \end{cases}$$

What is the largest value of K for which this could be a valid autocorrelation function?

14. Two stationary random zero-mean processes have variances of 25 each and a cross-correlation function of

$$R_{XY}(\tau) = \begin{cases} 25(2-\tau) & 0 \leq \tau \leq 2 \\ 0 & \text{elsewhere} \end{cases}$$

What value of the parameter K will minimize the mean square value of $Z(t) = X(t) - KY(t + 1)$?

15. Two stationary zero-mean random processes have variances of 100 each and a cross-correlation function of the form

$$R_{XY}(\tau) = 50e^{-(\tau-4)^2}$$

A random process $Z(t)$ is formed from $Z(t) = X(t) + Y(t + t_1)$. For what value of t_1 does $Z(t)$ have the largest variance?

16. Using MATHCAD or MATLAB, read the data labeled "nstatv."

   (a) Plot the data.

   (b) Compute the mean as a function of time but use adjacent as opposed to overlapped windows. Let the window length be 25 values long. Plot these data.

   (c) Compute the estimate of the median and the variance, using the same size window.

   Compare your result with those obtained in Computer Examples 4.1.

17. Consider the following function of time:

$$X(t) = 2e^{-t} \quad t \geq 0$$

   Refer to MATHCAD or MATLAB Computer Examples 4.3.

   (a) Obtain 128 digital values of the function of time every .02 s.

   (b) Plot the data.

   (c) Compute the autocorrelation function of the data, using a time domain technique. Use only 30 autocorrelation lags.

(d) Add now random noise to the function.

(e) Plot the data.

(f) Compute and plot the autocorrelation function.

(g) Compare this autocorrelation function with the one obtained earlier with pure data. Explain the differences.

18. Compute the cross-correlation function by a time domain technique. Refer to MATHCAD or MATLAB Computer Example 4.4. Consider the following functions:

$$X(t) = A\cos(\omega_0 t + \theta)$$

and

$$Y(t) = B\cos(\omega_0 t + \theta + \vartheta)$$

where A, B, f_0, and ϕ are constants and θ is uniformly distributed between 0 and 2π.

(a) Select at random a value for θ. Plot $X(t)$ and $Y(t)$.

(b) Compute the time domain cross-correlation function. Plot the results.

(c) Select a different value for θ. Plot $X(t)$ and $Y(t)$.

(d) Compute the time domain cross-correlation function. Plot the results.

(e) Compare the results.

19. Consider the following functions:

$$X(t) = A\cos(\omega_0 t + \theta)$$

and

$$Y(t) = B\cos(\omega_0 t + \theta + \vartheta) + N(t)$$

where $N(t)$ is random noise.

(a) Compute the cross-correlation between $X(t)$ and $Y(t)$. Plot the data.

(b) Increase the amplitude of the noise. Consider using increasing amounts of S/N ratio.

(c) Compute the cross-correlation between $X(t)$ and $Y(t)$. Plot the data.

(d) Compare the results of the cross-correlation function as the amount of noise increases.

20. Using MATHCAD or MATLAB, generate 100 values of noise. Call this data $X(t)$. Generate a second set of 100 values of noise. Call this data $Y(t)$.

(a) To $X(t)$ add a symmetrical triangular waveform centered about point number 25.

(b) To $Y(t)$ add a symmetrical triangular waveform centered about point number 55.

(c) Obtain the time domain cross-correlation between $X(t)$ and $Y(t)$.

(d) Comment on the results.

4.7 COMPUTER EXAMPLES

MATHCAD Computer Example 4.1
WIDE-SENSE STATIONARY versus NONSTATIONARY RANDOM PROCESSES . . . Some Insights

In order to show that a random process is stationary, we must show that all the higher-order moments do not change with time. This is difficult to prove, and for most applications the term *stationary* refers to the more relaxed requirement of wide-sense stationariness. A process which is stationary in the wide sense has a mean and variance which do not change as a function of time. For sampled data this means that if the estimated mean and variance change appreciably as a function of time through the sample, then the process is most likely nonstationary. (Refer to the text for a more detailed definition.) We will start by reading in some data sets and testing them for wide-sense stationariness by computing their mean and variance as a function of time.

$k := 0..199$

$X_k := \text{READ(nstatv)}$

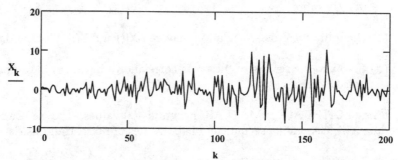

In order to compute the mean as a function of time, we will use a moving average with a window length of 50. This means we will compute the mean based on the first 50 points (1 , . . . , 50) then on the next 50 points (2 , . . . , 51), etc., out to the end of the sample, which for our case will be 149 since we have a 200-point sample to work with. This will give us some indication how the mean is changing with time.

$i := 0 .. 149$

$j := 0 .. 49$

$$M_i := \frac{1}{50} \cdot \sum_j X_{j+i}$$

Compute the moving average (mean as a function of time). A plot of the mean versus time is shown below.

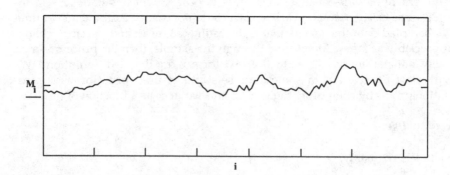

Now we will do the same thing with the variance, again using a moving average of window length 50. Since we already have the mean values M, the estimate for the variance is as follows:

$m := 0 .. 149$

$$l := 0 .. 49$$

$$V_m := \frac{1}{49} \cdot \sum_l \left(X_{l+m} - M_m \right)$$

And now we can plot the variance as a function of time.

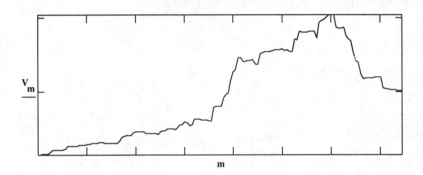

MATHCAD Computer Example 4.2
THE AUTOCORRELATION FUNCTION

The first part of this example computes the autocorrelation function of the same numbers up to lag 10.

Enter the number of points in the example.

$$M := 50$$

$$m := 0 .. M - 1$$

Define now the random numbers for the example:

c0=2.8, c1=7.8, c2=5.2, c3=6.6, c4=6.5, c5=5.4, c6=9.2, c7=4.3, c8=7.8, c9=5.1, c10=9.1, c11=9.4, c12=5.7, c13=4.9, c14=4.3, c15=6.3, c16=9.8, c17=9.8, c18=9.5, c19=9.0, c20=5.8, c21=7.1, c22=2.5, c23=4.8, c24=8.8, c25=3.3, c26=2.8, c27=3.2, c28=2.9, c29=9.3, c30=3.8, c31=3.9, c32=7.3, c33=2.8, c34=2.9, c35=9.3, c36=6.3, c37=8.5, c38=5.2, c39=4.8, c40=8.2, c41=4.8, c42=5.1, c43=8.8, c44=6.1, c45=5.7, c46=5.3, c47=8.2, c48=4.7, c49=8.5

Find the mean value. Note that there are 50 values of the random variable:

$$\text{mean} := \frac{\sum c}{50}$$

$$\text{mean} = 6.184$$

Subtract the mean value from each data value:

$$a_m := \left(c_m - \text{mean} \right)$$

Extend with zeros in order to avoid circular correlation:

$$a_{M+m} := 0$$

Lag values are needed:

$$\tau := 0 .. 10$$

Compute now the autocorrelation:

$$r_\tau := \frac{1}{M} \cdot \sum_m a_m \cdot a_{m+\tau}$$

$$n := 0 .. 5$$

$$r0 := r$$

These are the first values of the autocorrelation:

$\dfrac{r_n}{r0}$
1
0.031
0.061
0.086
- 0.177
0.17

r_n
5.064
0.157
0.309
0.433
- 0.897
0.863

Now let's use a cosinusoid with added random noise. In the present case we use a cosinusoid with a random phase. The random phase is uniformly distributed between 0 and 2π.

Generate 300 values and use an amplitude of 2.0:

$i := 1, 2 .. 300$

$\theta := rnd(2 \cdot \pi)$

$A := 2.0$

$M := 300$

Note that each time that the document is recalculated, the value of the phase changes:

$\theta = 0.008$

Define now the equation:

$v_i := A \cdot \cos(2 \cdot \pi \cdot 3.0 \cdot i \cdot 0.01 + \theta)$

Now add some noise to the random variable:

$x_i := v_i + 2 \cdot rnd(2) - 2$

Compute and subtract the mean:

$$mean := \frac{\sum x}{300}$$

$y_i := x_i - mea$

The number of autocorrelation values desired is

$\tau := 0 .. 100$

Extend with zeros:

$k := 0 .. 100$

$y_{M+k} := 0$

Compute the autocorrelation:

$$\mathbf{r}_\tau := \frac{1}{M} \cdot \sum_k \mathbf{y}_k \cdot \mathbf{y}_{\tau +}$$

Plot the autocorrelation:

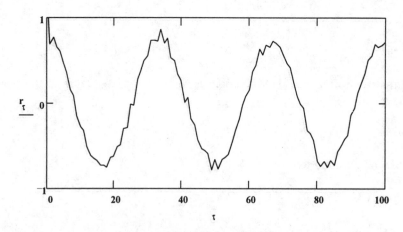

Noe explain the autocorrelation plot. Change the value of the amplitude of the noise and recompute. Plot the autocorrelation function and explain any changes in the graph. Add a dc component to the cosinusoid and explain the results.

MATHCAD Computer Example 4.3
COMPUTATION OF THE AUTOCORRELATION FUNCTION BY A TIME DOMAIN TECHNIQUE

The number of points is

$\mathbf{M} := \mathbf{128}$

$\mathbf{m} := \mathbf{0, 1 .. 127}$

Define now the equation:

$$\mathbf{x}_m := 2 \cdot \exp(-3 \cdot m \cdot 0.02)$$

Compute and subtract the dc value:

$$\mathbf{mean} := \frac{\sum \mathbf{x}}{128}$$

$y_m := x_m - mea$

Add zeros:

$y_{M+m} := 0$

The number of autocorrelation lags desired is

$\tau := 0 \ .. \ 128$

Compute now the autocorrelation:

$$r_\tau := \frac{1}{M} \cdot \sum_m y_m \cdot y_{\tau +}$$

$r0 := r$

Plot the data:

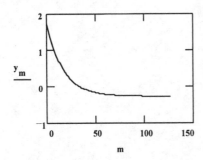

Plot the autocorrelation function:

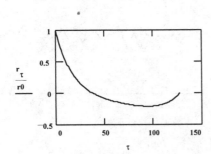

Now, let's compute the autocorrelation function of random noise. We generate the noise first:

$y_m := rnd(1)$

Compute and subtract the dc value:

$$mean := \frac{\sum y}{128}$$

$y_m := y_m - mea$

$M := 128$

Add zeros:

$y_{M+m} := 0$

The number of autocorrelation lags desired is

$\tau := 0 .. 50$

Compute now the autocorrelation:

$$r_\tau := \frac{1}{M} \cdot \sum_m y_m \cdot y_{\tau +}$$

$r0 := r$

Plot the autocorrelation function:

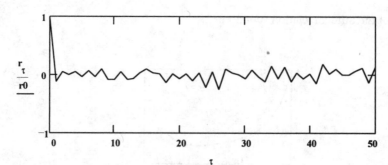

Is this the type of plot that you would expect? Why?

Add random noise to the function:

$$x_m := \left(x_m + \text{rnd}(1)\right) - 1$$

Compute and subtract the dc value:

$$\text{mean} := \frac{\sum x}{128}$$

$$x_m := x_m - \text{mea}$$

Plot the data:

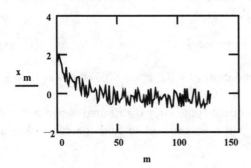

Add zeros:

$$x_{M+m} := 0$$

Compute and plot the autocorrelation function:

$$r_\tau := \frac{1}{M} \cdot \sum_m x_m \cdot x_{\tau +}$$

$$r0 := r$$

Plot the autocorrelation function:

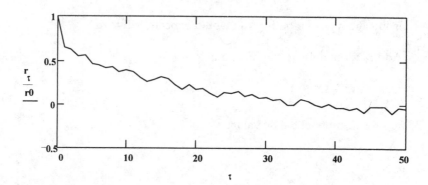

Compare this autocorrelation function with the one shown on page 245. Explain the differences.

MATHCAD Computer Example 4.4
COMPUTATION OF THE CROSS-CORRELATION FUNCTION BY A TIME DOMAIN TECHNIQUE

Consider the following processes:

$$X(t) = A \cos(\omega_0 t + \theta) \quad \text{and} \quad Y(t) = B \cos(\omega_0 t + \theta + \phi)$$

where θ is a random variable uniformly distributed between 0 and 2π and A, B, and ϕ are constants. We now proceed to find the cross-correlation function $R_{xy}(t)$.

First, define some of the constants:

$A := 1.0$
$B := 1.0$
$\phi := 1.6$

Generate 300 points for each of the two functions:

$i := 1 .. 300$

$M := 300$

$f_0 := 3.0$

Pick at random a phase angle:

$\theta := rnd(2 \cdot \pi)$

Define now the first equation:

$$x_i := A \cdot \cos\left(2 \cdot \pi \cdot f_0 \cdot i \cdot 0.01 + \theta\right)$$

Define now the second equation:

$$y_i := B \cdot \cos\left(2 \cdot \pi \cdot f_0 \cdot i \cdot 0.01 + \theta + \phi\right)$$

Plot the two functions:

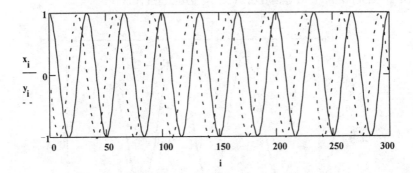

We wish to compute the entire cross-correlation between the two functions of time. We extend each of the arrays with zeros. Note that in this case both functions have the same length.

$$x_{M+i} := 0$$

$$y_{M+i} := 0$$

$$K :=$$

The lagged values desired are

$$\tau := -50, -49 .. 50$$

$$K = 300$$

$$k := 0 .. K - 1$$

$$z_\tau := \sum_k x_k \cdot y_{k+\tau}$$

z0 := z

The theoretical ACF is given by

$$r_\tau := \frac{A \cdot B}{2.0} \cdot \cos\left(2 \cdot \pi \cdot f_0 \cdot \tau \cdot 0.01 + \phi\right)$$

r0 := r

Plot the calculated and theoretical autocorrelation function:

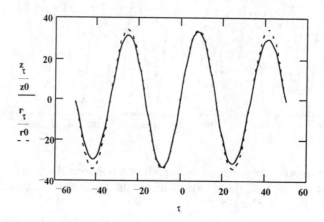

Please draw conclusions from this example.

MATHCAD Computer Example 4.5
USES OF THE CROSS-CORRELATION FUNCTION

Note: For this program to work properly, open the Math window and click on the Built-in Variables option. Change the origin option to -100.

In order to demonstrate one of the uses of the CCF, consider two square waves with random noise added.

$$R(n, N) \equiv (n \geq 0) \cdot (n \leq 19)$$

$$P(n, M) \equiv (n \geq 30) \cdot (n \leq 50)$$

$$N := 20$$

$$M := 20$$

$n := 0, 1 .. 299$

$x_n := 0.5 \cdot rnd(1) - 0.25$

$y_n := R(n, N) + x_n$

$z_n := P(n, M) + x_n$

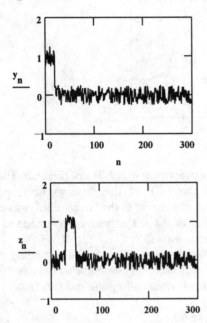

The only difference between the two functions is the position of the square wave. We now compute their CCF.

Extend each array with zeros:

$y_{300+n} := 0$

$z_{300+n} := 0$

$K := 300$

The lagged values desired are

$\tau := -100, -99 .. 100$

$$k := 0 .. K - 1$$

Compute and plot the cross-correlation:

$$yz_\tau := \sum_k y_k \cdot z_{k+\tau}$$

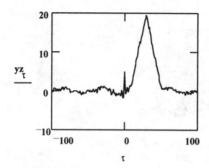

The interpretation of the preceding graph is not difficult. There is a peak at $t = 0$; this corresponds to the CCF of the random noise present in both waveforms. The second point of interest is the "triangular" wave present in the latter part of the CCF. The value of t at the peak corresponds to the separation in time between the two square waves.

In order to appreciate these results, return to the beginning of the program, and increase the amount of random noise present. Please remember to take into account the dc value introduced when calling the rnd function.

MATLAB Computer Example 4.1
WIDE SENSE STATIONARY VERSUS NONSTATIONARY RANDOM PROCESSES ... Some Insights

In order to show that a random process is stationary, we must show that all the higher-order moments do not change with time. This is difficult to prove, and for most applications the term *stationary* refers to the more relaxed requirement of wide-sense stationary. A process which is stationary in the wide sense has a mean and variance which do not change as a function of time. For sampled data this means that if the estimated mean and variance change appreciably as a function of time through the sample, then the process is most likely nonstationary. (Refer to the text for a more detailed definition.) We start by reading in some data sets and testing them for wide-sense stationariness by computing their mean and variance as a function of time.

```
%=============================================
% Input parameters
            signal1='nstatv.dat';
```

```
%===============================================

eval(['load ',signal1]);
limit=length(signal1)-4;
signal1=signal1(1:limit);
X_k=eval(signal1);
samples=length(X_k);
meanx=mean(X_k);
stdevx=std(X_k);
for i=1:1:samples-50,
 M_i(i)=mean(X_k(i:i+50));
 V_i(i)=std(X_k(i:i+50)).^2;
end
clg;s1=subplot(2,2,1);
plot(1:1:samples,X_k,'r');axisn(1:1:samples);
xlabel('k','FontSize',[10]);
ylabel('X_k','FontSize',[10]);
title('Original signal','FontSize',[10]);
set(s1,'FontSize',[10])
s2=subplot(2,2,2);
plot((1:1:samples-50),M_i,'r'); axisn(1:1:samples-50);
xlabel('i','FontSize',[10]);
ylabel('M_i','FontSize',[10]);
title('Mean value ( i )','FontSize',[10]);
set(s2,'FontSize',[10])
s3=subplot(2,2,3);
plot((1:1:samples-50),V_i,'r'); axisn(1:1:samples-50);
xlabel('i','FontSize',[10]);
ylabel('V_i','FontSize',[10])
title('Standard deviation ( i )','FontSize',[10]);
set(s3,'FontSize',[10])
s4=subplot(2,2,4);
axis('off');
set(s4,'Drawmode','fast'),
text(0,0.6,['Mean value: ',num2str(meanx)] ,'FontSize',[10]);
text(0, 0.4,['Standard deviation: ',num2str(stdevx)] ,'FontSize',[10]);
```

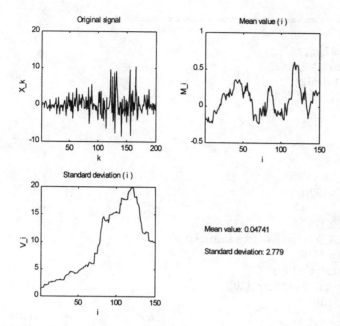

MATLAB Computer Example 4.2
THE AUTOCORRELATION FUNCTION

The first part of this example computes the autocorrelation function of the same numbers up to lag 10.

```
%===========================================
% Input parameter
           signal='dat51.dat';
%===========================================
eval(['load ',signal]);
limit=length(signal)-4;
signal=signal(1:limit);
c_k=eval(signal);
samples=length(c_k);
cmean=mean(c_k);
c_k=c_k-cmean;
r_k=xcorr(c_k,'biased');
clg;s1=subplot(2,2,1);
plot(1:1:samples,c_k,'r');axisn(1:1:samples);
xlabel('k','FontSize',[10]);
ylabel('c_k','FontSize',[10]);
title('Random variable','FontSize',[10]);
set(s1,'FontSize',[10])
```

```
s2=subplot(2,2,2);
plot(1:1:samples,c_k,'.r');axisn(1:1:samples);
xlabel('k','FontSize',[10]);
ylabel('c_k','FontSize',[10]);
set(s2,'FontSize',[10])
s3=subplot(2,2,3);
axis('off')
text(0.2,1,'Autocorrelation function:','FontSize',[10]);
for j=0:6,
 set(text(0.6,0.14*(6-j),['
',num2str(r_k(50+j))]),'HorizontalAlignment','right','FontSize',[10]);
end;
s4=subplot(2,2,4);
axis('off')
for j=0:6,
 text(0,1,'Normalized autocorrelation function: ','FontSize',[10]);
 set(text(0.6,0.14*(6-j),['
',num2str(r_k(50+j)/r_k(50))]),'HorizontalAlignment','right','FontSize',[10]);
end
```

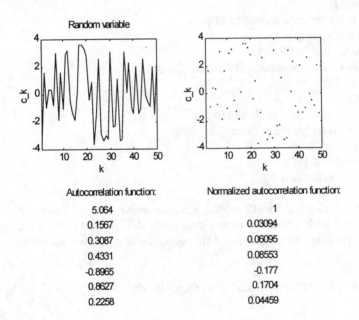

Autocorrelation function:	Normalized autocorrelation function:
5.064	1
0.1567	0.03094
0.3087	0.06095
0.4331	0.08553
-0.8965	-0.177
0.8627	0.1704
0.2258	0.04459

```
str=str2mat(...
```
    'Now let's use a cosinusoid with added random noise. We will use a
noisy cosinusoid signal with',...
    'a random phase uniformly distributed between 0 and 2*pi. We will
generate 300 signals of this',...

```
        'sequence and use an amplitude of 2.',...
        ' ',...
        '>> x(i)=A*cos(2*pi*f0*i*0.01+ theta) + A_r*(rand) - dc',...
        ' ');
disp(str);
%=============================================
% Input parameters
                A=2;
                A_r=4;
                dc=2;
                theta=0.008;
                f0=3.0;
                M=300;
%=============================================
i=0:1:M-1;
v_i=A*cos(2*pi*f0*i*0.01+theta);
x_i=v_i+A_r*rand(1,M)-dc;
y_i=x_i-mean(x_i);
r_k=xcorr(y_i,'biased');
clg;s1=subplot(211);
plot((0:1:100),y_i(1:101),'r'); axisn(0:1:100);
xlabel('i','FontSize',[10]);
ylabel('y_i','FontSize',[10]);
title('Noisy cosinusoid sequence','FontSize',[10]);
set(s1,'FontSize',[10])
s2=subplot(212);
plot((0:1:100),r_k(M:400),'r'); axisn(0:1:100);
title('Autocorrelation function','FontSize',[10]),
xlabel('k','FontSize',[10]);
ylabel('r_k','FontSize',[10]);
set(s2,'FontSize',[10]);clear;
```

Now let's use a cosinusoid with added random noise. We will use a noisy cosinusoid signal with a random phase uniformly distributed between 0 and 2*pi. We will generate 300 signals of this sequence and use an amplitude of 2.

```
>> x(i)=A*cos(2*pi*f0*i*0.01+ theta) + A_r*(rand) - dc
```

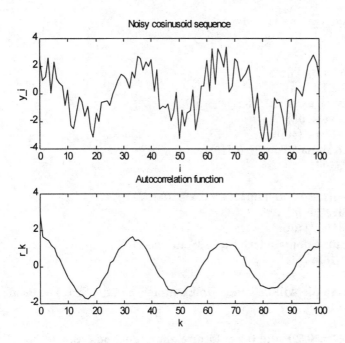

Change the value of the amplitude of the noise and recompute. Plot the autocorrelation function and explain any changes in the graph. Add a dc component to the cosinusoid and explain the results.

MATLAB Computer Example 4.3
COMPUTATION OF THE AUTOCORRELATION FUNCTION BY A TIME DOMAIN TECHNIQUE

```
%==============================================
% Input parameter
            M=128;
%==============================================
str=str2mat(...
    'The following example will be similar to Example 5.6 in the text. The
used signal is defined as',...
    ' ',...
    '>> x(m)=2*exp(-3*m*0.02)  and the autocorrelation function estimator
as:',...
    ' ',...
    '>> r(k)=(1/M)*sum(x(m)*x(m+k))');
disp(str);
m=0:1:M-1;
```

```
x_m=2*exp(-3*m*0.02);
mu=mean(x_m);
y_m=x_m-mu;
r_tau=xcorr(y_m,'biased');
clg;s1=subplot(2,1,1);
plot(0:1:M-1,x_m,'r'); axisn(0:1:M-1);
xlabel('m','FontSize',[10]);
ylabel('y_m','FontSize',[10]);
title('Exponential signal','FontSize',[10]);
set(s1,'FontSize',[10]);
s2=subplot(2,1,2);
plot(0:1:M-1,r_tau(128:255)/r_tau(128),'r');axisn(0:1:M-1);
xlabel('k','FontSize',[10]);
ylabel('r_k','FontSize',[10]);
title('ACF using time domain technique','FontSize',[10])
set(s2,'FontSize',[10]);
```

The following example will be similar to Example 5.6 in the text. The used signal is defined as

>> x(m)=2*exp(-3*m*0.02) and the autocorrelation function estimator as

>> r(k)=(1/M)*sum(x(m)*x(m+k))

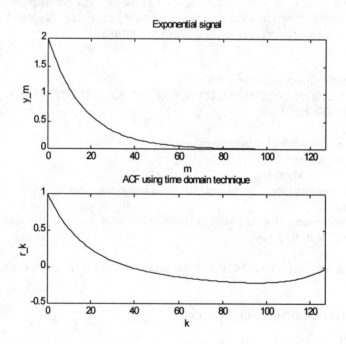

```
str=str2mat(...
     'Now, let's generate and compute the autocorrelation function of random
noise sequence. The mean',...
     'value is subtracted.');
disp(str);
y_m=rand(1,M);
ymean=mean(y_m);
y_m=y_m-ymean;
r_k=xcorr(y_m,'biased');
r_k=r_k/r_k(M);
s1=subplot(211);
plot(1:1:M,y_m,'r'); axisn(1:1:M);
xlabel('m','FontSize',[10]);
ylabel('y_m','FontSize',[10]);
title('Noise sequence','FontSize',[10])
set(s1,'FontSize',[10]);
s2=subplot(212);
plot(0:1:49,r_k(128:177),'r'); axisn(0:1:49);
xlabel('k','FontSize',[10])
ylabel('r_k','FontSize',[10]);
title('Normalized autocorrelation function','FontSize',[10]),
set(s2,'FontSize',[10]);
str=str2mat(...
     ' ',...
     'Is this the type of plot that you would expect. Why ??.');
disp(str);
```

Now, let's generate and compute the autocorrelation function of random noise
sequence. The mean value is subtracted.

Is this the type of plot that you would expect? Why?

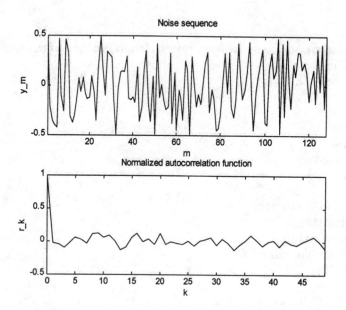

```
str=str2mat(...
    'Now, let's add the noise sequence to the exponential data and compute
the autocorrelation function of',...
    'this noisy sequence.');
disp(str);
x_m=x_m+rand(1,M)-1;
r_k=xcorr(x_m,'biased');
r_k=r_k/r_k(M);
k=0:1:59;
clg;s1=subplot(211);
plot((1:1:M),x_m,'r'); axisn((1:1:M));
xlabel('m','FontSize',[10]);
ylabel('x_m','FontSize',[10]);
title('Noisy sequence','FontSize',[10]);
set(s1,'FontSize',[10]);
s2=subplot(212);
plot(0:1:49,r_k(128:177),'r'); axisn(0:1:49);
xlabel('k','FontSize',[10]);
ylabel('r(k)','FontSize',[10]);
title('Normalized autocorrelation function','FontSize',[10]),
set(s2,'FontSize',[10]);clear
```

Now, let's add the noise sequence to the exponential data and compute the
autocorrelation function of this noisy sequence.

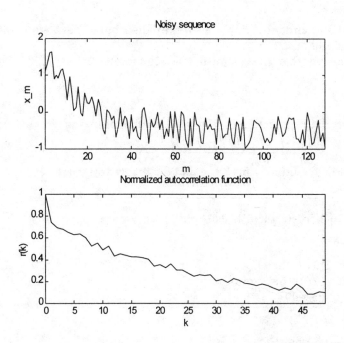

Compare this autocorrelation function with the one shown in the first figure. Explain the differences.

MATLAB Computer Example 4.4
COMPUTATION OF THE CROSS-CORRELATION FUNCTION BY A TIME DOMAIN TECHNIQUE

```
%===========================================
% Input Parameters
        A=1;
        B=1;
        phi=1.6;
        theta=0.008;
        f0=3;
        M=300;

%===========================================
str=str2mat(...
    'The processes used have the following form:',...
    ' ',...
    '>> x(i)=A*cos(2*pi*f0*i*0.01+ theta)',...
    '>> y(i+1)=B*cos(2*pi*f0*i*0.01+theta+phi)',...
    ' ',...
```

'Where theta is a random variable uniformly distributed between 0 and
2*pi and A, B, and phi are',...
   'constants. The length of the cosinusoids is 300 samples.');
disp(str);
str=str2mat(...
   ' ',...
   'The cross-correlation function is calculated via the following
estimator:',...
   ' ',...
   '>> z(k) = (1/M)*sum(x(m)*y(m+k))   and the theoretical cross-
correlation via:',...
   ' ',...
   '>> z(k) = (A*B/2)*cos(2*pi*f0*k*0.01+phi)');
disp(str);
i=0:1:299;
x_i=A*cos(2*pi*f0*i*0.01+theta);
y_i=B*cos(2*pi*f0*i*0.01+theta+phi);
r_xy=xcorr(x_i,y_i,'biased');
r_xy=r_xy/r_xy(300);
k=-50:1:50;
rt_xy = (A*B/2)*cos(2*pi*f0*k*0.01+phi);
rt_xy=rt_xy/rt_xy(51);
s1=subplot(211);
plot(0:1:299,x_i,'r', 0:1:299,y_i,':k'),xlabel('i','FontSize',[10]),ylabel('x _   y
...','FontSize',[10])
title('Cosinusoids','FontSize',[10]),axisn(i)
set(s1,'FontSize',[10])
s2=subplot(212);
plot(-50:1:50,r_xy(250:350),'r',k,rt_xy,'.k'),axisn(-50:1:50);grid;
xlabel('k','FontSize',[10]);
ylabel('Estimated _   Theoretical ....','FontSize',[10]);
title('Estimated and theoretical cross-correlation functions','FontSize',[10]);
set(s2,'FontSize',[10]),clear

The processes used have the following form:

```
>> x(i)=A*cos(2*pi*f0*i*0.01+ theta)
>> y(i+1)=B*cos(2*pi*f0*i*0.01+theta+phi)
```

Where theta is a random variable uniformly distributed between 0 and 2*pi and
A, B, and phi are constants. The length of the cosinusoids is 300 samples.

The cross-correlation function is calculated via the following estimator:

>> z(k) = (1/M)*sum(x(m)*y(m+k))   and the theoretical cross-correlation
via:

>> z(k) = (A*B/2)*cos(2*pi*f0*k*0.01+phi)

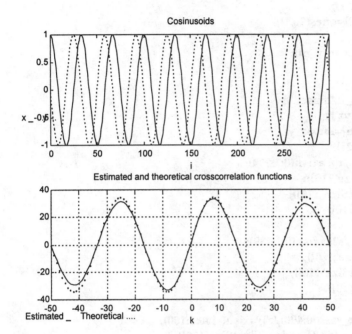

MATLAB Computer Example 4.5
USES OF THE CROSS-CORRELATION FUNCTION

In order to demonstrate one of the uses of the CCF, consider two square waves with random noise added.

```
%===============================================
% Input parameters
            amp=0.5;
            pos=20;
            M=300;
%===============================================
str=str2mat(...
    'In order to demonstrate one of the uses of the CCF, consider two square
waves with random noise',...
    'added. The only difference between the two functions is the position of
the square wave. We will',...
    'compute their cross-correlation function using the estimator:',...
    ' ',...
    '>> z(k) = sum(x(m)*y(m+k))');
```

```
disp(str);
R1=zeros(M,1);R2=R1;
R1(1:20)=ones(20,1);
R2(pos+1:pos+20)=ones(20,1);
x_n=rand(M,1);
x_n=amp*(x_n-mean(x_n));
y_n=R1+x_n;
z_n=R2+x_n;
r_yz=xcorr(z_n,y_n,'biased');
r_yz=M*fliplr(r_yz');
[value_p position]=max(r_yz);
clg;s1=subplot(221);
plot((0:1:299),y_n,'r'); axisn(0:1:299);
xlabel('n','FontSize',[10]);
ylabel('y_n','FontSize',[10]);
set(s1,'FontSize',[10])
s2=subplot(222);
plot((0:1:299),z_n,'r'); axisn(0:1:299);
xlabel('n','FontSize',[10]);
ylabel('z_n','FontSize',[10]);
set(s2,'FontSize',[10])
s3=subplot(223);
plot((-100:1:100),r_yz(200:400),'r'); axisn(-100:1:100);
title('Cross-correlation function','FontSize',[10]);
xlabel('k','FontSize',[10]);
ylabel('r_yz','FontSize',[10]);
set(s3,'FontSize',[10])
s4=subplot(2,2,4);
axis('off')
text(0,0.6,['Max. correlation value : ',num2str(value_p)] ,'FontSize',[10]);
text(0,0.4,['Position : ',num2str(position-300)] ,'FontSize',[10]);
if (amp==0.5 & pos==20)
 str=str2mat(...
    '',...
    'There is a peak at t = 0, this corresponds to the CCF of the random noise
present in both waveforms.',...
    'The second point of interest is the triangular wave present in the latter
part of the CCF. The value of t',...
    'at the peak corresponds to the separation in time between the two
square waves.');
 disp(str);
end;clear
```

In order to demonstrate one of the uses of the CCF, consider two square waves
with random noise added. The only difference between the two functions is the

position of the square wave. We will compute their cross-correlation function
using the estimator:

`>> z(k) = sum(x(m)*y(m+k))`

There is a peak at t = 0, this corresponds to the CCF of the random noise present
in both waveforms. The second point of interest is the triangular wave present
in the latter part of the CCF. The value of t at the peak corresponds to the
separation in time between the two square waves.

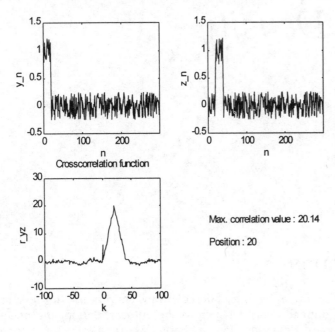

Max. correlation value : 20.14

Position : 20

CHAPTER

5 *Spectral Density*

5.0 INTRODUCTION

In this chapter, we continue the description of a random process. In previous chapters, we have defined concepts such as the autocorrelation, the mean value, and the standard deviation. In this chapter we seek ways to describe and estimate the properties of an observed random process in the frequency domain. It is important to point out that there are several methods by which we can estimate the spectral density of a random process. This is a challenge for students, as it may be difficult to imagine a problem with many outcomes depending on the assumptions made regarding the random process. We will describe, explain, and illustrate several of these techniques and offer theoretical and real-world examples in this chapter. Knowledge of Fourier transforms is important in understanding the frequency domain description of random processes. The appendix provides a self-contained description of Fourier transforms as well as practical computational techniques associated with frequency domain analysis.

5.1 BASIC DEFINITION OF THE SPECTRAL DENSITY OF A RANDOM PROCESS

The power spectral density, as the name indicates, is the density of average power given as a function of frequency. The power spectral density gives the frequency spectrum of the mean power of the random process. The spectrum of a deterministic signal $x(t)$ is given as its Fourier transform:

$$X(f) = \int_{-\infty}^{\infty} x(t)e^{-j2\pi ft}\, dt \qquad (5.1)$$

with the corresponding inverse transform

$$x(t) = \int_{-\infty}^{\infty} X(f)e^{j2\pi ft}\, df \qquad (5.2)$$

The conditions for the existence of Fourier transforms are discussed in Appendix A. Equation (5.1) cannot be computed for realistic sample functions of random processes. We can define the time-limited Fourier transform by restricting the limits of the integral in (5.1) to $\pm T$ by

$$X(f, T) = \int_{-T}^{T} x(t)e^{-j2\pi ft}\, dt \qquad (5.3)$$

Consider a stationary ergodic random process $X(t)$. Since the process is ergodic, we can consider a sample function $x(t)$ of length $2T$ in the interval $[-T, T]$. Please note that ergodicity is the first assumption we are making. Obviously, if the random process were not ergodic, different techniques of estimation would have to be used. The power spectral density may be expressed by the following relationship (Bendat and Piersol, 1986):

$$S_x(f) = \lim_{T \to \infty} \frac{1}{2T} E\left[|X(f, T)|^2\right] \qquad (5.4)$$

where $X(f, T)$ is the time-limited Fourier transform.

This expression conveys information not only about the theoretical definition, but also about possible approaches to estimating this result.

One common error in the interpretation of Equation (5.4) and in the estimation of spectral densities in general is the direct use of the fast Fourier transform (FFT). Direct use of the FFT often ignores the limit operation and the expectation operation of Equation (5.4). The limit operation will have to be ignored because we are dealing with time-limited functions of time. Ignoring the expectation operation may lead to spurious results. Ignoring both of these operations leads to the following estimate:

$$\hat{S}_x(f) = \frac{1}{2T} \cdot |X(f,T)|^2 \tag{5.5}$$

It can be proved (Bendat and Piersol, 1986) that the standard deviation of this estimate is as large as the mean of the quantity being estimated. This is an unacceptable error. It is therefore necessary to account in some manner for the expectation operation of Equation (5.4). To implement Equation (5.4), it will be necessary to employ some type of smoothing operation. This will lead to the Fourier-transform-based techniques to estimate the spectral density of a random process.

Another important definition that will permit us to understand and estimate the spectral density of a random process is the *Einstein-Wiener-Khinchine theorem*. This theorem states that the power spectral density of a random process equals the Fourier transform of its autocorrelation function. This theorem may be expressed by

$$S(f) = F\{R_X(\tau)\} \tag{5.6}$$

where $F\{\cdots\}$ is the Fourier transform operator. This equation equals

$$S(f) = \int_{-\infty}^{\infty} R_X(\tau) \exp[-j2\pi f\tau] \, d\tau \tag{5.7}$$

This equation can also be considered the definition of power spectral density.

There are a few caveats to the above theorem. First, we are dealing with at least a wide-sense stationary random process. Second, this theorem will allow us to estimate the power spectral density of a random process, provided that we carefully estimate the autocorrelation function of the random process.

Equation (5.7) may be expanded by using a well-known relationship

$$S(f) = \int_{-\infty}^{\infty} R_X(f)(\cos 2\pi f\tau - \sin 2\pi f\tau) \, d\tau \tag{5.8}$$

This last equation may be simplified as follows:

$$S(f) = \int_{-\infty}^{\infty} R_X(f)\cos 2\pi f\tau \, d\tau \tag{5.9}$$

Since $R_X(\tau)$ is an even function of time (for real random processes), the integral of an even function of time multiplied by an odd function of time equals zero. This result leads to the cosine transform which simplifies the computation of Equation (5.7).

Example 5.1 Consider the random process of Example 4.4:

$$X(t) = A \cos(\omega_0 t + \theta) \tag{5.10}$$

where A and ω_0 are constants and θ is a random variable. The random variable θ has a probability density function of the following type:

$$f(\theta) = \begin{cases} B & 0 \le \theta \le 2\pi \\ 0 & \text{elsewhere} \end{cases} \tag{5.11}$$

The value of B was found to be equal to $1/(2\pi)$. The autocorrelation function of the random process, as given by Equation (4.25), equals

$$R_X(\tau) = \frac{A^2}{2} \cos \omega_0 \tau \tag{5.12}$$

Invoking the Einstein-Wiener-Khinchine theorem yields

$$S(f) = \int_{-\infty}^{\infty} \frac{A^2}{2} \cos(\omega_0 \tau) \exp(-j2\pi f \tau) \, d\tau \tag{5.13}$$

which simplifies to

$$S(f) = \frac{A^2}{2} \int_{-\infty}^{\infty} \cos(\omega_0 \tau) \cos(2\pi f \tau) \, d\tau \tag{5.14}$$

From Fourier transform theory, we know that this last equation leads to

$$S(f) = \frac{A^2}{2} \cdot \left[\frac{1}{2} \delta(f - f_0) + \frac{1}{2} \delta(f + f_0) \right] \tag{5.15}$$

which equals

$$S(f) = \frac{A^2}{4} \cdot \delta(f - f_0) + \frac{A^2}{4} \cdot \delta(f + f_0) \tag{5.16}$$

which is an anticipated result.

Note that this is a purely theoretical result. Practically, if we were to estimate the spectral density of a time-limited cosinusoid, we would have to take into account a number of issues such as the sampling rate and truncation effects. These issues will be explored in greater detail later.

5.2 PROPERTIES

Equation (5.9) provides much insight into the properties of the spectral density of a random process. Note that the cosine transform and the auto-correlation function are even functions of time. Consequently, for a random process which is real and at least wide-sense stationary, the spectral density of a random process is a *real*, *positive*, and *even* function of frequency. These properties have important implications. For example, it is only necessary to compute the positive half of the spectral density, as the negative half will be identical. This is a consequence of the *even* property of the spectral density.

The Fourier transform pair corresponding to Equation (5.7) is the following:

$$R_X(\tau) = \int_{-\infty}^{\infty} S(f) \exp j2\pi f\tau \, df \tag{5.17}$$

Evaluating this equation at $\tau = 0$ yields

$$R_X(0) = \int_{-\infty}^{\infty} S(f) \, df = E\left[X^2\right] \tag{5.18}$$

This value is the mean square value of a random process. It is customary to refer to this quantity as the *average power* of the random process.

Example 5.2. Consider the random process used in Example 5.1:

$$X(t) = A \cos(\omega_0 t + \theta) \tag{5.19}$$

where A and ω_0 are constants and θ is a random variable. What is the average power of this process?

There are two ways of solving this process. The first is to evaluate the autocorrelation function at $\tau = 0$:

$$R_X(\tau)\Big|_{\tau=0} = \frac{A^2}{2} \cos \omega_0\tau \Big|_{\tau=0} = \frac{A^2}{2} \tag{5.20}$$

The second is to integrate the spectral density as shown in Equation (5.18):

$$R_X(0) = \int_{-\infty}^{\infty} S(f)\, df = \int_{-\infty}^{\infty} \left[\frac{A^2}{4} \delta(f - f_0) + \frac{A^2}{4} \delta(f + f_0) \right] df = \frac{A^2}{2} \qquad (5.21)$$

which yields the same result.

Example 5.3. In this example, a wide-sense stationary random process has the following type of autocorrelation function:

$$R(\tau) = 1.0 + \exp(-2|\tau|) + \cos 4\pi t \qquad (5.22)$$

As observed with the example, the autocorrelation function is made up of the sum of three distinct processes: the constant value of 1, an even exponential function, and a periodic waveform. Compute the spectral density of this process. To solve this problem, we will invoke the Einstein-Wiener-Khinchine theorem given by Equation (6.3):

$$S_X(f) = F\{1.0 + \exp(-2|\tau|) + \cos 4\pi t\} \qquad (5.23)$$

The standard Fourier transform pairs tables yield the following relationships:

$$\begin{cases} 1 & \Leftrightarrow & \delta(f) \\[2mm] \exp(-a|t|) \quad a > 0 & \Leftrightarrow & \dfrac{2a}{a^2 + (2\pi f)^2} \\[2mm] \cos 2\pi f_0 t & \Leftrightarrow & \dfrac{1}{2}\delta(f - f_0) + \dfrac{1}{2}\delta(f + f_0) \end{cases} \qquad (5.24)$$

When they are applied to the problem at hand, we obtain the following result:

$$\begin{cases} F\{1\} = \delta(f) \\[2mm] F\{\exp(-2|\tau|)\} = \dfrac{4}{4 + (2\pi f)^2} \\[2mm] F\{\cos 4\pi t\} = \dfrac{1}{2}\delta(f - 2) + \dfrac{1}{2}\delta(f + 2) \end{cases} \qquad (5.25)$$

The spectral density is then given by

$$S_X(f) = \delta(f) + \frac{4}{4 + (2\pi f)^2} + \frac{1}{2}\delta(f - 2) + \frac{1}{2}\delta(f + 2) \tag{5.26}$$

A plot of the spectral density is given in Figure 5.1.

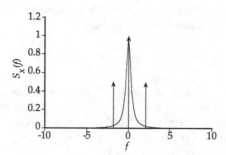

Figure 5.1 Plot of the spectral density of Example 5.3. Note that there are three delta functions, one at the origin and one each at $f_0 = \pm 2$.

It is important to realize the significance of these last two examples. These are *theoretical* examples, designed to provide an insight into the calculation of the spectral density of a wide-sense stationary random process. The autocorrelation function as depicted by Equation (5.19) ranges from $-\infty$ to $+\infty$. In a real physical situation, the autocorrelation function would be an *estimate* of the autocorrelation function of the random process in question, and it would be *time-limited*. These are important limitations that require careful attention to detail.

5.3 WHITE NOISE

The concept of white noise is an extremely important one. Applying white noise to the input of a linear system will permit us to explore the response of such a system to a variety of inputs. These concepts will be explored in detail in the next chapter.

By definition, *white noise* is random noise that has equal power at all frequencies of the spectrum. This definition allows us to define the spectral density and corresponding autocorrelation function of white noise:

$$S_X(f) = 1 \tag{5.27}$$

and

$$R_X(\tau) = \delta(\tau) \qquad (5.28)$$

The Fourier transform pair corresponding to these definitions is

$$\delta(\tau) \iff 1 \qquad (5.29)$$

Example 5.4 Let $X(t)$ be a random process having the following form:

$$X(t) = A\cos(\omega_0 t + \theta) + N(t) \qquad (5.30)$$

where A and ω_0 are constants, θ is a random variable, and $N(t)$ is independent white noise. From a physical point of view, the cosinusoid may be a noisy process we wish to observe. The random variable θ has a probability density function of the following type:

$$f(\theta) = \begin{cases} B & 0 \le \theta \le 2\pi \\ 0 & \text{elsewhere} \end{cases} \qquad (5.31)$$

The value of B is found to be equal to $1/(2\pi)$. Note that the only random variable in Equation (5.30) is θ. This example is similar to Example 5.1 except that there is white noise added to the system. Find the spectral density of this random process.

We proceed in the same way as we proceeded with Example 5.1. We first calculate the autocorrelation function and then compute its Fourier transform to arrive at the spectral density of the random process:

$$R_X(\tau) = E[X(t)X(t + \tau)] \qquad (5.32)$$

We let the cosinusoid term equal $G(t)$, and now we proceed to evaluate Equation (5.32):

$$R_X(\tau) = E\big[\{G(t) + N(t)\}\{G(t + \tau) + N(t + \tau)\}\big] \qquad (5.33)$$

which yields the following result:

$$R_X(\tau) = E[G(t)G(t + \tau) + G(t)N(t + \tau) \\ + N(t)G(t + \tau) + N(t)N(t + \tau)] \qquad (5.34)$$

Expanding this expression yields the following result:

$$R_X(\tau) = E[G(t)G(t+\tau)] + E[G(t)N(t+\tau)]$$
$$+ E[N(t)G(t+\tau)] + E[N(t)N(t+\tau)] \qquad (5.35)$$

By definition, white noise is a zero-mean uncorrelated random process. Therefore the cross terms go to zero. We are left with the following terms:

$$R_X(\tau) = R_G(\tau) + R_N(\tau) \qquad (5.36)$$

Note that

$$E[G(t)N(t+\tau)] = E[G(t)]E[N(t+\tau)] = 0 \qquad (5.37)$$

The autocorrelation term of the cosinusoidal term was found to be

$$R_G(\tau) = \frac{A^2}{2} \cos \omega_0 \tau \qquad (5.38)$$

and the autocorrelation function of the white noise is given by

$$R_N(\tau) = C\delta(\tau) \qquad (5.39)$$

where, for this example, C is an arbitrary constant.

The autocorrelation function of the process is therefore

$$R_X(\tau) = \frac{A^2}{2} \cos \omega_0 \tau + C\delta(\tau) \qquad (5.40)$$

We now compute the spectral density of the random process by taking the Fourier transform of the autocorrelation function and making use of the Fourier transform pair tables:

$$S_X(f) = \frac{A^2}{4} \delta(f - f_0) + \frac{A^2}{4} \delta(f + f_0) + C \qquad (5.41)$$

5.4 TIME DOMAIN AND FREQUENCY DOMAIN CORRESPONDENCE

Equation (5.18), repeated below for completeness,

$$R_X(0) = \int_{-\infty}^{\infty} S(f) \, df = E\left[X^2\right] \tag{5.42}$$

provides an interesting bridge between the frequency domain and the time domain. This equation relates the average power of a random process to the integral of its spectral density. Since the domain of the spectral density function is the frequency domain, one can then calculate the average power over a specific frequency band by integrating over the desired band. This implies that

$$2\int_{f_0}^{f_1} S(f) \, df = \text{average power in } \omega_0 \rightarrow \omega_1 \text{ band} \tag{5.43}$$

Note that the integral is multiplied by 2. This takes into account the positive and negative portions of the spectral density and is valid since the spectral density is a real, positive, and even function of frequency for a real signal.

5.5 ESTIMATE OF THE AUTOCORRELATION FUNCTION OF A DISCRETE FUNCTION OF TIME USING FREQUENCY DOMAIN TECHNIQUES

To understand the use of frequency domain techniques for the computation of the autocorrelation function of a sample function of a random process, it is necessary to compare the operations of convolution and correlation. Convolution in the time domain has a well-known counterpart in the frequency domain. This relationship may be expressed by the following Fourier transform pair:

$$x(t) * y(t) \quad \Leftrightarrow \quad X(f) \cdot Y(f) \tag{5.44}$$

This result is known as the *convolution theorem*. The Fourier transform of the function $x(t)$ is $X(f)$, and the Fourier transform of the function $y(t)$ is $Y(f)$. Equation (5.44) expresses the result that the convolution in the time domain of the two functions of time $x(t)$ and $y(t)$ is equivalent to the multiplication in the frequency domain of the Fourier transforms of the two time functions. This relationship permits the computation of the time convolution of two functions of time, via the frequency domain.

In the time domain the convolution integral is given by

$$z(t) = \int_{-\infty}^{\infty} x(\tau)y(t-\tau)\, d\tau = x(t) * y(t) \tag{5.45}$$

The time domain autocorrelation function is actually very similar in form to the convolution integral. This equation is presented again for ease of comparison.

$$R(\tau) = \lim_{T \to \infty} \frac{1}{2T} \int_{-T}^{T} x(t)x(t+\tau)\, dt \tag{5.46}$$

The only difference between the two integrals is that there is no reversal of one of the integrands in the correlation. It is possible to find a Fourier transform pair that will relate the time autocorrelation of two functions with some type of operation in the frequency domain. This relationship is now derived. The convolution of $y(t)$ in the time domain is expressed as follows:

$$y(t) = \int_{-\infty}^{\infty} x(\tau)x(t+\tau)\, d\tau \tag{5.47}$$

The limits in the integral reflect the fact that, for this derivation, the sample function is considered to be infinite. Now we evaluate the Fourier transform of Equation (5.47):

$$\int_{-\infty}^{\infty} y(t)\exp(-j2\pi ft)\, dt = \int_{-\infty}^{\infty} \left[\int_{-\infty}^{\infty} x(\tau)x(t+\tau)\, d\tau \right] \exp(-j2\pi ft)\, dt \tag{5.48}$$

Assuming that the order of the integration can be interchanged, we can rewrite this equation as follows:

$$Y(f) = \int_{-\infty}^{\infty} x(\tau) \left[\int_{-\infty}^{\infty} x(t+\tau)\exp(-j2\pi ft)\, dt \right] d\tau \tag{5.49}$$

Let $\rho = t + \tau$ and replace this term inside Equation (5.49):

$$\int_{-\infty}^{\infty} x(t+\tau)\exp(-j2\pi ft)\, dt = \int_{-\infty}^{\infty} x(\rho)\exp[-j2\pi f(\rho - \tau)]\, d\rho \tag{5.50}$$

This results in

$$\int_{-\infty}^{\infty} x(\rho)\exp[-j2\pi f(\rho-\tau)]\,d\rho = \int_{-\infty}^{\infty} x(\rho)\exp(-j2\pi f\,\rho)\exp(+j2\pi f\,\tau)\,d\rho$$

$$= \exp(+j2\pi f\,\tau)\int_{-\infty}^{\infty} x(\rho)\{\exp(-j2\pi f\,\rho)\}\,d\rho$$

$$= \exp(+j2\pi f\,\tau)\,X(f) \tag{5.51}$$

The term inside the brackets of Equation (5.49) is given by the last term of Equation (5.51). Making this substitution, we obtain the following result:

$$Y(f) = \int_{-\infty}^{\infty} x(\tau)\exp(+j2\pi f\,\tau)X(f)\,d\tau \tag{5.52}$$

This equation may be expressed as follows:

$$Y(f) = X(f)X^*(f) \tag{5.53}$$

Note that the term $X(f)$ may be taken outside the integral, and what remains of the integral is the complex conjugate of the Fourier transform of $x(\tau)$. Consequently, the Fourier transform of Equation (5.45) is given by the result of Equation (5.51). It is possible, then, to establish the following Fourier transform pair:

$$\int_{-\infty}^{\infty} x(\tau)x(t+\tau)\,d\tau \Leftrightarrow X(f)X^*(f) \tag{5.54}$$

This is an important relationship because it enables us (with some restrictions) to compute the autocorrelation of a sample function of a random process using the FFT. It is also possible to derive a similar relationship for the case of the cross-correlation function. We can then establish the following Fourier transform pair for the cross-correlation function:

$$\int_{-\infty}^{\infty} x(\tau)y(t+\tau)\,d\tau \Leftrightarrow X(f)Y^*(f) \tag{5.55}$$

In the rest of the chapter we use the fast Fourier transform algorithm to compute the samples of the Fourier transforms needed to apply the various frequency domain results to practical problems. A detailed discussion of the FFT and its application to random processes is provided in the Appendix. There are a number of issues associated with the use of the FFT for the computation of the autocorrelation that must be carefully considered prior to its use. We will illustrate these by an example.

Example 5.5. Consider the time function

$$x(t) = \begin{cases} \exp[-2t] & t \geq 0 \\ 0 & \text{elsewhere} \end{cases} \qquad (5.56)$$

This infinite function of time will be observed over approximately 2.5 s. The effective sampling rate of the function will be 50 samples per second. Figure 5.2 shows this function plotted up to 6 s.

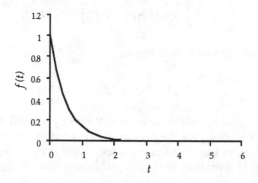

Figure 5.2 Plot of the function given by Equation (5.56)

The function is now time-limited to 2.56 s and therefore 128 points of data will be considered. Note that this value is a power of 2, as required by the FFT subroutine. Figure 5.3 shows this function time-limited to 2.56 s. Time-limiting the function may be done as follows:

$$x(t) = \begin{cases} \exp(-2t) & 0.0 \leq t \leq 2.56 \\ 0 & \text{elsewhere} \end{cases} \qquad (5.57)$$

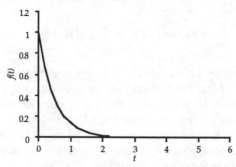

Figure 5.3 Plot of the function given by Equation (5.56) time-limited to 2.56 seconds.

The important implication here is that beyond 2.56 s and before 0.0 s, it is assumed that $x(t) = 0$. The function describing the discrete Fourier transform is given by

$$F(\omega) = \sum_{k=0}^{N-1} f(k\,\Delta t) \exp\left(\frac{-j2\pi nk}{N}\right) \tag{5.58}$$

where $n = 0, 1, \ldots, N-1$ and

$$\omega = \frac{2\pi n}{N\,\Delta t}$$

is the spacing between frequencies. The index range of this equation is $n = 0, 1, \ldots, N-1$. For values of n larger than $N-1$, the values of $F(\omega)$ repeat, due to the circular nature of the FFT. Consequently, when using the FFT to compute the autocorrelation function, we must add a number of zeros to the data to avoid this circular problem. Now we establish the general procedure for the computation of the autocorrelation function using the FFT:

(a) The function given by Equation (5.57) is first expressed in discrete form as follows:

$$x(k\,\Delta t) = \exp(-2k\,\Delta t) \quad 0 \le k \le 127 \tag{5.59}$$

(b) To avoid the circular convolution problem (associated with the FFT) we extend the discrete function of time with an equal number of zeros:

$$x(k\,\Delta t) = \begin{cases} \exp(-2k\,\Delta t) & 0 \le k \le 127 \\ 0 & 128 \le k \le 255 \end{cases} \tag{5.60}$$

This yields a total of $N = 256$ discrete values of data. Note that 256 is a power of 2.

(c) The Fourier transform is now computed using the FFT. Only the first half, or 129 points, needs to be computed, since the autocorrelation function is an even function of time.

(d) The Fourier transform is multiplied by its complex conjugate:

$$Y(f) = X(f)X^*(f) \tag{5.61}$$

(e) The inverse Fourier transform is computed using the FFT. This computation yields the positive lag values of the autocorrelation function. Figure 5.4 shows the positive half of the autocorrelation function.

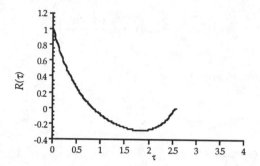

Figure 5.4 Plot of the autocorrelation function of Example 5.5. The FFT was used to arrive at these results.

5.6 ESTIMATION OF THE SPECTRAL DENSITY OF RECORDS OF LIMITED DURATION

The defining equation of the spectral density of a random process (5.4) has a number of limitations and requirements imposed on its use in practical situations. The most important requirement is that imposed by the expectation operation. As we use any of these techniques, we must remember that the result will always be an *estimate* of the spectral density of a time-limited sample of a random process. This estimate is intimately associated with any theoretical assumptions that have been made concerning the random process. For example, we may assume that the process is at least wide-sense stationary, that it is gaussian distributed, etc.

Transform of the Autocorrelation Function

The first estimate that we are going to consider is based on the Einstein-Wiener-Khinchine theorem. This theorem relates the spectral density of a random process to the autocorrelation function of the random process by the Fourier transform. This operation is given by the following equation [see Equation (5.6)]:

$$S(f) = F\{R_X(\tau)\} \tag{5.62}$$

To perform these operations, we will make extensive use of the FFT. The first step is to obtain a reliable estimate of the autocorrelation function of the time-limited sample of the random process. To make these operations easier to understand, we use an example to illustrate all the steps. Consider a time-limited sample function of a wide-sense stationary random process. Figure 5.5 shows this time-limited sample function. The sampling rate for this time-limited sample function was 0.01 s between samples, and the total sampling was 128 samples. Note that at this sampling rate and according to the sampling theorem, the maximum frequency that we can possibly discern through this process will be 50 Hz. Before we proceed with all the mathematical manipulations, it is a good idea to simply look at the waveform and try to analyze it visually. There is really no substitute for this type of analysis, as it oftens helps us check our results. Considering the waveform of Figure 5.5, we make the following observations:

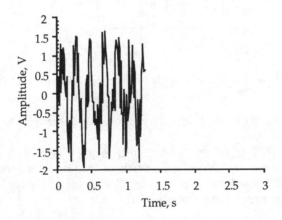

Figure 5.5. Plot of the time-limited sample function of a wide-sense stationary random process.

The data are periodic and have mean zero. The frequency is approximately 5. The data were generated using the following model:

$$x_i = A\cos(2\pi f_0 i\Delta t + \theta) + 2 \cdot \text{rnd}(1.0) - 1.0 \qquad (5.63)$$

where $A = 1.0$, $f_0 = 5.0$, θ is selected from a uniform distribution with range 0 to 2π, and rnd(1.0) is a random value uniformly distributed between 0 and 1.

These observations will make it easier for us to decide whether to believe the results of the operations that will follow. The initial step in the operation is to obtain an estimate of the autocorrelation function of the random process. This operation will be performed via the FFT. The general procedure established in Section 5.5 will be followed:

1. The sample function is extended with zeros in order to compute the FFT of the data and avoid the circular correlation problem. Figure 5.6 shows the data after they have been padded with zeros. Note that, in this case, adding zeros introduces an artificial end effect.

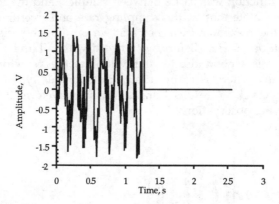

Figure 5.6 Plot of the time-limited sample function of a wide-sense stationary random process padded with zeros.

2. The FFT of the data is multiplied by its complex conjugate.

3. The inverse FFT is taken. This results in the *complete* autocorrelation function of the data. Only the first half of this result is necessary, as the autocorrelation function is an even function of time. Figure 5.7 shows the complete autocorrelation function of the data.

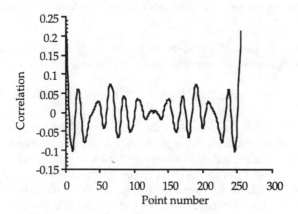

Figure 5.7 Autocorrelation function of the data.

It is interesting to consider the type of information given by Figure 5.7. To avoid the circular correlation problem, zeros were added to the data shown in Figure 5.5. There are a total of 128 points of data in the original sample function of the wide-sense stationary random process. The process of autocorrelation is such that as the number of lags increases, the number of data points used to compute the equation decreases proportionately. Figure 5.8 shows this phenomenon.

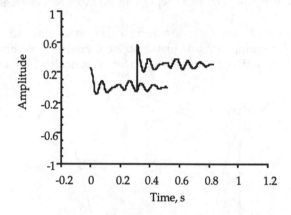

Figure 5.8 Example of the autocorrelation function at "long" lags.

A basic assumption made with time-limited sample functions of a random process is that outside the range, the sample function is equal to zero. This is why padding with zeros is needed when we proceed with this type of computation. It is then reasonable to assume that limits should be placed on how many lags one is expected to take and still have some statistical confidence in the results. Box (1976) recommends that the number of lags be defined as $K \leq N/4$, where N is the total number of points present in the time-limited sample function. In this case there are a total of 128 points; therefore, we should use only 32 values of the (positive half) autocorrelation function. The question now is, How can we effectively time-limit the estimate of the autocorrelation function to a number of lags $K \leq N/4$?

Window functions are often used to perform truncation operations. Window functions are extensively used in the context of estimating the power spectral density of random processes. If we apply a window to an estimated sample autocorrelation function, this results in a smoothing of the spectral estimate by virtue of convolution with the transform of the window. A wide variety of window functions are frequently used in the context of the power spectral density estimation. The Hamming window is one of the popular window functions. The defining equation for the Hamming window $w(t)$ is

$$w(t) = \begin{cases} .54 + .46\cos 2\pi t & |t| < .5 \\ 0 & \text{elsewhere} \end{cases} \tag{5.64}$$

Note that the Hamming window function given by Equation (5.64) can be scaled in time according to requirements. For this example, the Hamming window was selected and will be used to time-limit the estimate of the autocorrelation function. Figure 5.9 shows an example of a Hamming window.

4. The estimate of the autocorrelation function is now time-limited by using the Hamming window just described. Figure 5.10 shows the auto-correlation function before and after multiplication by the Hamming window.

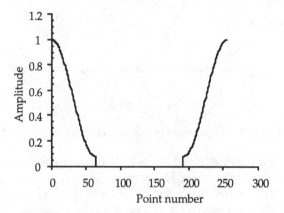

Figure 5.9 Hamming window used for the present example.

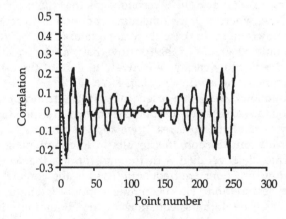

Figure 5.10 Estimate of the autocorrelation function before and after multiplying by the Hamming window.

There are several points that should be mentioned. First, the input array to the FFT must still be a power of 2. Second, there is an option as to how many zeros we add in the middle of the array. In this particular example, the total number of points used was 256. Third, the number of zeros added does not affect the maximum frequency resolution. This is completely determined by the sampling rate. In this case, this value is 50 Hz.

5. The FFT is now computed, and the absolute value of the answer is the desired estimate of the spectral density of the random process. Figure 5.11 shows this estimate. Note that the results from this procedure match what we found by *visual* inspection of the data.

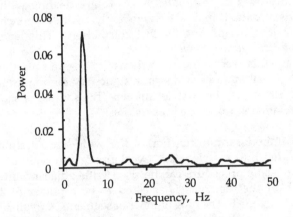

Figure 5.11 Estimate of the spectral density of the random process.

Units. Throughout these computations care must be taken to ensure that at the end we end up with the correct units. This is a "safety" precaution, because some FFT algorithms divide by the number of points in the data, and others do not. The first step we take is to compute the first value of the autocorrelation function. This is the value at zero lag and is computed from the following equation:

$$\hat{R}_X(0) = \frac{1}{N} \sum_{i=1}^{N} x_i^2 \tag{5.65}$$

For the data shown in Figure 5.5, this number is .406 (see the computer examples at the end of this chapter).

Upon returning from the procedure in which the autocorrelation function is computed via the FFT, we check the zero lagged value (dividing by the number of points). This is the second check. This value is also found to be .406. As a third and final step, we use Equation (5.42) and compute the zero lagged value of the autocorrelation function by integrating the estimate of the spectral density of the process. This number is found to be .407.

Periodogram Method

The periodogram method of spectral estimation closely resembles Equation (5.4). In this method, the sequence of observed values is split into shorter segments, the discrete Fourier transform of the short segments is computed, and then the results are averaged. The magnitude squared of the discrete Fourier transforms is called a *periodogram*. A single periodogram is not a good estimate of the spectral density of the process, but averaging several together significantly improves the estimate. Since we are dealing with short time-limited sections of data, it makes sense to apply a window function to each of the sections prior to performing the FFT calculations. This leads to what is called the *modified periodogram*.

To illustrate this technique, we will use as an example the same time-limited sample function of the wide-sense stationary random process that was used earlier to illustrate the first technique. This sample is shown in Figure 5.5. The computational procedure is as follows:

1. The time-limited sample function of the wide-sense stationary random process is divided into a series of K overlapping intervals of prespecified length L. (The same procedure can be used with adjacent intervals.) The length of each of the segments is selected to be a power of 2 for ease of computation via the FFT. The total number of segments K available for use by this technique will then be the integer part of $(N - L/2)/(L/2)$, where N is the total number of points in the observation. For our specific example, the data are $N = 128$ number of points long. We preselect $L = 32$ points as the power-of-2 number of points for each of the sections, and we also preselect 50 percent as the amount of overlap for each section. This leads to a total of $K = 7$ sections that will be used for the estimation. Figure 5.12 shows a plot of the data to be used and the relationship between the first three of the seven segments.

2. In order to avoid any end effects caused by the abrupt transition between segments, each segment is multiplied by a time window centered on the segment. For this example, we have selected a Hamming window.

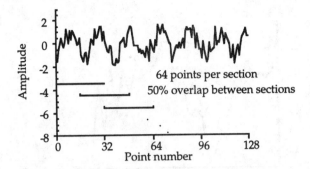

Figure 5.12 The time-limited observation and the first three of the seven segments to be used in the periodogram method of estimation of the spectral density of the random process.

Figure 5.13 shows the first section of data before and after multiplication by the Hamming window. In this particular case, the differences between the before and after data are not as obvious. Figure 5.14 shows the fourth section of the data. Again, the before and after data segments are shown. In this case, the differences are very obvious. The point number of each of the sections for this specific example is as follows:

Section 1: point 0 to point   31   Section 2: point 16 to point   47
Section 3: point 32 to point   63   Section 4: point 48 to point   79
Section 5: point 64 to point   95   Section 6: point 80 to point 111
Section 7: point 96 to point 127

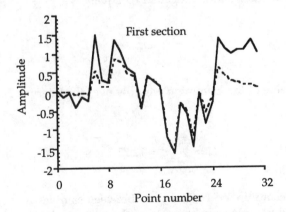

Figure 5.13 The first section of the data is 32 points long. The solid line is the data before multiplication of the segment by a Hamming window centered in the section. The dashed line is the data after multiplication by the window.

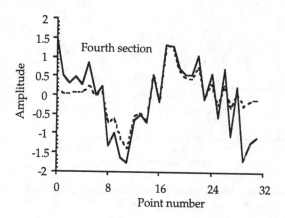

Figure 5.14 The fourth section of the data is also 32 points long. The solid line is the data before multiplication of the segment by a Hamming window centered in the section. The dashed line is the data after multiplication by the window.

3. For each one of the segments we compute the FFT and then the magnitude squared of the results. For the specific example at hand, using MATHCAD, each of the FFTs is multiplied by the number of points per section.

4. The modified periodograms are averaged to produce the estimate of the spectral density of the random process. To obtain an unbiased estimate, the results are divided by KU, where K is the number of segments that are being added together, U equals

$$U = \sum_{n=0}^{L-1} \omega_n^2 \tag{5.66}$$

and ω is the value of the Hamming window over the interval. For our case

$$U = \sum_{n=0}^{L-1} \omega_n^2 = 12.72 \tag{5.67}$$

The results are multiplied by the time between samples Δt. Figure 5.15 shows the modified periodogram estimate of the spectral density of the data in Figure 5.5. It is interesting to compare this estimate with the one obtained by the previous technique. There seems to be a fundamental frequency present around 5 Hz. With the previous technique there seems to be more spectral detail available than with the present technique.

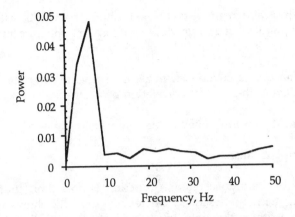

Figure 5.15 Modified periodogram estimate of the spectral density.

Example 5.6. In this example we will use a time-limited observation of the output of a physical system. The output of the system is considered to be a wide-sense stationary random process, and we are required to estimate the spectral density of the process.

Figure 5.16 shows this time-limited sample function. The sampling rate was 250 samples per second, or 0.004 seconds between time samples. The total time sampled was 0.5120 seconds for a total of 128 samples. Note that at this sampling rate, and according to the sampling theorem, the maximum frequency that can possibly be discerned is 125 Hz. A simple visual analysis of this waveform yields a fundamental frequency of about 13 Hz.

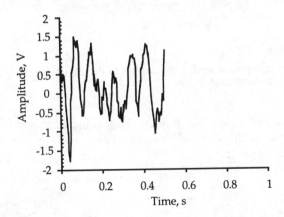

Figure 5.16 Plot of the time-limited sample function of the wide-sense stationary process under consideration.

The power spectral density of this time-limited data will be estimated using the two techniques just described. We will use first the transform of the autocorrelation function.

(a) The sample function is extended with zeros in order to compute the FFT of the data and avoid the circular correlation problem.

(b) The FFT of the data is multiplied by its complex conjugate.

(c) The inverse FFT is taken.

(d) The estimate of the autocorrelation function is now time-limited using the Hamming window described earlier. Figure 5.17 shows the auto-correlation function of the data before and after multiplication by the Hamming window.

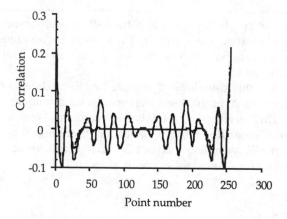

Figure 5.17 Estimate of the autocorrelation function of the data before (solid) and after (dashed) mutiplying by a Hamming window.

(e) The FFT is now computed, and the absolute value of the answer is the desired estimate of the spectral density of the random process. Figure 5.18 shows this estimate.

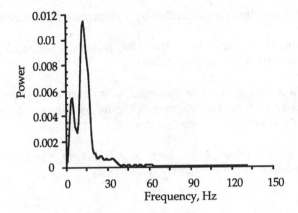

Figure 5.18 Estimate of the spectral density of the random process.

Analyzing the results from Figure 5.18 shows a large peak at about 13 to 15 Hz. This corresponds nicely to the visual analysis of the data that was done earlier.

We now use the periodogram method of estimating the power spectral density of the same time function. We proceed as before:

(a) The data are divided into K overlapping sections. The total data length is 128 points, and $L = 32$ points are arbitrarily preselected as the number of points to be analyzed for each of the K overlapping sections. We also select 50 percent as the overlap for each of the sections. These numbers yield a total of $K = 7$ sections. Figure 5.19 shows the data and the first of the three overlapping sections.

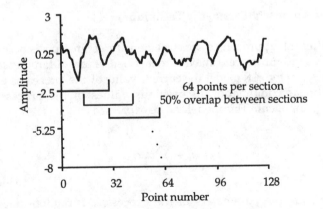

Figure 5.19 The time-limited observation and the first three of the $K = 7$ segments to be used in the periodogram method of estimation of the spectral density of the random process.

(b) Each of the time sections is multiplied by a Hamming time window.

(c) For each one of the segments, the FFT is computed, and then the magnitude squared of the results is found.

(d) The modified periodograms are averaged to produce the estimate of the spectral density of the random process. Figure 5.20 shows the modified periodogram estimate of the spectral density.

Comparison of this result with one shown in Figure 5.16 shows a close resemblance in location of the fundamental frequency.

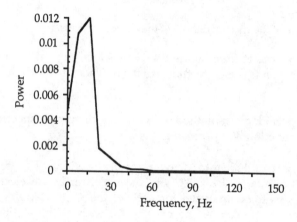

Figure 5.20 Modified periodogram estimate of the spectral density.

Model-Based Approach (Parametric Technique)

A totally different approach to spectral density estimation is based on the theory that a random process may be described as an *autoregressive* (AR) random process. In the AR model the current value of the process is expressed as a finite, linear combination of previous values of the process and an uncorrelated component. This model can be expressed as

$$x_t = a_t + \sum_{i=1}^{M} \alpha_i x_{t-i} \tag{5.68}$$

where the α's are the parameters of the process, a_t is random uncorrelated noise with zero mean value, and M is the order of the process. In Equation (5.68), we only need to estimate the α's and the order M of the model in order to describe the model.

Taking expected values, we find that the mean value of the process equals zero. The variance of the process is now

$$\sigma_X^2 = E[x_t x_t] \tag{5.69}$$

which equals

$$E[x_t x_t] = E\left[x_t a_t + x_t \sum_{i=1}^{M} \alpha_i x_{t-i} \right] \tag{5.70}$$

This is further simplified (Shanmugan, 1988) as

$$\sigma_X^2 = \sum_{i=1}^{M} \alpha_i R_x(i) + \sigma_N^2 \tag{5.71}$$

where σ_N^2 is the variance of the noise process.

By using Equation (5.70) and establishing a separation between the different x_t values, it is possible to obtain a set of equations commonly referred to as the *Yule-Walker equations*. In matrix form these equations can be expressed as follows:

$$
\begin{bmatrix}
R(1) \\
R(2) \\
\cdot \\
\cdot \\
\cdot \\
R(M)
\end{bmatrix}
=
\begin{bmatrix}
1 & R(1) & R(2) & \cdots & R(M-1) \\
R(1) & 1 & R(1) & \cdots & R(M-2) \\
\cdot & \cdot & \cdot & & \cdot \\
\cdot & \cdot & \cdot & & \cdot \\
\cdot & \cdot & \cdot & & \cdot \\
R(M-1) & R(M-2) & \cdots & & 1
\end{bmatrix}
\begin{bmatrix}
\alpha_1 \\
\alpha_2 \\
\cdot \\
\cdot \\
\cdot \\
\alpha_M
\end{bmatrix}
\tag{5.72}
$$

or

$$\bar{r} = \bar{R} \cdot \bar{\alpha} \tag{5.73}$$

where the overbars signify a vector and matrix, respectively.

The solution to this system of equations is by simple matrix inversion, although the symmetry which is obvious in this matrix will help in the solution of the system. This is a *Töeplitz matrix*. These matrices have the unique property that all diagonal terms are equal. For example, along the first off-diagonal, the terms all equal $R(1)$. *Levinson recursion* is one of the techniques often utilized for the solution of the system of equations. For now let us just pose a generalized solution as

$$\bar{\alpha} = \bar{R}^{-1} \cdot \bar{r} \tag{5.74}$$

The spectral density of x_t can now be shown to be

$$S(f) = \frac{\sigma_N^2}{\left|1 - \displaystyle\sum_{i=1}^{M} \alpha_i \exp(-j2\pi f)\right|^2} \quad |f| < .5 \tag{5.75}$$

There are two unknowns in the model expressed by Equation (5.68). They are the coefficients of the model and the order M of the model. The coefficients are found by the generalized solution given in Equation (5.74). Before we describe a method to find the order of the model, let us study a few simple random processes that are described by an autoregressive model.

To estimate the spectral density based on this type of model, the autocorrelation function of the process must be estimated. This leads to Equation (5.72) being rewritten as

$$
\begin{bmatrix}
\hat{R}(1) \\
\hat{R}(2) \\
\cdot \\
\cdot \\
\cdot \\
\hat{R}(M)
\end{bmatrix}
=
\begin{bmatrix}
1 & \hat{R}(1) & \hat{R}(2) & \cdots & \hat{R}(M-1) \\
\hat{R}(1) & 1 & \hat{R}(1) & \cdots & \hat{R}(M-2) \\
\cdot & \cdot & \cdot & & \cdot \\
\cdot & \cdot & \cdot & \cdots & \cdot \\
\cdot & \cdot & \cdot & \cdots & \cdot \\
\hat{R}(M-1) & \hat{R}(M-2) & \cdots & \cdots & 1
\end{bmatrix}
\begin{bmatrix}
\hat{\alpha}_1 \\
\hat{\alpha}_2 \\
\cdot \\
\cdot \\
\cdot \\
\hat{\alpha}_M
\end{bmatrix}
\tag{5.76}
$$

where the $\hat{R}$'s are the estimates of the autocorrelation function of the process. Likewise, the variance of the noise process must be estimated. An estimate may be found by

$$\hat{\sigma}_N^2 = \frac{N-M}{N-2M-1}\left[\hat{R}(0) + \sum_{i=1}^{M} \hat{\alpha}_i \hat{R}_x(i)\right] \tag{5.77}$$

where N is the number of observations of the random process and M is the order of the model.

Autoregressive Model of Order 1

Consider the model:

$$x_t = \alpha_1 x_{t-1} + a_t \tag{5.78}$$

where a_t defines random white noise. This is the *Markov process*. For this process to be stationary, the parameter α_1 must satisfy the condition $-1 < \alpha_1 < 1$.

Pick arbitrarily $\alpha_1 = .7$ and $a_t = 1$. Figure 5.21 shows one realization of this random process.

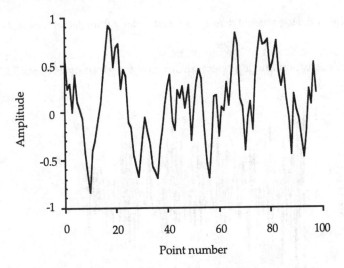

Figure 5.21 Realization of a first-order autoregressive model.

For the given model, the spectral density has the following form:

$$S(f) = \frac{\sigma_N^2}{\left|1 - \alpha_1 \exp(-j2\pi f)\right|^2} \quad |f| < .5 \tag{5.79}$$

which equals:

$$S(f) = \frac{\sigma_N^2}{1 - \alpha_1 \cos(2\pi f) + \alpha_1^2} \quad |f| < 0.5 \tag{5.80}$$

To obtain a plot of the spectral density, it will be assumed that the variance of the noise equals 1. A plot is given in Figure 5.22.

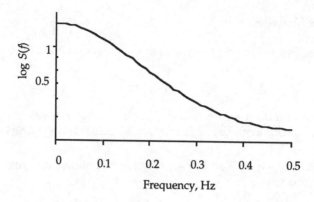

Figure 5.22 Plot of the log spectral density of a first-order AR model with $\alpha_1 = .7$.

Now let the coefficient of the model be $\alpha_1 = -.7$. A realization of this new model and its corresponding log spectral density are given by Figures 5.23 and 5.24 respectively.

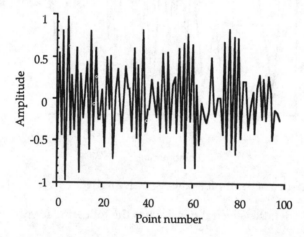

Figure 5.23 Realization of a first-order autoregressive model with $\alpha_1 = -.7$.

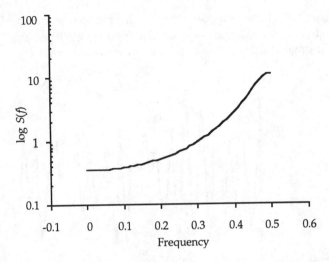

Figure 5.24 Plot of the log spectral density of a first order autoregressive model with $\alpha_1 = -.7$.

Note that a simple change in the sign of the coefficient of the model has drastically changed the type of random process we obtain. This is the model of a high-frequency system with the spectrum increasing in power as a function of the frequency.

Autoregressive Model of Order 2

The second-order autoregressive model is expressed by the following equation:

$$x_t = \alpha_1 x_{t-1} + \alpha_2 x_{t-2} + a_t \tag{5.81}$$

For stationariness, the parameters α_1 and α_2 must lie in the triangular region

$$\begin{aligned}
\alpha_1 + \alpha_2 &< 1 \\
\alpha_2 - \alpha_1 &< 1 \\
-1 < \alpha_2 &< 1
\end{aligned} \tag{5.82}$$

For the given model, the spectral density has the following form:

$$S(f) = \frac{\sigma_N^2}{1 + \alpha_1^2 + \alpha_2^2 - 2\alpha_1(1 - \alpha_2)\cos 2\pi f - 2\alpha_2 \cos 4\pi f} \quad |f| < .5 \tag{5.83}$$

Figure 5.25 shows one realization of this random process for the arbitrary values of $\alpha_1 = .7$ and $\alpha_2 = -.5$, and Figure 5.26 shows the corresponding spectral density as computed by Equation (5.83).

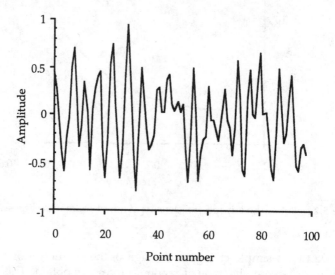

Figure 5.25 Realization of a second-order autoregressive model with $\alpha_1 = .7$, and $\alpha_2 = -.5$.

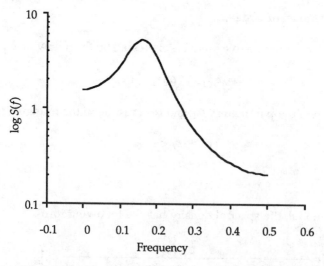

Figure 5.26 Plot of the log spectral density of a second order autoregressive model.

Estimation of the Order of the Model

The order of the model is an unknown parameter for which appropriate values have to be deduced from the observed data. One of the most utilized methods of estimating the order of the process is due to Akaike (1969) and is called the *final prediction error*, or FPE. The FPE is given by

$$\text{FPE}(M) = \frac{N+M}{N-M} \hat{\sigma}_N^2 \qquad (5.84)$$

where N is the number of observations and M is the order of the model. An estimate of σ_N^2 may be obtained by the use of Equation (5.77). Plotting FPE(M) against M should show a definite minimum which will indicate the estimate of the order of the model under consideration.

Example 5.7 Consider the autoregressive model expressed by the following equation:

$$x_t = .7x_{t-1} - .5x_{t-2} + a_t \qquad (5.85)$$

This is the same model formulated by Equation (5.81). A realization of this model is shown in Figure 5.25. Suppose that we now begin with a time-limited realization, as shown in Figure 5.25 and we then apply Equations (5.74), (5.77), and (5.84) to find the order and coefficients of the model. The following steps were followed.

(a) The autocorrelation function of the data is computed up to lag 10. We used the following method:

$$r_\tau = \frac{1}{N} \cdot \sum_j x_j \cdot x_{j+\tau} \qquad (5.86)$$

(b) Equation (5.74) is now solved for successively higher orders of autocorrelation matrices.
   For order 1, the following set is solved:

$$\overline{R} = r_0$$
$$\overline{r} = r_1 \qquad (5.87)$$
$$\overline{\alpha} = \overline{R}^{-1} \cdot \overline{r}$$

For order 2, the following set is solved:

$$\overline{R} = \begin{bmatrix} r_0 & r_1 \\ r_1 & r_0 \end{bmatrix}$$

$$\overline{r} = \begin{bmatrix} r_1 \\ r_2 \end{bmatrix} \qquad (5.88)$$

$$\overline{\alpha} = \overline{R}^{-1} \cdot \overline{r}$$

For order 3, the following set is solved:

$$\overline{R} = \begin{bmatrix} r_0 & r_1 & r_2 \\ r_1 & r_0 & r_1 \\ r_2 & r_1 & r_0 \end{bmatrix}$$

$$\overline{r} = \begin{bmatrix} r_1 \\ r_2 \\ r_2 \end{bmatrix} \qquad (5.89)$$

$$\overline{\alpha} = \overline{R}^{-1} \cdot \overline{r}$$

For this example, the computations were continued until the R matrix had a size of 6 x 6.

(c) Equation (5.77) was then computed for each of the cases given:

$$\hat{\sigma}_N^2 = \frac{N-M}{N-2M-1} \left[\hat{R}(0) + \sum_{i=1}^{M} \hat{\alpha}_i \cdot \hat{R}_x(i) \right] \qquad (5.90)$$

To compare the results obtained under each case, the coefficients estimated under each of the cases considered are shown below:

True model $\begin{cases} \text{Order} = 2 \\ \alpha_1 = .7 \quad \alpha_2 = -.5 \end{cases}$

Order 1 $\begin{cases} \hat{\alpha}_1 = .431 \\ \hat{\sigma}_1^2 = .169 \end{cases}$

Order 2 $\begin{cases} \hat{\alpha}_1 = .617 \quad \hat{\alpha}_2 = -.433 \\ \hat{\sigma}_2^2 = .193 \end{cases}$

Order 3 $\begin{cases} \hat\alpha_1 = .484 \quad \hat\alpha_2 = -.244 \quad \hat\alpha_3 = -.307 \\ \hat\sigma_3^2 = .197 \end{cases}$

Order 4 $\begin{cases} \hat\alpha_1 = .586 \quad \hat\alpha_2 = -.352 \quad \hat\alpha_3 = -.128 \quad \hat\alpha_4 = .072 \\ \hat\sigma_4^2 = .198 \end{cases}$

Order 5 $\begin{cases} \hat\alpha_1 = .587 \quad \hat\alpha_2 = -.354 \quad \hat\alpha_3 = -.133 \quad \hat\alpha_4 = .08 \quad \hat\alpha_5 = -.014 \\ \hat\sigma_5^2 = .201 \end{cases}$

Order 6 $\begin{cases} \hat\alpha_1 = .588 \quad \hat\alpha_2 = -.362 \quad \hat\alpha_3 = -.12 \quad \hat\alpha_4 = .115 \quad \hat\alpha_5 = -.072 \\ \hat\alpha_6 = .099 \\ \hat\sigma_6^2 = .202 \end{cases}$

(d) Finally, the final prediction error is computed for each case according to Equation (5.84). A plot of the FPE versus the order is shown in Figure 5.27.

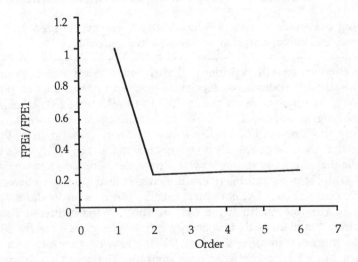

Figure 5.27 Final prediction error (FPE) for Example 5.6.

Note that the first minimum for the FPE occurs at order 2. Thus the

Akaike criterion has correctly selected the order for this process. All that remains now is to compute the estimate of the spectral density, using Equation (5.75), and compare the results to those obtained earlier and shown in Figure 5.26.

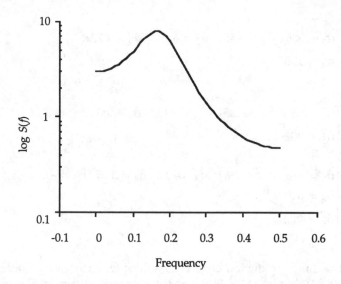

Figure 5.28 Plot of the estimate of the log spectral density for the data from Example 5.7.

Note that the results shown in Figure 5.28, are essentially identical with the theoretical values obtained and shown in Figure 5.26.

Example 5.8 Now we will consider a slightly more complicated example. Consider the data of Example 5.6. Figure 5.29 reproduces this data as used in Example 5.6. The original data were sampled at the rate of 250 samples per second, resulting in a Nyquist rate of 125 Hz. The spectral density, as computed by the Einstein-Wiener-Khinchine theorem and shown in Figure 5.18, shows that there is essentially no energy beyond 40 Hz. It behooves us to undersample the original data; otherwise, the data sampled at the original rate of 250 samples per second is oversampled and leads to high correlation between adjacent values. Extreme care must be taken when undersampling data, since aliasing may occur if the data are not low-pass filtered first. In this case, since it is evident that no energy exists over approximately 30 Hz, we are safe in undersampling the data from a rate of 4 ms between time samples to a rate of 16 ms between time samples. The new Nyquist rate is computed as

$$f_0 = \frac{1}{2 \cdot \Delta t} = \frac{1}{2 \cdot .0016} = 31.25 \text{ Hz} \qquad (5.91)$$

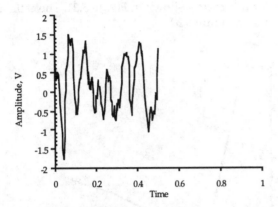

Figure 5.29 Plot of the time-limited sample function of the wide-sense stationary process used in Example 5.6.

The undersampled data are shown in Figure 5.30.

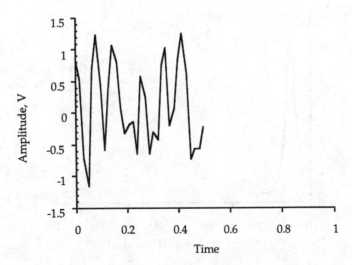

Figure 5.30 Plot of the time-limited sample function of the wide-sense stationary process used in Example 5.6 undersampled to one sample every 16 ms.

Once the data have been undersampled, the same process that was followed for Example 5.6 is now followed for this example. The plot of the final prediction error for this case is shown in Figure 5.31. The estimate of the spectral density is shown in Figure 5.32.

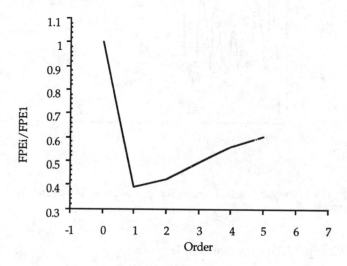

Figure 5.31 Final prediction error for Example 5.8.

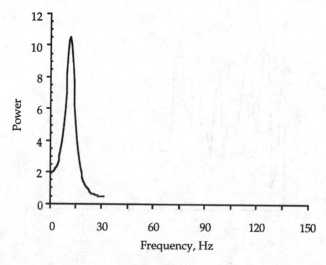

Figure 5.32 Estimate of the spectral density of the random process of Example 5.8.

Comparing this result to the one in Figure 5.18 reveals that the fundamental frequencies coincide by both methods. The results in Figure 5.18 show a smaller power peak at about 6 Hz. This is not shown in Figure 5.32.

5.7 CROSS-SPECTRAL DENSITY

The cross-spectral density between two random processes X and Y may be defined by the following equation:

$$S_{XY}(f) = \int_{-\infty}^{\infty} R_{XY}(\tau) \exp(-j2\pi f\tau)\, d\tau \tag{5.92}$$

Cross-spectral density can also be defined as the limit of the time-limited Fourier transform similar to Equation (5.4) as

$$S_{XY}(f) = \lim_{T\to\infty} \frac{1}{2T} E\left[X^*(f,T)Y(f,T)\right] \tag{5.93}$$

where $X(f, T)$ and $Y(f, T)$ are the time-limited Fourier transforms of random processes X and Y.

Note that the cross-spectral density carries the same information as the cross-correlation function, except that the information is conveyed in the frequency domain. It is important to note that since the cross-correlation function is *not* an even function of time, the resulting Fourier transform will be complex. Therefore, we must take into account not only the *magnitude* of the result but also the *phase* of the resulting calculations. The magnitude of the cross-spectral density is defined as

$$\left|S_{XY}(f)\right| = \sqrt{\mathrm{Re}^2\left[S_{XY}(f)\right] + \mathrm{Im}^2\left[S_{XY}(f)\right]} \tag{5.94}$$

and the phase of the cross-spectral density is defined as

$$\theta_{XY}(f) = \tan^{-1}\left\{-\frac{\mathrm{Im}\left[S_{XY}(f)\right]}{\mathrm{Re}\left[S_{XY}(f)\right]}\right\} \tag{5.95}$$

where $\mathrm{Re}(\cdots)$ and $\mathrm{Im}(\cdots)$ are the real part and the imaginary part of the complex cross-spectrum, respectively. Equation (5.92) provides insight into the properties of the cross-spectral density between two random processes X and Y. Based on the properties of the cross-correlation function and Fourier transforms, we can state the following properties of the cross-spectral density:

$$S_{XY}(f) = S_{YX}(-f)$$
$$S_{XY}(f) = S_{YX}^*(f)$$

To obtain a better understanding of these concepts, consider the following examples.

Example 5.9 Let the sample functions of two random processes be defined as

$$x_i = \cos(2\pi f_0 i\,\Delta t) + \cos(2\pi f_1 i\,\Delta t) + a_i \tag{5.96}$$

and

$$y_i = \cos(2\pi f_1 i\,\Delta t) + a_i \tag{5.97}$$

where a_t is random noise. Figure 5.33 shows a plot x_t and Figure 5.34 a plot of y_t.

For this example, the following parameters were selected:

$$\Delta t = .01 \text{ s}$$
$$f_0 = 5 \text{ Hz}$$
$$f_1 = 14 \text{ Hz}$$

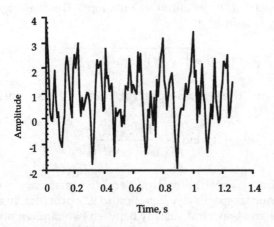

Figure 5.33 Plot of x_t for Example 5.9.

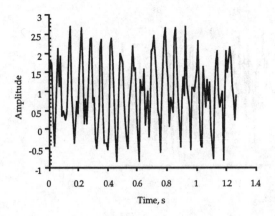

Figure 5.34 Plot of y_t *for Example 5.9.*

Note that f_1 is common to both sample functions. The magnitude of the cross-spectrum between these two sample functions is shown in Figure 5.35.

Note that there is a large peak at 15 Hz, indicating the frequency at which both sample functions have common values. The phase of the cross-spectrum is shown in Figure 5.36.

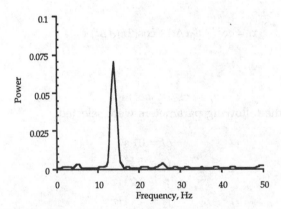

Figure 5.35 Magnitude of the cross-spectrum of x_t and y_t.

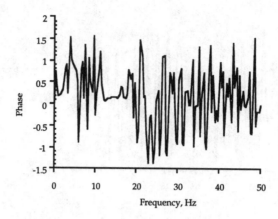

Figure 5.36 Phase of the cross-spectrum of x_t and y_t.

The phase information is difficult to explain. One observation that may be made is that the phase at the frequency where the magnitude of the cross-spectrum is largest is essentially zero.

The next example is similar, but there will not be a common frequency between both sample functions.

Example 5.10 Let the sample functions of two random processes be defined as follows:

$$x_i = \cos(2\pi f_0 i\, \Delta t) + \cos(2\pi f_1 i\, \Delta t) + a_i \tag{5.98}$$

and

$$y_i = \cos(2\pi f_2 i\, \Delta t) + a_i \tag{5.99}$$

For this example, the following parameters were selected:

$$\Delta t = .01 \text{ s}$$
$$f_0 = 5 \text{ Hz}$$
$$f_1 = 14 \text{ Hz}$$
$$f_2 = 21 \text{ Hz}$$

The magnitude of the cross-spectrum between these two sample functions is shown in Figure 5.37. This has been plotted to the same scale as Figure 5.35.

Note that in this example there are no frequencies in common between the two sample functions. At best, there is a certain amount of random noise introduced in both functions, causing the low level of power present in the plot. The phase of the cross-spectrum is shown in Figure 5.38. In this case, there appear to be no portions of the phase cross-spectrum, where the value is consistently equal to zero. As mentioned earlier, the interpretation of the phase of the cross-spectrum is, at best, difficult.

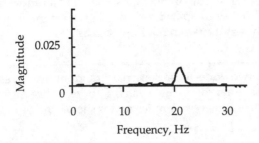

Figure 5.37 Magnitude of the cross-spectrum of x_t and y_t for Example 5.10.

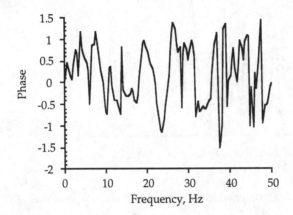

Figure 5.38 Phase of the cross-spectrum of x_t and y_t for Example 5.10.

5.8 ESTIMATION OF CROSS-SPECTRAL DENSITY OF RECORDS OF LIMITED DURATION

The process to be followed to estimate the cross-spectral density between two random processes is very similar to the one outlined in Section 5.6. The stages in the estimation process are outlined by following these steps:

1. The direct current is removed from both sample functions.
2. Extend the data with zeros. In this case, the number of data values was 128; the data are, therefore, extended to the next power of 2, or 256 points.
3. Now compute the FFT of both sample functions, x and y, that have been extended with zeros.
4. Multiply the Fourier transform of x by the complex conjugate of the Fourier transform of y.

5. Compute the inverse FFT of this product. The result is the complete cross-correlation function (CCF) between x and y.
6. The estimate of the CCF is time-limited by using a Hamming window.
7. The FFT of the CCF is now computed, resulting in the estimate of the cross-spectral density between the two random processes.

Example 5.11 Consider two wide-sense stationary random processes X and Y. A time-limited sample function from each of these processes has been sampled at a rate of .004 s between samples and is shown in Figures 5.39 and 5.40.

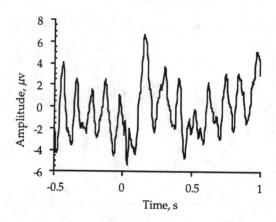

Figure 5.39 Sample function of wide-sense stationary random process X.

The random processes from which these sample functions were recorded are brain waves from similar areas of the brain. Figure 5.39 shows data recorded from the left auditory area of the brain, whereas Figure 5.40 shows data recorded from the right auditory area of the brain. A total of 1.5 s of data were sampled. Time equal to 0 s corresponds to the time when a loud click was presented to each ear. Both data sets were recorded simultaneously. Figure 5.41 shows the complete cross-correlation between the two sample functions. This is the output of step 5. Next, the cross-correlation function is time-limited by using a Hamming window.

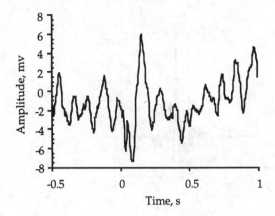

Figure 5.40 Sample function of wide-sense stationary random process Y.

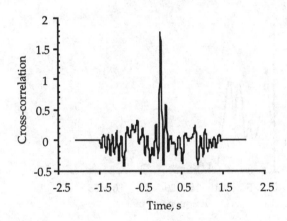

Figure 5.41 Complete cross-correlation between the two sample functions.

Figure 5.42 shows the cross-correlation function after it has been time-limited by a Hamming window.

The next step in the process is the estimation of the magnitude of the cross-spectral density using the FFT on the estimate of the cross-correlation function. This estimate is shown in Figure 5.43.

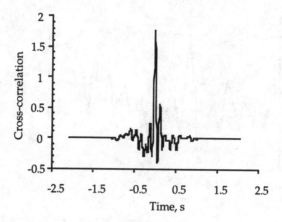

Figure 5.42 Complete cross-correlation between the two sample functions after time-limiting using a Hamming window.

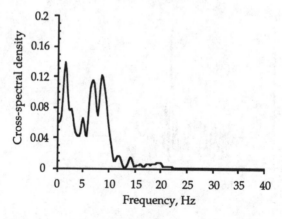

Figure 5.43 Estimate of the magnitude of the cross-spectral density between the two random processes.

The cross-spectral density shows that the two random processes have a strong common component at 1.5 Hz and two other common components at 6.8 and 8.6 Hz. Upon close observation of the data it may be observed that there is a low-frequency phenomenon common to both sample functions. This phenomenon probably is not as significant as the fact that there are two frequencies, 6.8 and 8.6 Hz, common to both random processes. The estimate of the phase angle of the cross-spectrum is shown in Figure 5.44.

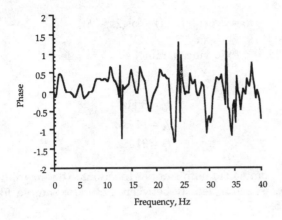

Figure 5.44 Estimate of the phase of the cross-spectral density between the two random processes.

5.9 COHERENCE

The coherence function is often used to express the relationship between two random processes. This function is defined as follows:

$$\rho_{XY}^2(f) = \frac{|S_{XY}(f)|^2}{S_{XX}(f)S_{YY}(f)} \leq 1 \tag{5.100}$$

Coherence can be viewed as the correlation between the two random processes X and Y at frequency f. When the value of the function equals 1 at a particular frequency or range of frequencies, it is said that the random processes are *fully coherent* at those frequencies. When the function equals zero at a particular frequency, then it is said that the random processes are *incoherent*. If the two processes are statistically independent, then the coherence equals zero.

On the surface it would seem an easy task to compute the results of Equation (5.100). One of the problems associated with this definition is that we are dividing by quantities that may go to zero. In these cases erroneous results will be obtained.

Example 5.12 Let the sample functions of two random processes be defined as follows:

$$x_i = \cos(2\pi f_0 i \,\Delta t) + \cos(2\pi f_1 i \,\Delta t) \tag{5.101}$$

and

$$y_i = 2\cos(2\pi f_1 i\,\Delta t) + \cos(2\pi f_2 i\,\Delta t) \qquad (5.102)$$

For this example, the following parameters were selected:

$$\Delta t = .01\text{ s}$$
$$f_0 = 5\text{ Hz}$$
$$f_1 = 14\text{ Hz}$$
$$f_2 = 21\text{ Hz}$$

Note that f_1 is common to both sample functions. The magnitude of the estimate of the cross-spectral density between these two sample functions is shown in Figure 5.45.

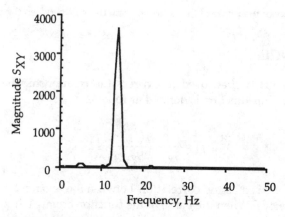

Figure 5.45 Magnitude of the cross-spectral density.

The magnitude of the spectral densities is shown in Figures 5.46 and 5.47.

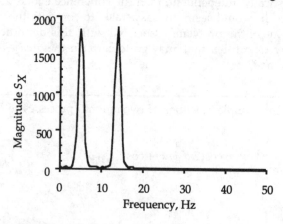

Figure 5.46 Magnitude of the spectral density S_X.

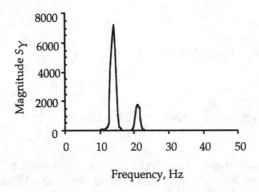

Figure 5.47 Magnitude of the spectral density S_Y.

The cross-spectral density and the two spectral densities follow patterns that we would expect. In other words they show consistency with the model used. The estimate of the coherence between these two sample functions is now shown in Figure 5.48.

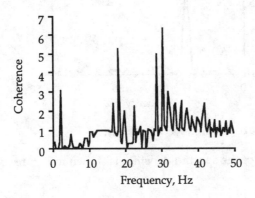

Figure 5.48 Coherence

The interesting fact about Figure 5.48 is that it does not appear to bear any resemblance to what we would expect the coherence between these two functions to look like. First, the values of the coherence are greater than 1, which violates the basic definition as shown in Equation (5.100). Second, we anticipate that the coherence would be greater at the area close to 14 Hz, since this is the frequency where the cross-spectral density is largest.

The reason the coherence, as shown, is not correct is that the values are artificially large due to the division by numbers that are close to zero.

Example 5.13 Let us reconsider the same example, adding some random noise to each of the functions:

$$x_i = \cos(2\pi f_0 i\, \Delta t) + \cos(2\pi f_1 i\, \Delta t) + a_i \qquad (5.103)$$

and

$$y_i = 2\cos(2\pi f_1 i\, \Delta t) + \cos(2\pi f_2 i\, \Delta t) + b_i \qquad (5.104)$$

We will use the same parameters that were used for Example 5.12. The only difference is that each of the two functions has random noise added to it. The magnitude of the estimate of the cross-spectral density between these two sample functions is shown in Figure 5.49.

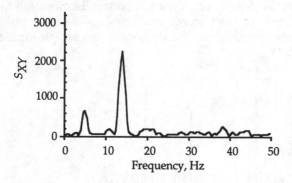

Figure 5.49 Magnitude of the cross-spectral density.

The magnitude of the spectral densities is shown in Figures 5.50 and 5.51.

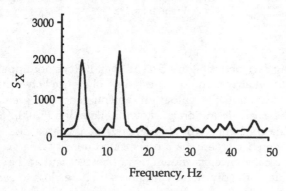

Figure 5.50 Magnitude of the spectral density S_X.

The several spectral plots use the same scales for ease of comparison.

Note the effect of adding noise on the different spectral densities.

The coherence is plotted in Figure 5.52. Note that now the coherence is distributed between 0 and 1. The largest peak present is centered at approximately 14 Hz, which is the value of the common frequencies.

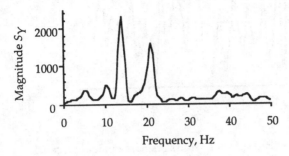

Figure 5.51 Magnitude of the spectral density S_Y.

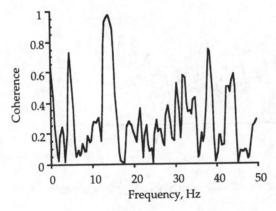

Figure 5.52 Coherence.

Example 5.14 Consider the data used for Example 5.11. The data are repeated below for completeness. The two time-limited sample functions of different random processes were sampled at a rate of 0.004 s between samples. The data are shown in Figures 5.53 and 5.54. The random processes from which these sample functions were recorded are brain waves from similar areas of the brain. The data shown were recorded from the left and right auditory areas of the brain. A total of 1.5 s of data was sampled. Time equal to 0 s corresponds to the time when a loud click was presented to each ear. Both data sets were recorded simultaneously. Both data sets are very similar, so one would anticipate a great deal of coherence between these two time-limited waveforms. Figure 5.55 shows the mag-nitude of the cross-spectral density between both waveforms, and Figure 5.56 shows the coherence.

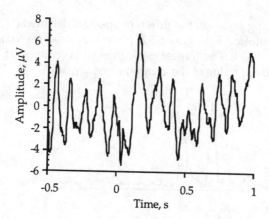

Figure 5.53 Sample function of wide-sense stationary random process X.

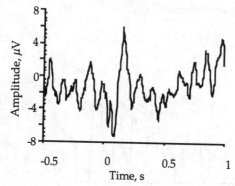

Figure 5.54 Sample function of wide-sense stationary random process Y.

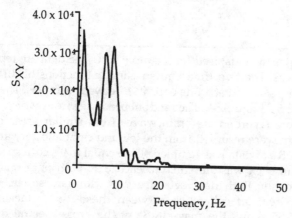

Figure 5.55 Magnitude of the cross-spectral density S_{XY}.

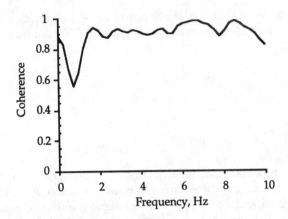

Figure 5.56 Coherence between the two sample functions.

The coherence has been computed up to 10 Hz because this is where most of the energy is concentrated for the sample functions. As expected, the amount of coherent energy in this range is high.

Before leaving this topic, the author would like to caution the reader again about using the coherence function in a judicious manner. It is possible to obtain results which may be tainted by unknown relationships between the random process under study.

5.10 CHAPTER SUMMARY

The power spectral density of a random process may be expressed by the following relationship:

$$S_x(f) = \lim_{T \to \infty} \frac{1}{2T} E\left[|X(f, T)|^2\right]$$

The Einstein-Wiener-Khinchine theorem states that the power spectral density of a random process equals the Fourier transform of its autocorrelation function. This theorem may be expressed by

$$S(f) = F\{R_X(\tau)\}$$

For a random process which is real and at least wide-sense stationary, the spectral density of a random process is a real, positive, and even function of frequency.

White noise is random noise that has equal power at all frequencies of the spectrum. The spectral density and corresponding autocorrelation function of white noise are given by

$$S_X(f) = 1$$

and

$$R_X(\tau) = \delta(\tau)$$

respectively.

One of the most important sections of this chapter deals with the estimation of the spectral density of records of limited duration. A number of techniques were explained and used. The first technique involved computing the transform of the autocorrelation function were the FFT is used extensively. The second one entailed computing the periodogram. In this method, the sequence of observed values is split into shorter segments, the discrete Fourier transform of the short segments is computed, and then the results are averaged. The last method explained was the model-based approach. In this approach the estimation is based on the theory that a random process may be described as an autoregressive (AR) random process. In the AR model the current value of the process is expressed as a finite, linear combination of previous values and an uncorrelated component.

The cross-spectral density between two random processes X and Y may be defined by the following equation:

$$S_{XY}(f) = \int_{-\infty}^{\infty} R_{XY}(\tau) \exp(-j2\pi f \tau) \, d\tau$$

The magnitude of the cross-spectral density is defined as

$$|S_{XY}(f)| = \sqrt{\text{Re}^2[S_{XY}(f)] + \text{Im}^2[S_{XY}(f)]}$$

and the phase of the cross-spectral density is defined as

$$\theta_{XY}(f) = \tan^{-1}\left\{-\frac{\text{Im}[S_{XY}(f)]}{\text{Re}[S_{XY}(f)]}\right\}$$

where $\text{Re}(\cdots)$ and $\text{Im}(\cdots)$ are the real part and the imaginary part of the complex cross-spectrum, respectively.

The coherence function is often used to express the relationship between two random processes. This function is defined as follows:

$$\rho_{XY}^2(f) = \frac{|S_{XY}(f)|^2}{S_{XX}(f)S_{YY}(f)} \le 1$$

Coherence can be viewed as the correlation between the two random processes X and Y at frequency f. When the value of the function equals 1 at a particular frequency or range of frequencies, it is said that the random processes are fully coherent at those frequencies. When the function equals zero at a particular frequency, then it is said that the random processes are incoherent. If the two processes are statistically independent, then the coherence equals zero.

5.11 PROBLEMS

1. A truncated sample function $X_T(t)$ from a stationary random process has a Fourier transform of

$$F[X_T(t)] = \sqrt{\frac{2(1-\cos 2\omega)T}{\omega^2}} \cos(\omega T + \theta)$$

where θ is a random variable uniformly distributed between 0 and π. Find the spectral density of this process.

2. A stationary random process has the spectral density shown below. Find the mean square value of this process.

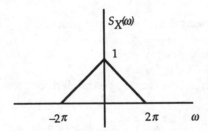

3. An ergodic random process has an autocorrelation function of the form

$$R_X(\tau) = 50\delta(\tau) + 36$$

What is the value of the spectral density of this process at $\omega = 100$?

4. A stationary random process has an autocorrelation function of the form

$$R_X(\tau) = \begin{cases} 4 & |\tau| > 3 \\ 4 + \dfrac{2}{3}\left(1 - \dfrac{|\tau|}{3}\right) & |\tau| < 3 \end{cases}$$

Compute the spectral density of the process.

5. For the following functions of ω, state whether it is a valid expression for the spectral density of a random process. If it is not valid, state the reason.

a) $\dfrac{\omega^2 + 4}{\omega^4 + 10\omega^2 + 9}$

(b) $\dfrac{\omega^2}{\omega^4 - 10\omega^2 + 9}$

(c) $\dfrac{\omega^2 + 8\omega + 2}{\omega^6 + 10\omega^4 + 4\omega^2 + 4}$

(d) $\dfrac{1 - \cos 2\omega}{2\omega^2}$

(e) $\dfrac{\sqrt{\omega^2 - 1}}{\omega^4 + 6\omega^2 + 9}$

(f) $\dfrac{\omega + 6}{\omega^3 + 6\omega^2 + \omega + 6}$

6. A stationary random process has an autocorrelation function given by

$$R_X(\tau) = \begin{cases} \left(1 - \dfrac{|\tau|}{T}\right)\cos \omega_0 \tau & |\tau| \le T \\ 0 & |\tau| > T \end{cases}$$

(a) Determine the spectral density of this process.

(b) Sketch the spectral density.

7. A stationary random process has a spectral density given by

$$S_X(\omega) = \begin{cases} A^2\left[1 - \dfrac{|\omega|}{2\pi W}\right] & |\omega| \le 2\pi W \\ 0 & |\omega| > 2\pi W \end{cases}$$

(a) Determine the autocorrelation function of this process.

(b) Sketch the autocorrelation function.

8. The power spectrum of a random process is given by

$$S_X(\omega) = \frac{\omega^4 + 7\omega^2 + 9}{\omega^4 + 5\omega^2 + 5}[1 + 2\pi\delta(\omega)]$$

(a) Compute the mean value of $X(t)$.

(b) Compute the spectral value S_0 for the white noise portion of $X(t)$.

(c) Compute the mean square value of $X(t)$.

(d) Compute the autocorrelation function of $X(t)$.

9. A stationary random process has the spectral density shown below. Find the mean square value of this process.

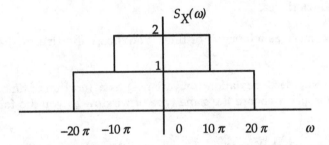

10. Consider the time function

$$x(t) = \begin{cases} \exp(-2t) + n(t) & t \geq 0 \\ 0 & \text{elsewhere} \end{cases}$$

This is similar to Example 5.5 except that a random noise term has been added to the function of time. Compute the autocorrelation of this function, using the FFT.

11. Consider the time function

$$x_i = A\cos(2\pi f_0 i\,\Delta t + \theta) + 2\,\text{rnd}(1.0) - 1.0$$

where $A = 1.0$, $f_0 = 10$, θ is selected from a uniform distribution with range 0 to 2π, and the function rnd is simply a random number generator. Estimate the spectral density of this function. Use the autocorrelation method (refer to Section 5.6) using the FFT.

(a) Increase the amount of random noise by using larger numbers in the multiplier.

(b) Comment on the effects of the amount of random noise on the spectral density.

12. Repeat Problem 11, but now use the periodogram technique.

13. Consider the data KOBE.DAT supplied to you. These are data recorded from a seismograph (vertical acceleration, nm/s^2) of the Kobe earthquake recorded at Tasmania University, Hobart, Australia, on January 16, 1995, beginning at 20:56:51 Greenwich mean time (GMT) and continuing for 51 minutes at 1-s intervals.

(a) Plot the data. Please describe the data regarding their characteristics: mean as a function of time, variance as a function of time, spectral characteristics.

(b) Estimate the mean as a function of time. You choose the size of the window.

(c) Estimate the standard deviation (or variance) as a function of time. You should probably choose the same type of window as you did for part (a).

(d) Estimate the spectral density in at least two different ways. Please explain carefully the assumptions you are making for each of the two ways.

14. Consider the data METHANE.DAT supplied to you. These are data of methane input into a gas furnace in cubic feet per minute. Sampling interval is 9 s. Repeat parts (a) to (d) as given in Problem 13.

15. Consider the data E921.DAT supplied to you. The data are 200 simulated values of an AR process. Please estimate the order of the process, the parameters, and the spectral density.

16. Consider the OSHORTS.DAT data supplied to you. The data correspond to 57 consecutive daily overshorts from an underground gasoline tank at a filling station in Colorado. If $y(t)$ is the measured amount of fuel in the tank at time t and $a(t)$ is the measured amount sold minus the amount delivered during the course of the day t, then the overshort at the end of the day t is defined as $y(t) - y(t-1) + a(t)$. Source: Brockwell and Davis.

(a) Plot the data.

(b) Is this an autoregressive process? If so, confirm the order of the process.

(c) Estimate the spectral density of the process.

17. Consider the data CO2.DAT supplied to you. These data are the carbon dioxide output from a gas furnace in percent of output gas. Sampling interval is 9 s. (Source: Box and Jenkins.) Repeat parts (a) to (d) as given in Problem 13.

18. Consider the data SUNSPLUS.DAT supplied to you. These are the monthly means of daily relative sunspot numbers, from January 1749 to March 1977. (Source: Andrews and Herzbe.)

(a) Plot the data. Please describe the data characteristics: mean as a function of time, variance as a function of time, spectral characteristics.

(b) Estimate the mean as a function of time. You choose the size of the window.

(c) Estimate the standard deviation (or variance) as a function of time. You should probably choose the same type of window as you did for part (a).

(d) Estimate the spectral density in at least two different ways. Please explain carefully the assumptions you are making for each of the two ways.

19. Estimate the coherence between the data CO2.DAT and the data METHANE.DAT.

20. In Problem 13 you described the characteristics of an earthquake. Using the Internet, retrieve the data from another earthquake and repeat the process. Provide your instructor with the Internet address of the new earthquake.

5.12 MATHCAD COMPUTER EXAMPLES

MATHCAD Computer Example 5.1
COMPUTATION OF THE AUTOCORRELATION FUNCTION USING THE FFT

Number of points:

M := 128

m := 0 , 1 .. 127

Define now the equation:

$x_m := 2 \cdot \exp(-3 \cdot m \cdot 0.02)$

Compute the DC value:

$mean := \dfrac{\sum x}{128}$

Subtract the DC value:

$y_m := x_m - mea$

Add zeros:

$y_{M+m} := 0$

Compute now the FFT:

$f := FFT(y)$

Determine the number of autocorrelation lags desired:

j := 0 .. 128

$\tau := 0 .. 128$

$h_j := f_j \cdot \overline{f}$

Compute now the inverse FFT:

$z := IFFT(h)$

$z0 := z$

Compute now the autocorrelation, using a time domain technique:

$$r_\tau := \frac{1}{M} \cdot \sum_m y_m \cdot y_{\tau +}$$

$$r0 := r$$

Plot of the ACF using the FFT.

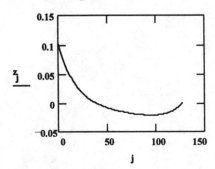

Plot of the ACF using time domain technique.

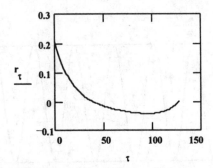

MATHCAD Computer Example 5.2
**COMPUTATION OF THE CROSS-CORRELATION FUNCTION VIA
THE FFT**

The purpose of this computer program is to demonstrate the computation of
the CCF via the FFT.

$$A := 1.0$$

$$B := 1.0$$

$$\phi := 1.5$$

$i := 0 .. 300$

$M := 300$

$f_0 := 3.0$

Pick at random a phase angle:

$\theta := \mathbf{rnd}(2 \cdot \pi)$

$\theta = 0.008$

Define the first equation:

$x_i := A \cdot \cos\left(2 \cdot \pi \cdot f_0 \cdot i \cdot 0.01 + \theta\right)$

Define now the second equation:

$y_i := B \cdot \cos\left(2 \cdot \pi \cdot f_0 \cdot i \cdot 0.01 + \theta + \phi\right)$

Plot the data:

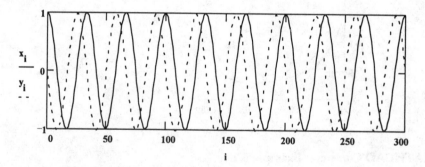

Extend with zeros up to a power of 2:

$m := 1 .. 211$

$j := 0 .. 256$

$x_{M+m} := 0$

$y_{M+m} := 0$

Check the length of data vectors:

length(x) = 512

Compute the FFTs:

xx := FFT(x)

yy := FFT(y)

$$\mathbf{zz_j := xx_j \cdot \overline{yy}}$$

z := IFFT(zz)

j := 0 .. 50

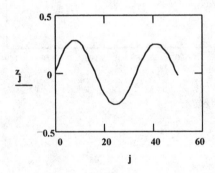

MATHCAD Computer Example 5.3
ESTIMATE OF THE SPECTRAL DENSITY OF A RANDOM PROCESS

In this program we will compute the estimate of the spectral density of a random process, using the first. We will use a known time-limited sample function from a wide-sense stationary random process.

i := 0 .. 127

A := 1.0

$$\mathbf{f_0 := 5.0}$$

deltat := 0.01

M := 128

$\theta := \mathbf{rnd}(2 \cdot \pi)$

$\theta = 0.008$

$x_i := A \cdot \mathbf{cos}\left(2 \cdot \pi \cdot f_0 \cdot i \cdot \mathbf{deltat} + \theta\right) + 4.0 \cdot \mathbf{rnd}(1.0) - 1.0$

$\mathbf{WRITEPRN(\ cosdat)} := x_i$

The data we have just generated are a cosinusoid of amplitude 1.0, frequency fixed at 3 Hz, with a random phase added, and random noise added to the entire process. A plot of the data is shown below.

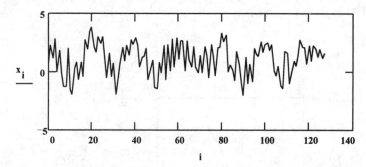

Remove the dc component of the data:

$$a := \frac{\sum\limits_{i} x_i}{M}$$

$$x_i := x_i -$$

Perform an autocorrelation check:

$$r_0 := \frac{\sum\limits_{i} \left(x_i\right)^2}{256}$$

$r_0 = 0.9$

Note that we are dividing by 256 since these are the number of points we will have after we add zeros to the data in order to avoid circular correlation problems.

Extend the data with zeros.

$j := 0..255$

$x_{M+i} := 0$

$length(x) = 256$

WRITEPRN(cosdat0) := x

Note that we are using the complex form of the FFT in MATHCAD for completeness. MATHCAD uses a different definition for the computation of the FFT.

$f := CFFT(x) \cdot length(x)$

Multiply the FFT by the absolute value squared:

$h_j := \left(\left| f_j \right| \right)$

Compute now the inverse FFT (IFFT). Note that we have to divide by the number of points used, since MATHCAD uses a different definition for the IFFT.

$z := \dfrac{ICFFT(h)}{length(x)}$

The result is the complete autocorrelation of the original data. This includes *all* lagged values. Check to see if the value obtained earlier for the zero lagged value of the autocorrelation function corresponds.

$\dfrac{z_0}{256} = 0.914$

It corresponds!

Divide each value of the ACF by the number of points. This completes the computation of the ACF for all lagged values.

$z_j := \dfrac{z_j}{256}$

WRITEPRN(cosacf) $:= z$

Plot the data:

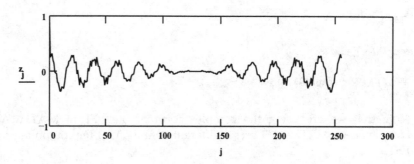

To estimate the spectral density using the autocorrelation function, we need   to multiply the estimate of the autocorrelation function by a time domain window (such as the Hamming window). We are only going to accept between 20 and 25 percent of the lagged values of the ACF.

$N \equiv 256$

$k := 0 .. (N) \cdot (0.25)$

$kk := N \cdot (0.25) - 1$

$kk = 63$

$jj := 1 .. (N) \cdot (0.25)$

Now we generate the Hamming window:

$$w_k := 0.54 + \left(0.46 \cdot \cos \left(\pi \cdot \frac{k}{kk} \right) \right)$$

$kkk := 0 .. 255$

$y_{kkk} := 0$

$y_k := w$

$y_{256 - jj} := w_j$

Plot of the Hamming window.

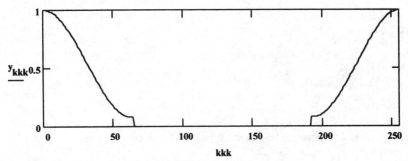

Multiply the estimate of the ACF by the time window:

$zz_{kkk} := 0$

$zz_{kkk} := z_{kkk} \cdot y_{kk}$

WRITEPRN(cosham) $:= zz_{kk}$

Plot the ACF before and after multiplication by the Hamming window.

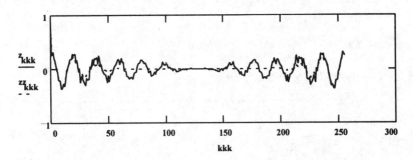

The solid line is the original, and the dotted line is the ACF after multiplication by the Hamming window.

We now compute the FFT and thus an estimate of the spectral density of the data. Note that we are multiplying the results by the number of points to compensate for the FFT definition used by MATHCAD.

$g := \text{CFFT}(zz) \cdot \text{length}(x)$

We need to compute the absolute value of g and multiply by Δt, the spacing between time samples. Note that earlier we made $\Delta t = .01$. We are going to take only the positive half of the spectrum since the negative half is identical.

$$\mathbf{MM} := \frac{\text{length}(\mathbf{g})}{2} + 1$$

$$\mathbf{MM} = 129$$

$$\mathbf{s} := 0 .. \mathbf{MM} - 1$$

$$\mathbf{ff_s} := \left| \mathbf{g_s} \right| \cdot \mathbf{delta}$$

We now need to compute the spacing between the frequency:

$$\mathbf{deltaf} := \frac{1}{\mathbf{N \cdot delta}}$$

$$\mathbf{deltaf} = 0.3$$

Compute now the last check of the units. We multiply the summation by 2 because we are dealing with only the positive side of the spectrum.

$$\mathbf{check} := \left(\sum_{\mathbf{s}} \mathbf{ff_s} \right) \cdot 2 \cdot \mathbf{delta}$$

$$\mathbf{check} = 0.9$$

Plot now the estimate of the spectra:

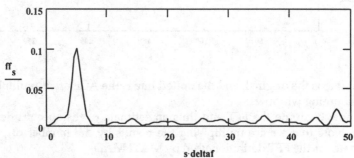

$$\mathbf{WRITEPRN} \; (\; \mathbf{cospec} \;) := \mathbf{ff_s}$$

MATHCAD Computer Example 5.4
ESTIMATE OF THE SPECTRAL DENSITY OF REAL DATA (1st case)

Read the data file:

$i := 0..127$

$M := 128$

$deltat = 0.004$

$x_i := READ(\,tonecz\,)$

Remove the dc of the data:

$$a := \frac{\displaystyle\sum_i x_i}{M}$$

$x_i := x_i -$

$a = 0.159$

Plot the data:

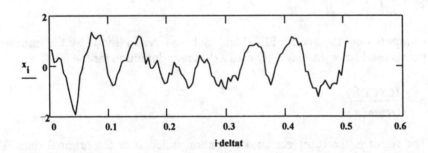

i·deltat

This is the check used later to make sure that the units of the ACF as returned by the FFT are correct. Note that we have to divide by the number of points and multiply by Δt to obtain the correct units. The number of points we are using includes the number of zeros we are going to add later.

$$r_0 := \frac{\displaystyle\sum_i \left(x_i\right)^2}{256}$$

$r_0 = 0.2$

Before computing the FFT, we need to extend the array with zeros. This is necessary since we are going to compute the complete ACF of the data.

$j := 0, 1 .. 255$

$x_{M+i} := 0$

$length(x) = 256$

This checks the length of x.

Compute now the FFT of the data:

$f := CFFT(x) \cdot length(x)$

Note that we are using the complex form of the FFT in MATHCAD for completeness. MATHCAD uses a different definition for the computation of the FFT.

Multiply the FFT by the absolute value squared:

$$h_j := \left(\left| f_j \right| \right)$$

Compute now the inverse FFT. Note that we have to divide by the number of points used since MATHCAD uses a different definition for the IFFT.

$$z := \frac{ICFFT(h)}{length(x)}$$

The result is the complete autocorrelation function of the original data. This includes *all* lagged values. Check to see if the value obtained corresponds to the direct computation performed above.

$$\frac{z_0}{256} = 0.2$$

$WRITEPRN(toneacf) := z$

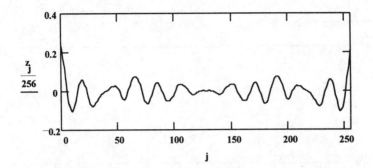

Divide by the number of points to conform to the definition of the ACF.

$$z_j := \frac{z_j}{256}$$

To estimate the spectral density using the autocorrelation function we need to multiply the estimate of the autocorrelation function by a time-domain window. We are only going to accept between 20 and 25 percent of the lagged values of the ACF.

$N \equiv 256$

Generate a Hamming window:

$k := 0 .. (N) \cdot (0.25)$

$kk := N \cdot 0.25 - 1$

$kk = 63$

$jj := 1 .. (N) \cdot (0.25)$

$$w_k := 0.54 + \left(0.46 \cdot \cos\left(\pi \cdot \frac{k}{kk}\right)\right)$$

$kkk := 0 .. 255$

$y_{kkk} := 0$

$y_k := w$

$y_{256-jj} := w_j$

Plot the Hamming window:

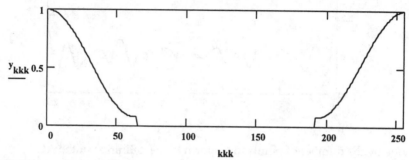

WRITEPRN(hamm) := y_{kk}

Note: Typically, 25 percent of the length of the data is used (see Box and Jenkins, 1976, p. 33). We now multiply the estimate of the autocorrelation function by the time window:

$$zz_{kkk} := 0$$

$$zz_{kkk} := z_{kkk} \cdot y_{kk}$$

WRITEPRN(acfham) := zz_{kk}

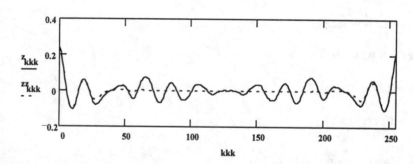

The solid line is the original estimate, and the dotted line is the estimate after multiplying by the time-limited Hamming window. We now compute the FFT and thus an estimate of the spectral density of the data. Note that we multiply the result by the number of points.

$$g := CFFT(zz) \cdot length(x)$$

We now need to compute the absolute value of g and multiply by Δt. We are going to take only the positive half of the spectrum since the negative half is identical.

$$MM := \frac{length(g)}{2} + 1$$

$$MM = 129$$

$$s := 0 .. MM - 1$$

$$ff_s := |g_s| \cdot delta$$

$$length(g) = 256$$

$$deltaf := \frac{1}{N \cdot delta}$$

$$check := \left(\sum_s ff_s \right) \cdot 2 \cdot delta$$

$$check = 0.2$$

Note that the summation is multiplied by 2 because we are only dealing with the positive side of the spectrum and the function is an even function of frequency. The estimate of the spectral density is plotted in a regular linear scale.

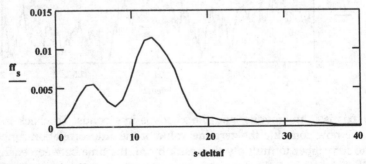

$$WRITEPRN(spec) := ff_s$$

MATHCAD Computer Example 5.5
ESTIMATE OF THE SPECTRAL DENSITY OF REAL DATA (2nd case)

Prepare to read in the data:

$i := 0 .. 499$

$M := 500$

$deltat := 0.004$

$x_i := READ(spec2)$

Remove the dc component of the data first:

$$a := \frac{\sum_i x_i}{M}$$

$$x_i := x_i -$$

$a = -2.147$

Plot the data:

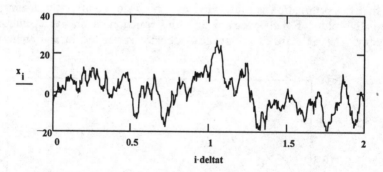

The y axis is in microvolts, and the x axis is in seconds. To check for proper units, we now compute the zero lag value of the autocorrelation function. We have to remember to multiply the result by Δt, the time between each discrete sample. This will be indeed the true approximation to a continuous integral.

$$r_0 := \frac{\sum_i (x_i)^2}{1024}$$

$$r_0 = 38.$$

A caveat about this computation. We are dividing by N. In this case N is not 500 but 1024, since we are going to extend the data with that many zeros. Now we extend the data with zeros up to 1024 to avoid circular correlation problems

$$l := 0 .. 523$$

$$x_{M+l} := 0$$

$$\text{length}(x) = 1.0$$

This checks the length of x. We now compute the ACF via the FFT. Compute the complex FFT first.

$$f := CFFT(x) \cdot \text{length}(x)$$

Multiply the FFT by its complex conjugate squared:

$$j := 0 .. 1023$$

$$h_j := \left(|f_j| \right)$$

Compute now the inverse FFT:

$$z := \frac{ICFFT(h)}{\text{length}(x)}$$

The result should be the complete (positive and negative) autocorrelation function of the original data. This includes *all* lagged values. Check now the first value to compare with value obtained above. As before, we have to multiply by Δt.

$$r_0 := \frac{z_0}{\text{length}(x)}$$

$r_0 = 38.$

The comparison checks!

We are now ready to accept only 25 percent of the lagged values of the autocorrelation function. This is done by using a Hamming window.

$$z_j := \frac{z_j}{\text{length}(x)}$$

We now generate the Hamming window:

$N \equiv 1024$

$k := 0 \,..\, (N) \cdot (0.25)$

$kk := N \cdot 0.25 - 1$

$kk = 255$

$jj := 1 \,..\, (N) \cdot (0.25)$

$$w_k := 0.54 + \left(0.46 \cdot \cos\left(\pi \cdot \frac{k}{kk} \right) \right)$$

$kkk := 0 \,..\, 1023$

$y_{kkk} := 0$

$y_k := w \qquad y_{1024 - jj} := w_j$

Plot the Hamming window:

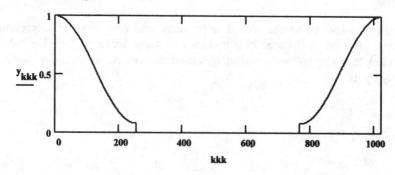

We now multiply the estimate of the autocorrelation function by the time window:

$$zz_{kkk} := 0$$

$$zz_{kkk} := z_{kkk} \cdot y_{kk}$$

Plot now the data before and after multiplication by the time window.

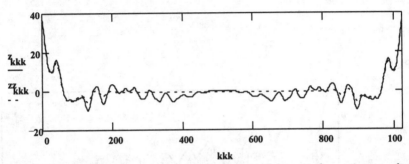

The solid line is the original estimate, and the dotted line is the estimate after multiplying by the time-limited Hamming window. We now compute the FFT and thus an estimate of the spectral density of the data. Note that we multiply the result by the number of points.

$$g := CFFT(zz) \cdot length(x)$$

We now need to compute the absolute value of g and multiply by deltat. We are going to take only the positive half of the spectrum.

$$MM := \frac{length(g)}{2} + 1$$

$$MM = 513$$

$$s := 0 .. MM - 1$$

$$ff_s := |g_s| \cdot deltat$$

$$deltaf := \frac{1}{N \cdot deltat}$$

$$\text{check} \ := \ \left(\sum_{s} \mathbf{ff}_s \right) \cdot 2 \cdot \mathbf{deltaf}$$

check = 39.

We plot the results on a decibel (dB) scale. (Note that we multiply by 10 since we are already working with power quantities.)

$$\mathbf{p}_s \ := \ 10 \cdot \log \left(\mathbf{ff}_s \right)$$

$\mathbf{p}_0 = 6.8$

Plot:

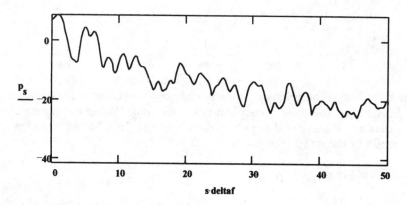

s·deltaf

MATHCAD Computer Example 5.6
ESTIMATE OF THE SPECTRAL DENSITY OF A RANDOM PROCESS. Use of the Periodogram

In this program we will compute the estimate of the spectral density of a random process using the periodogram method. We will use a known time-limited sample function from a wide-sense stationary random process.

$i \ := \ 0 .. 127$

$A \ := \ 1.0$

$f_0 \ := \ 5.0$

$\mathbf{deltat} \ := \ 0.01$

$M \ := \ 128$

$\theta := \mathbf{rnd}(2 \cdot \pi)$

$\theta = 0.008$

$x_i := A \cdot \cos\left(2 \cdot \pi \cdot f_0 \cdot i \cdot \mathbf{deltat} + \theta\right) + 2.0 \cdot \mathbf{rnd}(1.0) - 1.0$

$\mathbf{WRITEPRN}(\text{cosdat}) := x_i$

The data we have just generated are a cosinusoid of amplitude 1.0, frequency fixed at 3 Hz, with a random phase added, and random noise added to the entire process. A plot of the data is shown:

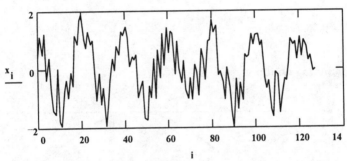

Remove the dc component of the data:

$$a := \frac{\sum_i x_i}{M}$$

$x_i := x_i -$

Number of points per section and overlap:

$\mathbf{npsec} := 32$

$\mathbf{overl} := 0.5$

$\mathbf{knew} := 16$

Compute now the number of sections.

$$\text{kcompl} := \text{floor}\left(\frac{M}{\text{npsec}}\right) \cdot \text{npsec}$$

$$\text{kcompl} = 128$$

$$\text{isect} := \text{floor}\left[\frac{\text{kcompl} - (\text{npsec} - \text{knew})}{\text{knew}}\right]$$

$$\text{isect} = 7$$

Generate a Hamming window over the number of points per section:

$$i := 0 \ldots 31$$

$$w_i := 0.54 - 0.46 \cdot \cos\left[\frac{2 \cdot \pi \cdot (i)}{32}\right]$$

Plot the Hamming window:

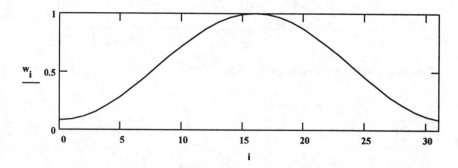

There are seven overlapping sections. The beginning and endpoint numbers of each section (note the 50 percent overlap): 0-31, 16-47, 32-63, 48-79, 64-95, 80-111, 96-127.

Compute the normalizing factor due to the window we are using:

$$u := \left[\sum_i (w_i)^2\right]$$

$$u = 12.71$$

Begin processing the different sections:

Section 1:

$ii := 0..31 \quad y_{ii} := x_{ii}$

$z_{ii} := y_{ii} \cdot w_{ii} \quad f := CFFT(z) \cdot npsec$

$jj := 0..16$

$g_{jj} := 0.0$

$g_{jj} := g_{jj} + \left(\left|f_{jj}\right|\right)$

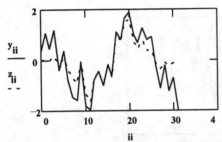

The solid line is the data prior to windowing, and the dashed line is after multiplication by the Hamming window. These are the first 64 points of the data.

$WRITEPRN(perbef) := y_{ii} \quad WRITEPRN(peraft) := z_{ii}$

Section 2:

$y_{ii} := x_{ii+1} \quad z_{ii} := y_{ii} \cdot w_{ii} \quad f := CFFT(z) \cdot npsec$

$g_{jj} := g_{jj} + \left(\left|f_{jj}\right|\right)$

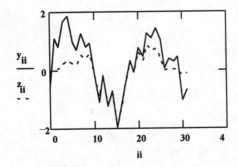

Section 3:

$$y_{ii} := x_{ii+3}$$

$$z_{ii} := y_{ii} \cdot w_{ii}$$

$$f := CFFT(z) \cdot npsec$$

$$g_{jj} := g_{jj} + \left(\left| f_{jj} \right| \right)$$

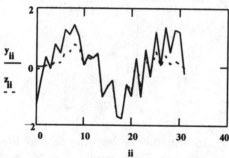

Section 4:

$$y_{ii} := x_{ii+4}$$

$$z_{ii} := y_{ii} \cdot w_{ii}$$

$$f := CFFT(z) \cdot npsec$$

$$WRITEPRN(\ perbef1\ ) := y_{ii}$$
$$WRITEPRN\ (\ peraft2\ ) := z_{ii}$$

Section 5:

$$y_{ii} := x_{ii+6}$$

$$z_{ii} := y_{ii} \cdot w_{ii}$$

$$f := CFFT(z) \cdot npsec$$

$$g_{jj} := g_{jj} + \left(\left| f_{jj} \right| \right)$$

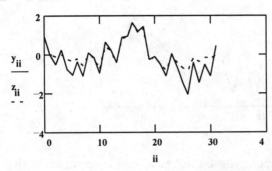

Section 6:

$$y_{ii} := x_{ii+8}$$

$$z_{ii} := y_{ii} \cdot w_{ii}$$

$$f := CFFT(z) \cdot npsec$$

$$g_{jj} := g_{jj} + \left(\left| f_{jj} \right| \right)$$

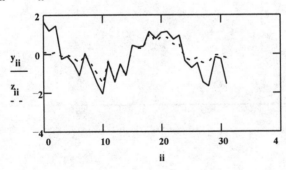

Section 7:

$$y_{ii} := x_{ii+9}$$

$$z_{ii} := y_{ii} \cdot w_{ii}$$

$$f := CFFT(z) \cdot npsec$$

$$g_{jj} := g_{jj} + \left(\left| f_{jj} \right| \right)$$

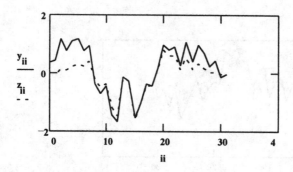

We are now ready to compute the estimate of the spectral density via the periodogram.

$$g_{jj} := \frac{g_{jj} \cdot delta}{u \cdot isect}$$

$$deltaf := \frac{1}{npsec \cdot delta}$$

$deltaf = 3.1$

$\mathbf{WRITEPRN(\ perspc\ )} := g_j$

$length(\ g\ ) = 17$

Plot the data:

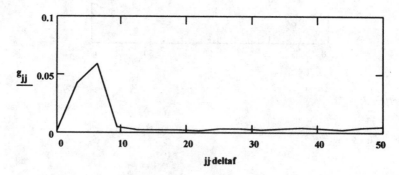

$deltat = 0.01$

$npsec = 32$

MATHCAD Computer Example 5.7
ESTIMATE OF THE SPECTRAL DENSITY OF REAL DATA. Use of the
Periodogram (1st case)

Read in the data file:

$i := 0 .. 127$ $M := 128$

$deltat \equiv 0.004$ $x_i := READ(tonecz)$

Remove the dc component of the data:

$$a := \frac{\sum_i x_i}{M}$$

$x_i := x_i -$ $a = 0.159$

Plot the data:

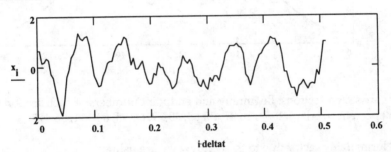

i·deltat

Number of points per section and overlap:

$npsec := 32$

$overl := 0.5$

$knew := 16$

Compute now the number of sections:

$$kcompl := floor\left(\frac{M}{npsec}\right) \cdot npsec$$

$kcompl = 128$

$$\text{isect} := \text{floor}\left[\frac{\text{kcompl} - (\text{npsec} - \text{knew})}{\text{knew}}\right]$$

$\text{isect} = 7$

Generate a Hamming window over the number of points per section:

$i := 0 .. 31$

$$w_i := 0.54 - 0.46 \cdot \cos\left[\frac{2 \cdot \pi \cdot (i)}{32}\right]$$

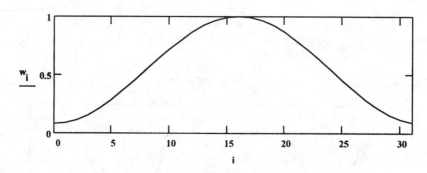

There are seven sections. Beginning and endpoint numbers of each section (note the 50 percent overlap): 0-31, 16-47, 32-63, 48-79, 64-95, 80-111, 96-127.

Normalizing factor due to the window we are using:

$$u := \left[\sum_i (w_i)^2\right]$$

$u = 12.$

Begin processing the different sections:

Section 1:

$ii := 0 .. 31$

$y_{ii} := x_{ii}$

$z_{ii} := y_{ii} \cdot w_{ii}$

$f := CFFT(z) \cdot npsec$

$jj := 0 .. 16$

$g_{jj} := 0.0$

$g_{jj} := g_{jj} + \left(\left| f_{jj} \right| \right)$

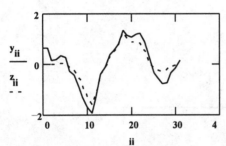

The solid line is the data prior to windowing, and the dashed line is after multiplication by the Hamming window. These are the first 64 points of the data.

$WRITEPRN(perbef) := y_{ii}$

$WRITEPRN(peraft) := z_{ii}$

Section 2:

$y_{ii} := x_{ii+1} \quad z_{ii} := y_{ii} \cdot w_{ii}$

$f := CFFT(z) \cdot npsec \quad g_{jj} := g_{jj} + \left(\left| f_{jj} \right| \right)$

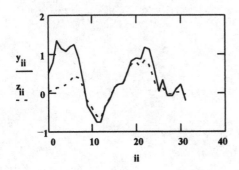

ii

Section 3:

$$y_{ii} := x_{ii+3} \quad z_{ii} := y_{ii} \cdot w_{ii}$$

$$f := CFFT(z) \cdot npsec \quad g_{jj} := g_{jj} + \left(\left| f_{jj} \right| \right)$$

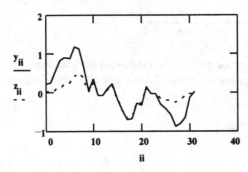

ii

Section 4:

$$y_{ii} := x_{ii+4} \quad z_{ii} := y_{ii} \cdot w_{ii}$$

$$f := CFFT(z) \cdot npsec \quad g_{jj} := g_{jj} + \left(\left| f_{jj} \right| \right)$$

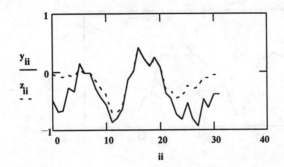

WRITEPRN(perbef1) := y_{ii} **WRITEPRN(peraft2)** := z_{ii}

Section 5:

$y_{ii} := x_{ii+6}$ $z_{ii} := y_{ii} \cdot w_{ii}$

$f := CFFT(z) \cdot npsec$ $g_{jj} := g_{jj} + \left(\left| f_{jj} \right| \right)$

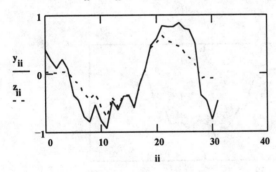

Section 6:

$y_{ii} := x_{ii+8}$ $z_{ii} := y_{ii} \cdot w_{ii}$

$f := CFFT(z) \cdot npsec$ $g_{jj} := g_{jj} + \left(\left| f_{jj} \right| \right)$

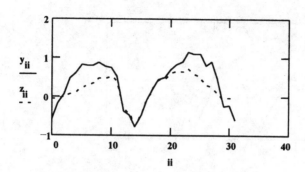

Section 7:

$$y_{ii} := x_{ii+9}$$

$$z_{ii} := y_{ii} \cdot w_{ii}$$

$$f := CFFT(z) \cdot npsec$$

$$g_{jj} := g_{jj} + \left(\left| f_{jj} \right| \right)$$

2
1
$\dfrac{y_{ii}}{}$ 0
z_{ii}
- - −1
−2
   0   10   20   30   40
                      ii

We are now ready to compute and plot the estimate of the spectral density via the periodogram.

$$g_{jj} := \frac{g_{jj} \cdot delta}{u \cdot isect}$$

$$deltaf := \frac{1}{npsec \cdot delta}$$

$$deltaf = 7.813$$

WRITEPRN(perspc) $:=$ **g**$_\text{j}$

length(g) = 17

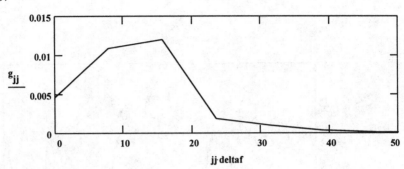

$$jj \cdot deltaf$$

MATHCAD Computer Example 5.8
ESTIMATE OF THE SPECTRAL DENSITY OF REAL DATA: Use of the
Periodogram (2nd case)

We are going to compute the estimate of the spectral density using the periodogram method. Read the data first.

$i := 0 .. 499$

$M := 500$

$deltat := 0.004$

Amount of overlap equals 50 percent:

$overl := 0.5$

$x_i := READ(spec2)$

Number of points per section:

$npsec := 64$

Remove the dc component of the data first:

$$a := \frac{\sum_i x_i}{M}$$

$$x_i := x_i -$$

$$a = {}^-2.147$$

Plot the data:

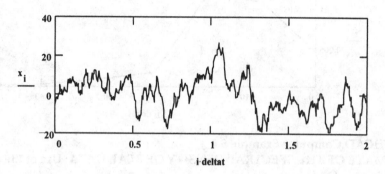

Determine now the number of complete sections with 50 percent overlap and 64 points per section.

$$\mathbf{knew} := \mathbf{npsec} - (\mathbf{npsec \cdot overl})$$

$$\mathbf{knew} = 32$$

$$\mathbf{kcompl} := \mathbf{floor}\left(\frac{\mathbf{M}}{\mathbf{npsec}}\right) \cdot \mathbf{npsec}$$

$$\mathbf{kcompl} = 448$$

$$\mathbf{isect} := \mathbf{floor}\left[\frac{\mathbf{kcompl} - (\mathbf{npsec} - \mathbf{knew})}{\mathbf{knew}}\right]$$

Number of complete sections:

$$\mathbf{isect} = 13$$

So, for the present data, which are 500 points long, we use sections which are 64 points long, an overlap of sections of 50 percent, and there will be a total of 14 sections.

Generate a Hamming window which will be 64 points long:

$$\mathbf{i} := \mathbf{0 \dots 63}$$

$$w_i := 0.54 - 0.46 \cdot \cos\left[\frac{2 \cdot \pi \cdot (i)}{63}\right]$$

Plot the Hamming window:

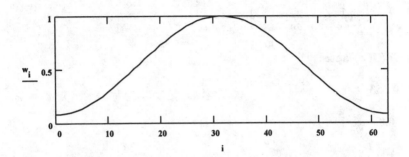

Every section of data will be multiplied by the Hamming window prior to computing its FFT. Normalizing factor due to the window we are using:

$$u := \sum_i (w_i)$$

$u = 25.0$

$u := u \cdot isect$

$u = 325$

   Now we start processing the sections of data. Unfortunately, this is going to be performed one section at a time.

Section 1:

It consists of the first 64 points of data.

$ii := 0 .. 63$

$y_{ii} := x_{ii}$

$z_{ii} := y_{ii} \cdot w_{ii}$

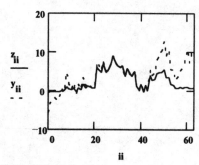

$$f := CFFT(z) \cdot npsec$$

$$jj := 0..33$$

$$g_{jj} := 0.0$$

$$g_{jj} := g_{jj} + \left(\left| f_{jj} \right| \right)$$

Section 2:

It consists of points 32 to 95   (Note that there is a 50 percent overlap of windows)

$$y_{ii} := x_{ii+3}$$

$$z_{ii} := y_{ii} \cdot w_{ii}$$

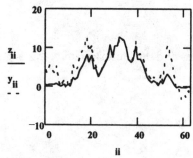

$$f := CFFT(z) \cdot npsec$$

$$g_{jj} := g_{jj} + \left(\left| f_{jj} \right| \right)$$

$$g_0 = 8.3$$

Section 3:

It consists of points 64 to 127.

$$y_{ii} := x_{ii+6}$$

$$z_{ii} := y_{ii} \cdot w_{ii}$$

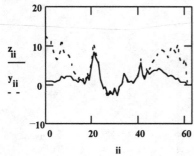

$$f := CFFT(z) \cdot npsec$$

$$length(f) = 64$$

$$g_{jj} := g_{jj} + \left(\left| f_{jj} \right| \right)$$

$$g_0 = 9.5$$

Section 4:

It consists of points 96 to 159.

$$y_{ii} := x_{ii+9}$$

$$z_{ii} := y_{ii} \cdot w_{ii}$$

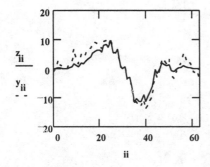

$f := CFFT(z) \cdot npsec \quad length(f) = 64$

$g_{jj} := g_{jj} + \left(\left| f_{jj} \right| \right) \quad g_0 = 9.5$

Section 5:

It consists of points 128 to 191.

$y_{ii} := x_{ii+12} \quad z_{ii} := y_{ii} \cdot w_{ii}$

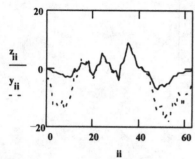

$f := CFFT(z) \cdot npsec \quad length(f) = 64$

$g_{jj} := g_{jj} + \left(\left| f_{jj} \right| \right) \quad g_0 = 9.6$

Section 6:

It consists of points 160 to 223.

$y_{ii} := x_{ii+16} \quad z_{ii} := y_{ii} \cdot w_{ii}$

$f := CFFT(z) \cdot npsec$

$length(f) = 64$

$$g_{jj} := g_{jj} + \left(\left| f_{jj} \right| \right)$$

$$g_0 = 1.16$$

Section 7:

It consists of points 192 to 255.

$$y_{ii} := x_{ii+19}$$

$$z_{ii} := y_{ii} \cdot w_{ii}$$

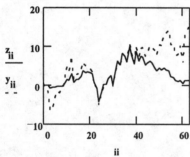

$$f := CFFT(z) \cdot npsec$$

$$length(f) = 64$$

$$g_{jj} := g_{jj} + \left(\left| f_{jj} \right| \right)$$

$$g_0 = 1.48$$

Section 8:

It consists of points 224 to 287.

$$y_{ii} := x_{ii+22}$$

$$z_{ii} := y_{ii} \cdot w_{ii}$$

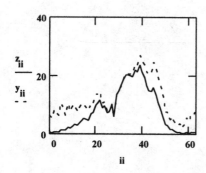

$f := CFFT(z) \cdot npsec$

$length(f) = 64$

$$g_{jj} := g_{jj} + \left(\left| f_{jj} \right| \right)$$

$g_0 = 4.0$

Section 9:

It consists of points 256 to 319.

$$y_{ii} := x_{ii+25}$$

$$z_{ii} := y_{ii} \cdot w_{ii}$$

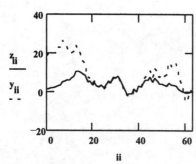

$f := CFFT(z) \cdot npsec$

$length(f) = 64$

$$g_{jj} := g_{jj} + \left(\left| f_{jj} \right| \right)$$

$g_0 = 4.7$

Section 10:

It consists of points 288 to 351.

$y_{ii} := x_{ii+28}$

$z_{ii} := y_{ii} \cdot w_{ii}$

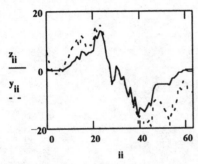

$f := CFFT(z) \cdot npsec$

$length(f) = 64$

$g_{jj} := g_{jj} + \left(\left| f_{jj} \right| \right)$

$g_0 = 4.7$

Section 11:

It consists of points 320 to 383.

$y_{ii} := x_{ii+32}$

$z_{ii} := y_{ii} \cdot w_{ii}$

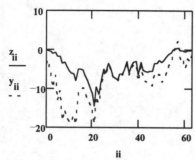

$$f := CFFT(z) \cdot npsec$$

$$length(f) = 64$$

$$g_{jj} := g_{jj} + \left(\left| f_{jj} \right|\right)$$

$$g_0 = 5.5$$

Section 12:

It consists of points 352 to 415.

$$y_{ii} := x_{ii+35}$$

$$z_{ii} := y_{ii} \cdot w_{ii}$$

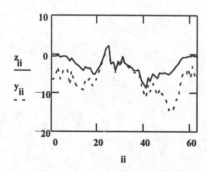

$$f := CFFT(z) \cdot npsec$$

$$length(f) = 64$$

$$g_{jj} := g_{jj} + \left(\left| f_{jj} \right|\right)$$

$g_0 = 5.8$

Section 13:

It consists of points 384 to 447.

$$y_{ii} := x_{ii+38}$$

$$z_{ii} := y_{ii} \cdot w_{ii}$$

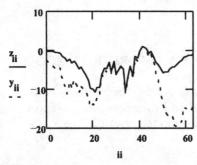

$$f := CFFT(z) \cdot npsec$$

$$length(f) = 64$$

$$g_{jj} := g_{jj} + \left(\left| f_{jj} \right| \right)$$

$$g_0 = 6.4$$

We have now computed all complete sections. We take the average of the array *g* divide by the normalization factor and multiply by Δt.

$$g_{jj} := \frac{g_{jj} \cdot delta}{u}$$

$$deltaf := \frac{1}{npsec \cdot delta}$$

$$p_{jj} := 10 \cdot \log\left(g_{jj}\right)$$

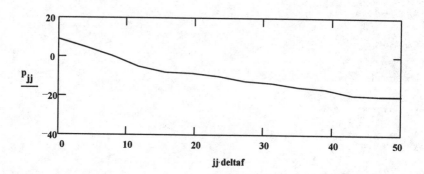

MATHCAD Computer Example 5.9
AUTOREGRESSIVE ESTIMATION OF A THEORETICAL MODEL

In this example, we are going to estimate by "brute force" the order and coefficients of an autoregressive model and then compute the spectral density based on the order and coefficients of the model.

$i := 2..100 \quad M := 101$

$j := 0..100 \quad z_0 := 1$

$z_i := 0.75 \cdot z_{i-1} - 0.5 \cdot z_{i-2} + rnd(1) - 0.5 \quad z_{M+j} := 0$

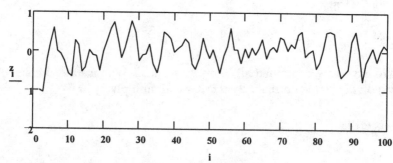

$\tau := 0..10 \quad r_\tau := \dfrac{1}{101} \cdot \sum_j z_j \cdot z_{j+\tau}$

$k := 5..100 \quad A := r$

$x := r \quad \alpha := A^{-1}.$

$\alpha = 0.431$

$$\mathbf{B} := \begin{pmatrix} r_0 & r_1 \\ r_1 & r_0 \end{pmatrix} \qquad \mathbf{y} := \begin{pmatrix} r_1 \\ r_2 \end{pmatrix}$$

$$\beta := \mathbf{B}^{-1}.$$

$$\beta = \begin{pmatrix} 0.617 \\ -0.433 \end{pmatrix} \qquad \mathbf{C} := \begin{pmatrix} r_0 & r_1 & r_2 \\ r_1 & r_0 & r_1 \\ r_2 & r_2 & r_0 \end{pmatrix}$$

$$\mathbf{z} := \begin{pmatrix} r_1 \\ r_2 \\ r_3 \end{pmatrix} \qquad \gamma := \mathbf{C}^{-1} \cdot \mathbf{z}$$

$$\gamma = \begin{pmatrix} 0.484 \\ -0.244 \\ -0.307 \end{pmatrix} \qquad \mathbf{D} := \begin{bmatrix} r_0 & r_1 & r_2 & r_3 \\ r_1 & r_0 & r_1 & r_2 \\ r_2 & r_1 & r_0 & r_1 \\ r_3 & r_2 & r_1 & r_0 \end{bmatrix}$$

$$\mathbf{yy} := \begin{bmatrix} r_1 \\ r_2 \\ r_3 \\ r_4 \end{bmatrix} \qquad \delta := \mathbf{D}^{-1} \cdot \mathbf{y}$$

$$\delta = \begin{bmatrix} 0.586 \\ -0.352 \\ -0.128 \\ 0.072 \end{bmatrix} \qquad E := \begin{bmatrix} r_0 & r_1 & r_2 & r_3 & r_4 \\ r_1 & r_0 & r_1 & r_2 & r_3 \\ r_2 & r_1 & r_0 & r_1 & r_2 \\ r_3 & r_2 & r_1 & r_0 & r_1 \\ r_4 & r_3 & r_2 & r_1 & r_0 \end{bmatrix}$$

$$zz := \begin{bmatrix} r_1 \\ r_2 \\ r_3 \\ r_4 \\ r_5 \end{bmatrix} \qquad \varepsilon := E^{-1} \cdot zz$$

$$\varepsilon = \begin{bmatrix} 0.587 \\ -0.354 \\ -0.133 \\ 0.08 \\ -0.014 \end{bmatrix} \qquad zzz := \begin{bmatrix} r_1 \\ r_2 \\ r_3 \\ r_4 \\ r_5 \\ r_6 \end{bmatrix}$$

$$\mathbf{F} := \begin{bmatrix} r_0 & r_1 & r_2 & r_3 & r_4 & r_5 \\ r_1 & r_0 & r_1 & r_2 & r_3 & r_4 \\ r_2 & r_1 & r_0 & r_1 & r_2 & r_3 \\ r_3 & r_2 & r_1 & r_0 & r_1 & r_2 \\ r_4 & r_3 & r_2 & r_1 & r_0 & r_1 \\ r_5 & r_4 & r_3 & r_2 & r_1 & r_0 \end{bmatrix} \qquad \zeta := \mathbf{F}^{-1} \cdot \mathbf{zzz}$$

$$\zeta = \begin{bmatrix} 0.588 \\ -0.362 \\ -0.12 \\ 0.115 \\ -0.072 \\ 0.099 \end{bmatrix} \qquad N := 101$$

$M := 1$

$$\sigma_1 := \left(\frac{N - M}{N - 2 \cdot M - 1} \right) \cdot (r_0 + \alpha \cdot r_1) \qquad \sigma_1 = 0.169$$

$$\sigma_2 := \left(\frac{N - M}{N - 2 \cdot M - 1} \right) \cdot \left[r_0 + (\beta_0 \cdot r_1 + \beta_1 \cdot r_2) \right] \qquad \sigma_2 = 0.191$$

$$\sigma_3 := \left(\frac{N - M}{N - 2 \cdot M - 1} \right) \cdot \left[r_0 + (\gamma_0 \cdot r_1 + \gamma_1 \cdot r_2 + \gamma_2 \cdot r_3) \right] \qquad \sigma_3 = 0.193$$

$$\sigma_4 := \left(\frac{N - M}{N - 2 \cdot M - 1} \right) \cdot \left[r_0 + \left[\delta_0 \cdot r_1 + \delta_1 \cdot r_2 + (\delta_2 \cdot r_3 + \delta_3 \cdot r_4) \right] \right] \qquad \sigma_4 = 0.192$$

$$\sigma_5 := \left(\frac{N - M}{N - 2 \cdot M - 1} \right) \cdot \left[r_0 + \left[\varepsilon_0 \cdot r_1 + \varepsilon_1 \cdot r_2 + \left[\varepsilon_2 \cdot r_3 + (\varepsilon_3 \cdot r_4 + \varepsilon_4 \cdot r_5) \right] \right] \right]$$
$$\sigma_5 = 0.192$$

$$\sigma_6 := \left(\frac{N - M}{N - 2 \cdot M - 1} \right) \cdot \left[r_0 + \left[\zeta_0 \cdot r_1 + \zeta_1 \cdot r_2 + \left[\zeta_2 \cdot r_3 + \left[\zeta_3 \cdot r_4 + (\zeta_4 \cdot r_5 + \zeta_5 \cdot r_5) \right] \right] \right] \right]$$

$\sigma_6 = 0.191$

$$FPE_1 := \left(\frac{N + M}{N - M}\right) \cdot \sigma \qquad FPE_2 := \left(\frac{N + M}{N - M}\right) \cdot \sigma$$

$$FPE_3 := \left(\frac{N + M}{N - M}\right) \cdot \sigma \qquad FPE_4 := \left(\frac{N + M}{N - M}\right) \cdot \sigma$$

$$FPE_5 := \left(\frac{N + M}{N - M}\right) \cdot \sigma \qquad FPE_6 := \left(\frac{N + M}{N - M}\right) \cdot \sigma$$

$i := 1 .. 6$

$$FPE_i := \left(\frac{FPE_i}{FPE_1}\right)$$

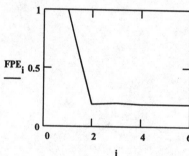

Calculate the spectrum:

$ij := 0, 0.1 .. 0.5$

$$s(ij) := \frac{2}{\left[1 + \left(\beta_0\right)^2 + \left(\beta_1\right)^2\right] - 2 \cdot \beta_0 \cdot \left(1 - \beta_1\right) \cdot \cos(2 \cdot \pi \cdot ij) - 2 \cdot \beta_1 \cdot \cos(4 \cdot \pi \cdot ij)}$$

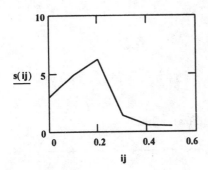

MATHCAD Computer Example 5.10
ESTIMATION OF THE CROSS-SPECTRAL DENSITY

$i := 0 .. 127$ $A := 1.0$ $f_0 := 5.0$ $deltat := 0.01$

$M := 128$ $f_2 := 21.$ $\theta := rnd(2 \cdot \pi)$ $\theta = 0.008$

$\phi := 1.0$ $B := 0.0$ $f_1 := 21.$ $C := 0$

$x_i := \cos\left(2 \cdot \pi \cdot f_1 \cdot i \cdot deltat\right) \cdot 1.0 + \cos\left(2 \cdot \pi \cdot f_0 \cdot i \cdot deltat\right) + rnd(1) \cdot 2$

$y_i := \cos\left(2 \cdot \pi \cdot f_2 \cdot i \cdot deltat\right)$

In this particular case, x and y have one frequency in common.

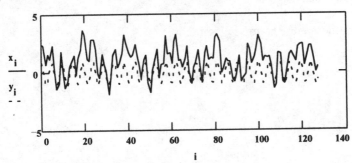

Remove now the dc component of the data:

$$a := \frac{\displaystyle\sum_i x_i}{M} \qquad x_i := x_i - \qquad b := \frac{\displaystyle\sum_i y_i}{M} \qquad y_i := y_i -$$

Extend the data with zeros:

$j := 0..255$ $x_{M+i} := 0$ $\text{length}(x) = 256$ $K :=$

$y_{M+i} := 0$ $\text{length}(y) = 256$ $xx := \text{CFFT}(x)$ $yy := \text{CFFT}(y)$

$zz_j := xx_j \cdot \overline{yy}$ $z := \text{ICFFT}(zz)$ $\text{length}(z) = 256$

To estimate the spectral density using the autocorrelation function, we need to multiply the estimate of the autocorrelation function by a time domain window (such as the Hamming window). We are only going to accept between 20 and 25 percent of the lagged values of the ACF.

$N \equiv 256$ $k := 0..(N) \cdot (0.25)$ $kk := N \cdot (0.25) - 1$ $kk = 63$

$jj := 1..(N) \cdot (0.25)$

Now we generate the Hamming window:

$$w_k := 0.54 + \left(0.46 \cdot \cos\left(\pi \cdot \frac{k}{kk}\right)\right)$$ $kkk := 0..255$ $y_{kkk} := 0$

$y_k := w$ $y_{256-jj} := w_j$

Plot of the Hamming window:

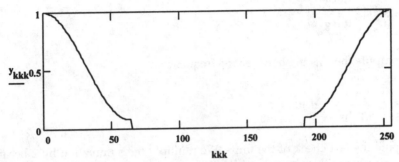

Now multiply the estimate of the ACF by the time window:

$$zz_{kkk} := 0 \qquad zz_{kkk} := z_{kkk} \cdot y_{kk}$$

Plot the ACF before and after multiplication by the Hamming window:

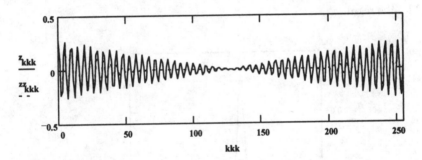

The solid line is the original, and the dotted line is the ACF after multiplication by the Hamming window.

We compute the FFT and thus an estimate of the spectral density of the data. Note that we are multiplying the results by the number of points to compensate for the FFT definition used by MATHCAD.

$$\text{length}(zz) = 256 \quad g := \text{CFFT}(zz) \cdot \text{length}(x) \quad \text{length}(g) = 256$$

We now need to compute the absolute value of g and multiply by Δt, the spacing between time samples. Note that earlier we made $\Delta t = 0.01$. We are going to take only the positive half of the spectrum since the negative half is identical.

$$MM := \frac{\text{length}(g)}{2} + 1 \quad MM = 129 \quad s := 0 .. MM - 1 \quad ff_s := |g_s| \cdot delta$$

$$ph_s := atan\left(-\frac{Im(g_s)}{Re(g_s)}\right)$$

We compute the spacing between the frequency:

$$deltaf := \frac{1}{N \cdot delta} \qquad deltaf = 0.3$$

Compute the last check of the units. We multiply the summation by 2 because we are dealing with only the positive side of the spectrum.

$$check := \left(\sum_s ff_s\right) \cdot 2 \cdot delta \qquad check = 0.29$$

Plot the results:

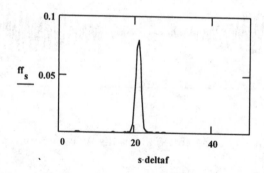

WRITEPRN(cospc) := ff_s WRITEPRN(coph) := ph_s

MATHCAD Computer Example 5.11
CROSS-SPECTRAL DENSITY ESTIMATION

$$i := 0 .. 374$$

$$M := 375$$

$$deltat := 0.004$$

$$x_i := READ(click3)$$

$$y_i := READ(click4)$$

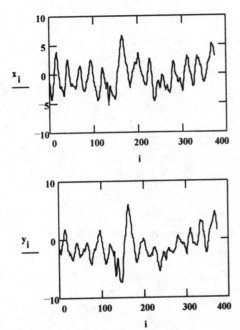

Remove the dc component of the data:

$$a := \dfrac{\displaystyle\sum_i x_i}{M}$$

$$x_i := x_i -$$

$$b := \dfrac{\displaystyle\sum_i y_i}{M}$$

$$y_i := y_i -$$

Extend the data with zeros:

$$j := 0 .. 648$$

$$x_{M+j} := 0$$

length$(x) = 1.024 \cdot 10^3$

jj $:= 0 .. 1023$

$y_{M+j} := 0$

length$(y) = 1.024 \cdot 10^3$

Compute the FFTs.

xx $:=$ **CFFT**(x)

yy $:=$ **CFFT**(y)

Multiply the FFT of x by the complex conjugate of the FFT of y:

$zz_{jj} := xx_{jj} \cdot \overline{yy_j}$

$z :=$ **ICFFT**(zz)

length$(z) = 1.0$

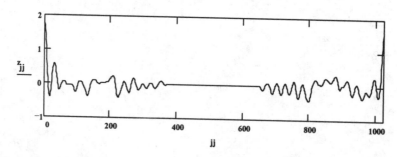

WRITEPRN$(ccf) := z_j$

In order to estimate the spectral density using the autocorrelation function we need to multiply the estimate of the autocorrelation function by a time-domain window (such as the Hamming window). We accept between 20 and 25 percent of the lagged values of the ACF.

$N \equiv 1024 \quad k := 0 .. (N) \cdot (0.25)$

$kk := N \cdot (0.25) - 1 \quad kk = 255$

$jj := 1 .. (N) \cdot (0.25)$

Now we generate the Hamming window:

$$w_k := 0.54 + \left(0.46 \cdot \cos \left(\pi \cdot \frac{k}{kk} \right) \right) \qquad kkk := 0 .. 1023$$

$y_{kkk} := 0 \quad y_k := w$

$y_{1024 - jj} := w_j$

Plot the Hamming window:

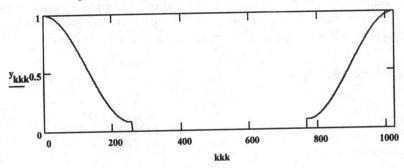

kkk

Now multiply the estimate of the ACF by the time window:

$zz_{kkk} := 0$

$zz_{kkk} := z_{kkk} \cdot y_{kk}$

Plot the ACF before and after multiplication by the Hamming window:

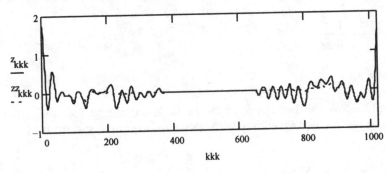

kkk

WRITEPRN(ccfham) $:= zz_{kk}$

The solid line is the original, and the dotted line is the ACF after multiplication by the Hamming window. We now compute the FFT and thus an estimate of the spectral density of the data. Note that we are multiplying the results by the number of points to compensate for the FFT definition used by MATHCAD.

$length(zz) = 1.0$

$g := CFFT(zz) \cdot length(x)$

$length(g) = 1.0$

We compute the absolute value of g and multiply by Δt, the spacing between time samples. Note that earlier we made $\Delta t = 0.01$. We are going to take only the positive half of the spectrum since the negative half is identical.

$$MM := \frac{length(g)}{2} + 1$$

$MM = 513$

$s := 0 .. MM - 1$

$ff_s := \left| g_s \right| \cdot delta$

$$ph_s := atan\left(-\frac{Im\left(g_s \right)}{Re\left(g_s \right)} \right)$$

We compute the spacing between the frequency:

$$deltaf := \frac{1}{N \cdot delta}$$

$deltaf = 0.2$

Compute now the last check of the units. We multiply the summation by 2 because we are dealing with only the positive side of the spectrum.

$$check := \left(\sum_s ff_s \right) \cdot 2 \cdot delta$$

check = 1.8

Plot now the estimate of the spectra:

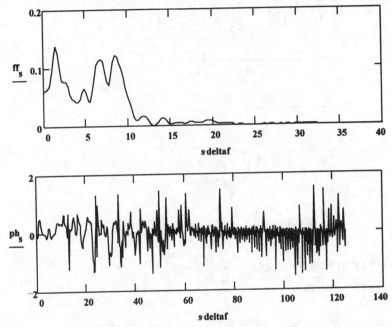

WRITEPRN (cospc88) := **ff**$_s$
WRITEPRN (copha88) := **ph**$_s$

MATHCAD Computer Example 5.12
COHERENCE ESTIMATION

i := **0 .. 374**

M := **375**

deltat := **0.004**

x$_i$:= **READ(click3)**

y$_i$:= **READ(click4)**

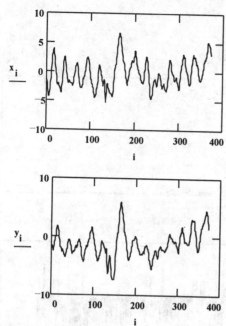

WRITEPRN(click3) := x_i

WRITEPRN(click4) := y_i

Remove the dc component of the data.

$$a := \frac{\sum_i x_i}{M}$$

$$x_i := x_i -$$

$$b := \frac{\sum_i y_i}{M}$$

$$y_i := y_i -$$

Extend the data with zeros.

$$j := 0 .. 648$$

$x_{M+j} := 0$

$jj := 0 .. 1023$

$y_{M+j} := 0$

Compute the FFTs.

$xx := CFFT(x) \cdot length(x)$

$yy := CFFT(y) \cdot length(y)$

Multiply the FFT of x by the complex conjugate of the FFT of y:

$g_{jj} := \left(\left| xx_{jj} \right| \right)$

$h_{jj} := \left(\left| yy_{jj} \right| \right)$

$zz_{jj} := xx_{jj} \cdot \overline{yy_j}$

$rx := \dfrac{ICFFT(g)}{length(x)}$

$ry := \dfrac{ICFFT(h)}{length(y)}$

$rxy := \dfrac{ICFFT(zz)}{length(x)}$

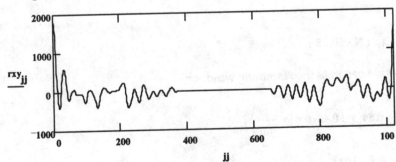

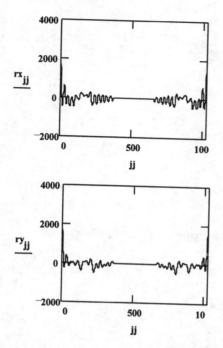

To estimate the spectral density using the autocorrelation function, we multiply the estimate of the autocorrelation function by a time domain window (such as the Hamming window). We accept between 20 and 25 percent of the lagged values of the ACF.

$N \equiv 1024$

$k := 0 .. (N) \cdot (0.25)$

$kk := N \cdot (0.25) - 1$

$kk = 255$

$jj := 1 .. (N) \cdot (0.25)$

Now we generate the Hamming window:

$$w_k := 0.54 + \left(0.46 \cdot \cos \left(\pi \cdot \frac{k}{kk} \right) \right)$$

$kkk := 0 .. 1023$

$y_{kkk} := 0$

$y_k := w$

$y_{1024-jj} := w_j$

Multiply the estimate of the ACF by the time window:

$zz_{kkk} := 0$

$zz_{kkk} := rxy_{kkk} \cdot y_{kk}$

$rxx_{kkk} := 0$

$rxx_{kkk} := rx_{kkk} \cdot y_{kk}$

$ryy_{kkk} := 0$

$ryy_{kkk} := ry_{kkk} \cdot y_{kk}$

We now compute the FFT and thus an estimate of the spectral density of the data. Note that we are multiplying the results by the number of points to compensate for the FFT definition used by MATHCAD.

$length(zz) = 1.0$

$ii := 0 .. 511$

$g := CFFT(zz) \cdot length(x)$

$length(g) = 1.0$

$gx := CFFT(rxx) \cdot length(x)$

$deltaf := \dfrac{1}{N \cdot delta}$

$gy := CFFT(ryy) \cdot length(x)$

$deltaf = 0.2$

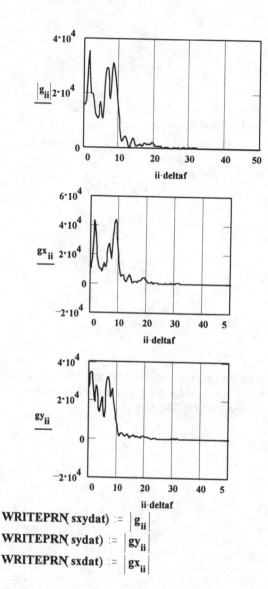

$$\text{WRITEPRN}(\text{sxydat}) := \left| g_{ii} \right|$$

$$\text{WRITEPRN}(\text{sydat}) := \left| gy_{ii} \right|$$

$$\text{WRITEPRN}(\text{sxdat}) := \left| gx_{ii} \right|$$

We now need to compute the absolute value of g and multiply by Δt, the spacing between time samples. Note that earlier we made $\Delta t = 0.01$. We are going to take only the positive half of the spectrum since the negative half is identical.

$$MM := \frac{\text{length}(g)}{2} + 1$$

$$MM = 513$$

$s := 0 .. 41$

$ff_s := \left(\left| g_s \right| \right)$

$fx_s := \left| gx_s \right|$

$fy_s := \left| gy_s \right|$

$coh_s := \dfrac{ff_s}{fx_s \cdot fy_s}$

Plot now the estimate of the coherence:

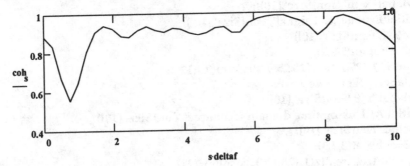

$$s \cdot deltaf$$

$\text{WRITEPRN} \, (\, cohdat \,) := coh_s$

MATLAB Computer Example 5.1
COMPUTATION OF THE AUTOCORRELATION FUNCTION USING THE FFT

```
%==========================================
str=str2mat(...
    'This program obtains the autocorrelation function of the signal defined
as:',...
    ' ',...
    '>> x(m)=2*exp(-3*m*0.02)',...
    ' ',...
    'We obtain the autocorrelation function via the estimator:',...
    ' ',...
    '>> 1/M*sum(y(m)y(m+k)',...
    ' ',...
    'and via FFT of the padded signal.' );
```

```
disp(str);
M=128;
m=0:1:127;j=0:1:127;
x_m=2*exp(-3*m*0.02);
mu=mean(x_m);
stdx=std(x_m);
y_m=x_m-mean(x_m);
r_tau=xcorr(y_m,'biased');
y_z=zeros(size([m m]));
y_z(1:128)=y_m;
f=fft(y_z,256);
h_j=f.*conj(f);
z=ifft(h_j)/256;
clg;s1=subplot(2,2,1);
plot(0:1:127,x_m,'r');axisn(0:1:127);
xlabel('m','FontSize',[10]);
ylabel('x_m','FontSize',[10]);
title('Exponential signal','FontSize',[10]);
set(s1,'FontSize',[10])
s2=subplot(2,2,2);
plot(0:1:127,r_tau(128:255),'r');axisn(0:1:127);
xlabel('j','FontSize',[10]);
ylabel('r_j','FontSize',[10]);
title('ACF using time domain technique','FontSize',[10])
set(s2,'FontSize',[10])
s3=subplot(2,2,3);
plot(0:1:127,real(z(1:128)),'r');axisn(0:1:127);
title('ACF using the FFT','FontSize',[10]);
xlabel('k','FontSize',[10]);
ylabel('z_k','FontSize',[10]);
set(s3,'FontSize',[10])
s4=subplot(2,2,4);
axis('off')
text(0,0.6,['Mean value: ',num2str(mu)],'FontSize',[10]);
text(0,0.4,['standard deviation: ',num2str(stdx)],'FontSize',[10]);
clear;
```

This program obtains the autocorrelation function of the signal defined as

```
>> x(m)=2*exp(-3*m*0.02)
```

We obtain the autocorrelation function via the estimator

```
>> 1/M*sum(y(m)y(m+k)
```

and via the FFT of the padded signal.

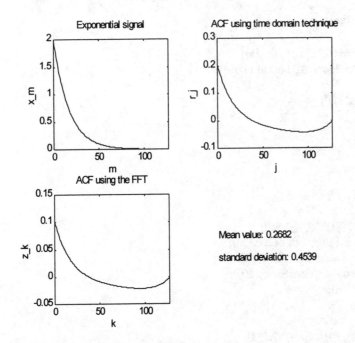

Mean value: 0.2682

standard deviation: 0.4539

MATLAB Computer Program 5.2
COMPUTATION OF THE CROSSCORRELATION FUNCTION VIA THE FFT

The purpose of this computer program is to demonstrate the computation of the CCF via the FFT.

```
%==========================================
% Input parameters
                A=1.0;
                B=1.0;
                phi=0.;
                f0=3.0;
                M=301;
%==========================================
str=str2mat(...
    'This program obtains the cross-correlation function via FFT of the
extended signals defined as:',...
    ' ',...
    '>> x(i)=A*cos(2*pi*f0*i*0.01+ theta)',...
    ' ',...
    '>> y(i+1)=B*cos(2*pi*f0*i*0.01+theta+phi)');
disp(str);
theta=2*pi*rand;
x_i=zeros(size(1:1:301));
```

```
y_i=zeros(size(1:1:301));
for i=0:1:300,
x_i(i+1)=A*cos(2*pi*f0*i*0.01+theta);
y_i(i+1)=B*cos(2*pi*f0*i*0.01+theta+phi);
end
vz=zeros(size(1:1:211));
x_e=[x_i vz];
y_e=[y_i vz];
xx=fft(x_e,512);
yy=fft(y_e,512);
zz=xx.*conj(yy);
z=ifft(zz)/length(x_e);
clg;s1=subplot(2,2,1);
plot(0:1:300,x_i,'r', 0:1:300,y_i,':k');
xlabel('i','FontSize',[10]); axisn(0:1:300);
ylabel('x_i _      y_i ...','FontSize',[10])
title('Original Signals.','FontSize',[10]);
set(s1,'FontSize',[10]);
s2=subplot(2,2,2);
t=(0:1:511);
plot(0:1:511,real(z),'r'); axisn(0:1:511);
xlabel('j','FontSize',[10]);
ylabel('z_j','FontSize',[10]);
title('Cross-correlation function.','FontSize',[10]);
set(s2,'FontSize',[10])
s4=subplot(2,2,4);
axis('off')
text(0.1,0.6,['Theta: ',num2str(theta)],'FontSize',[10]);
text(0.1,0.4,['Length of the data: ',num2str(M)],'FontSize',[10]);
text(0.1,0.2,['Length of padded data: ',num2str(length(z))],'FontSize',[10]);
set(s4,'FontSize',[10]);clear;
```

This program obtains the cross-correlation function via FFT of the extended
signals defined as:

```
>> x(i)=A*cos(2*pi*f0*i*0.01+ theta)
```

```
>> y(i+1)=B*cos(2*pi*f0*i*0.01+theta+phi)
```

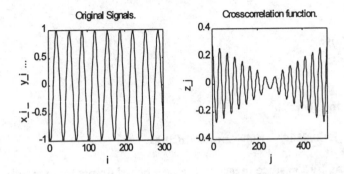

Theta: 2.436

Length of the data: 301

Length of padded data: 512

MATLAB Computer Example 5.3
ESTIMATE OF THE SPECTRAL DENSITY OF A RANDOM PROCESS

In this program we will compute the estimate of the spectral density of a random process using the first method. We will use a known time-limited sample function from a wide-sense stationary random process.

```
%=============================================
% Input parameters
                A=1;
                f0=5;
                deltat=0.01;
                theta=0.008;
                M=128;
%=============================================
str=str2mat(...
    'The program uses a cosinusoid of amplitude 1.0, frequency fixed at 3 Hz,
with a random phase, and',...
    'random noise added to the entire process.',...
    ' ',...
    '>> x_i=A*cos(2*pi*f0*i*deltat+theta)+4*rand(size(i))-1;');
disp(str);
i=0:1:M-1;
x_i=A*cos(2*pi*f0*i*deltat+theta)+4*rand(size(i))-1;
xmean=mean(x_i);
```

```
x_i=x_i-xmean;
xstd=std(x_i);
lengthx=length(x_i);
r0=sum(x_i.^2)/256;
str=str2mat(...
    ' ',...
    'Autocorrelation check:',...
    ' ',...
    ['>> r0=sum(x_i.^2)/256;',' r(0)= ',num2str(r0)],...
    ' ',...
    'We are dividing by 256 since these are the number of points that we will
have after we add zeros to',...
    'the data in order to avoid circular correlation problems.');
disp(str);
clg;s1=subplot(2,1,1);
plot(0:1:M-1,x_i,'r'); axisn(0:1:M-1);
xlabel('i','FontSize',[10]);
ylabel('x_i','FontSize',[10]);
title('Cosinusoid plus noise','FontSize',[10]);
set(s1,'FontSize',[10])
s2=subplot(2,1,2);
axis('off')
text(0.3,0.8,['Mean value: ',num2str(xmean)] ,'FontSize',[10]);
text(0.3,0.6,['Standard deviation: ',num2str(xstd)] ,'FontSize',[10]);
text(0.3,0.4,['Length of data: ',num2str(lengthx)] ,'FontSize',[10]);
%=========================================
```

The program uses a cosinusoid of amplitude 1.0, frequency fixed at 3 Hz, with a random phase, and random noise added to the entire process.

```
>> x_i=A*cos(2*pi*f0*i*deltat+theta)+4*rand(size(i))-1;
```

Autocorrelation check:

```
>> r0=sum(x_i.^2)/256; r(0)= 0.8365
```

We are dividing by 256 since these are the number of points that we will have after we add zeros to the data in order to avoid circular correlation problems.

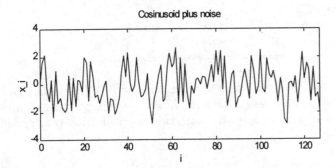

Mean value: 1.062

Standard deviation: 1.299

Length of data: 128

```
%===========================================
str=str2mat(....
    'We now compute the autocorrelation function via the FFT.',...
    ' ',...
    'The result is the complete autocorrelation of the original data. Check to
see if the value obtained',...
    'earlier for the zero lagged value of the autocorrelation function
corresponds.');
disp(str);
x_izeros=zeros(size(0:1:255));
x_izeros(1:M)=x_i;
f=fft(x_izeros,length(x_izeros)).*length(x_izeros);
h_j=f.*conj(f);
z=ifft(h_j)./(length(x_izeros)^2);
r0=real(z(1))/length(x_izeros);
z_j=real(z)/length(x_izeros);
str=str2mat(...
    ' ',...
    'Autocorrelation check:',...
    ' ',...
    ['>> r0=sum(x_i.^2)/256;','   r(0)= ',num2str(r0),'    It corresponds !!!']);
disp(str);
s1=subplot(1,1,1);
plot(0:1:255,z_j,'r'); axisn(0:1:255);
xlabel('j','FontSize',[10]);
```

```
ylabel('z_j','FontSize',[10]);
title('Autocorrelation function','FontSize',[10])
set(s1,'FontSize',[10])
%=============================================
```

We now compute the autocorrelation function via the FFT. The result is the complete autocorrelation of the original data. Check to see if the value obtained earlier for the zero lagged value of the autocorrelation function corresponds. Autocorrelation check:

>> r0=sum(x_i.^2)/256; r(0)= 0.8365   It corresponds!

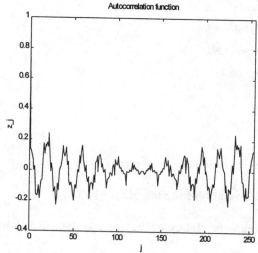

```
%=============================================
str=str2mat(...
    'We are only going to accept 25% of the lagged values of the
autocorrelation function. This is done',...
    'using a Hamming window.',...
    ' ',...
    '>> hamming(0.5*length(x_izeros));',...
    ' ',...
    'We multiply the estimate of the ACF by the time Hamming window.');
disp(str);
w=hamming(0.5*length(x_izeros));
w_zeros=zeros(size(1:1:256));
w_zeros(1:64)=w(65:128);
w_zeros(193:256)=w(1:64);
y_kkk=w_zeros;
t=(0:1:255);
zz_kkk=z_j.*y_kkk;
clg;s1=subplot(2,1,1);
```

```
plot(0:1:255,y_kkk);axisn(0:1:255);
xlabel('kkk','FontSize',[10]);
ylabel('y_kkk','FontSize',[10]);
title('Hamming window','FontSize',[10]);
set(s1,'FontSize',[10])
s2=subplot(2,1,2);
plot(0:1:255,zz_kkk,'r', 0:1:255,z_j,':k');axisn(0:1:255);
xlabel('kkk','FontSize',[10]);
ylabel('z_j ...    zz_kkk _','FontSize',[10])
title('Rxx: unwindowed(black) - windowed(red)','FontSize',[10]);
set(s2,'FontSize',[10])
%=========================================
```

We are only going to accept 25% of the lagged values of the autocorrelation function. This is done using a Hammnig window.

>> hamming(0.5*length(x_izeros));

We multiply the estimate of the ACF by the time Hamming window.

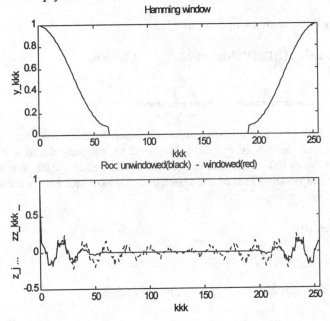

```
%=========================================
str=str2mat(...
      'Now compute the FFT of the windowed autocorrelation function and
thus an estimate of the spectral',...
```

'density of the data. We need to compute the absolute value and multiply
by deltat. Note that earlier',...
'we made deltat = 0.01. We are going to take only the positive half of the
spectrum. We now need to',...
'compute the frequency spacing.',...
' ',...
'>> deltaf=1/(N*deltat);');

```
disp(str);
g=fft(zz_kkk,length(zz_kkk)).*length(zz_kkk);
ff_s=abs(g)*deltat/256;
deltaf=1/(256*deltat);
f=deltaf*(0:1:length(g)/2-1);
s1=subplot(1,1,1);
plot(f,ff_s(1:128),'r'); axisn(f);
xlabel('s*deltaf','FontSize',[10]);
ylabel('ff_s','FontSize',[10]);
title('Spectrum via Rxx','FontSize',[10]);
set(s1,'FontSize',[10])
check=sum(ff_s(1:128))*2*deltaf;
str=str2mat(...
    'Compute now the last check of the units.',...
    ' ',...
    ['>> Check=sum(ff_s(1:128))*2*deltaf/N;',''          Check=
',num2str(check)]);
disp(str);clear;
  %============================================
```

Now compute the FFT of the windowed autocorrelation function and thus an
estimate of the spectral density of the data. We need to compute the absolute
value and multiply by deltat. Note that earlier we made deltat = 0.01. We are
going to take only the positive half of the spectrum. We now need to compute
the frequency spacing.

>> deltaf=1/(N*deltat);

Compute now the last check of the units.

>> Check=sum(ff_s(1:128))*2*deltaf/N;   Check= 0.8346

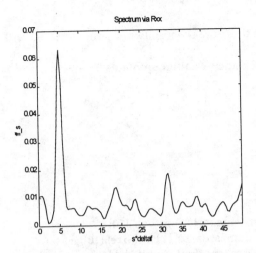

MATLAB Computer Example 5.4
ESTIMATE OF THE SPECTRAL DENSITY OF REAL DATA

```
%===========================================
% Input parameter
                signal='tonecz.dat';
%===========================================
eval(['load ',signal]);
limit=length(signal)-4;
signal=signal(1:limit);
x_i=eval(signal);
lengthx=length(x_i);
if (lengthx ~=128)
str=str2mat(...
    '           The length of the data must be 128 !!!');
clear signal;
disp(figNumber,str);
end
i=0:1:127;M=128;whitebg
deltat=0.004;
t=0:deltat:(M-1)*deltat;
xmean=mean(x_i);
x_i=x_i-xmean;
xstd=std(x_i);
lengthx=length(x_i);
r0=sum(x_i.^2)/256;
str=str2mat(...
    'Autocorrelation check:',...
    ' ',...
    ['>> r0=sum(x_i.^2)/256;',' r(0)= ',num2str(r0)]....
```

```
'',...
'We are dividing by 256 since these are the number of points that we will
have after we add zeros to',...
'the data in order to avoid circular correlation problems.');
disp(str);
clg;s1=subplot(2,1,1);
plot(t,x_i,'r');axisn(t);
xlabel('i*deltat','FontSize',[10]);
ylabel('x_i','FontSize',[10]);
title(num2str(signal),'FontSize',[10]);
set(s1,'FontSize',[10]);
s2=subplot(2,1,2);
axis('off')
text(0.3,0.8,['Mean signal: ',num2str(xmean)] ,'FontSize',[10]);
text(0.3,0.6,['Standard deviation: ',num2str(xstd)] ,'FontSize',[10]);
text(0.3,0.4,['Length of data: ',num2str(lengthx)] ,'FontSize',[10]);
text(0.3,0.2,['Deltat: ',num2str(deltat)] ,'FontSize',[10]);
set(s2,'FontSize',[10])
%=============================================
```

Autocorrelation check:

```
>> r0=sum(x_i.^2)/256; r(0)= 0.2411
```

We are dividing by 256 since these are the number of points that we will have
after we add zeroes to the data in order to avoid circular correlation problems.

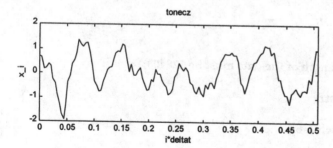

Mean signal: 0.159

Standard deviation: 0.6972

Length of data: 128

Deltat: 0.004

```
%============================================
str=str2mat(...
    'We now compute the autocorrelation function via the FFT.',...
    ' ',...
    'The result is the complete autocorrelation of the original data. Check to
see if the signal obtained',...
    'earlier for the zero lagged signal of the autocorrelation function
corresponds.');
disp(str);
x_izeros=zeros(size(0:1:255));
x_izeros(1:M)=x_i;
f=fft(x_izeros,length(x_izeros)).*length(x_izeros);
h_j=f.*conj(f);
z=ifft(h_j)./(length(x_izeros)^2);
r0=real(z(1))/length(x_izeros);
z_j=real(z)/length(x_izeros);
str=str2mat(...
    ' ',...
    'Autocorrelation check:',...
    ' ',...
    ['>> r0=sum(x_i.^2)/256;',' r(0)= ',num2str(r0)]);
disp(str);
clg;s1=subplot(1,1,1);
plot(0:1:255,z_j,'r');axisn(0:1:255);
xlabel('j','FontSize',[10]);
ylabel('z_j','FontSize',[10]);
title('Autocorrelation function','FontSize',[10])
set(s1,'FontSize',[10])
%============================================
```

We now compute the autocorrelation function via the FFT.The result is the
complete autocorrelation of the original data. Check to see if the signal obtained
earlier for the zero lagged signal of the autocorrelation function corresponds.

Autocorrelation check:

>> r0=sum(x_i.^2)/256; r(0)= 0.2411

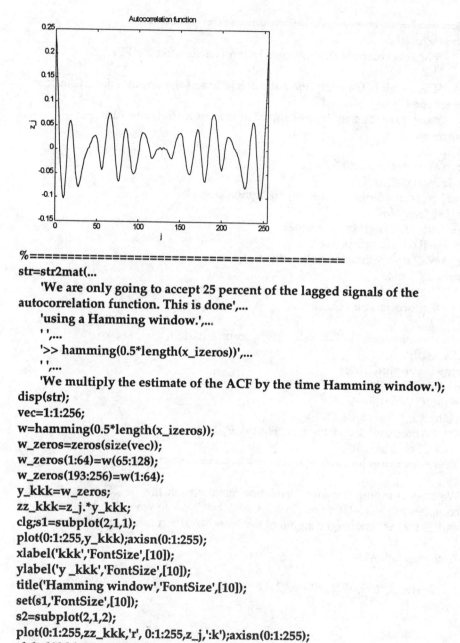

```
%===============================================
str=str2mat(...
    'We are only going to accept 25 percent of the lagged signals of the
autocorrelation function. This is done',...
    'using a Hamming window.',...
    ' ',...
    '>> hamming(0.5*length(x_izeros))',...
    ' ',...
    'We multiply the estimate of the ACF by the time Hamming window.');
disp(str);
vec=1:1:256;
w=hamming(0.5*length(x_izeros));
w_zeros=zeros(size(vec));
w_zeros(1:64)=w(65:128);
w_zeros(193:256)=w(1:64);
y_kkk=w_zeros;
zz_kkk=z_j.*y_kkk;
clg;s1=subplot(2,1,1);
plot(0:1:255,y_kkk);axisn(0:1:255);
xlabel('kkk','FontSize',[10]);
ylabel('y _kkk','FontSize',[10]);
title('Hamming window','FontSize',[10]);
set(s1,'FontSize',[10]);
s2=subplot(2,1,2);
plot(0:1:255,zz_kkk,'r', 0:1:255,z_j,':k');axisn(0:1:255);
xlabel('kkk','FontSize',[10]);
ylabel('z_j ...      zz_kkk _','FontSize',[10]);
title('Rxx: unwindowed(black)  -  windowed(red)','FontSize',[10]);
set(s2,'FontSize',[10])
%===============================================
```

We are only going to accept 25% of the lagged signals of the autocorrelation function. This is done using a Hammnig window.

>> hamming(0.5*length(x_izeros))

We multiply the estimate of the ACF by the time Hamming window.

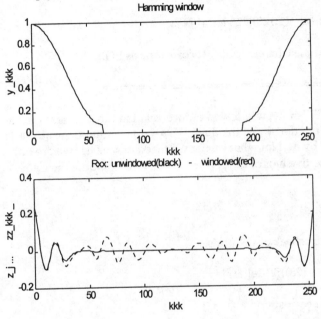

```
%=============================================
str=str2mat(...
    'Now compute the FFT of the windowed autocorrelation function and
thus an estimate of the spectral',...
    'density of the data. We need to compute the absolute signal and
multiply by deltat. Note that earlier we',...
    'made deltat = 0.004. We are only going to take the positive half of the
spectrum. We now compute the',...
    'frequency spacing.',...
    '',...
    '>> deltaf=1/(N*deltat)');
disp(str);
g=fft(zz_kkk,length(zz_kkk)).*length(zz_kkk);
ff_s=abs(g)*deltat/256;
deltaf=1/(256*deltat);
check=sum(ff_s(1:128))*2*deltaf;
f=deltaf*(0:1:length(g)/2-1);
clg;s1=subplot(1,1,1);
plot(f(1:1:40),ff_s(1:40),'r'); axisn(f(1:1:40));
```

```
xlabel('s*deltaf','FontSize',[10]);
ylabel('ff_s','FontSize',[10]);
set(s1,'FontSize',[10])
str=str2mat(...
    ' ',...
    'Compute now the last check of the units.',...
    ' ',...
    '>> Check=sum(ff_s(1:128))*2*deltaf/N',...
    ' ',...
    ['    Check: ',num2str(check),'        It corresponds !!!']);
disp(str);clear;
%=============================================
```

Now compute the FFT of the windowed autocorrelation function and thus an estimate of the spectral density of the data. We need to compute the absolute signal and multiply by Δt. Note that earlier we made $\Delta t = 0.004$. We are only going to take the positive half of the spectrum. We now compute the frequency spacing.

>> deltaf=1/(N*deltat)

Compute now the last check of the units.

>> Check=sum(ff_s(1:128))*2*deltaf/N

Check: 0.2417   It corresponds!

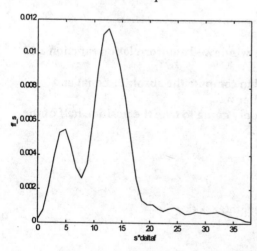

MATLAB Computer Example 5.5
ESTIMATE OF THE SPECTRAL DENSITY OF REAL DATA (2nd case)

```
%===========================================
% Input parameter
                signal='spec2.dat';
                deltat=0.004;
                M=500;
%===========================================
str=str2mat(...
    'The program uses the data file called spec2.dat. You can read in other
data file with 500 samples.');
disp(str);
eval(['load ',signal]);
limit=length(signal)-4;
signal=signal(1:limit);
x_i=eval(signal);
lengthx=length(x_i);
if (lengthx~=500)
str=str2mat(...
    '            The length of the data must be 500 ');
clear signal;
disp(str);
end
i=0:1:M-1;t=0:deltat:(M-1)*deltat;
xmean=mean(x_i);
x_i=x_i-xmean;
xstd=std(x_i);
lengthx=length(x_i);
r0=sum(x_i.^2)/1024;
str=str2mat(...
    ' ',...
    'Autocorrelation check:',...
    ' ',...
    ['>> r0=sum(x_i.^2)/256;','   r(0)= ',num2str(r0)],...
    ' ',...
'We are dividing by 256 since these are the number of points that we will
have after we add zeros to',...
    'the data in order to avoid circular correlation problems.');
disp(str);
clg;s1=subplot(2,1,1);
plot(t,x_i,'r');axisn(t);
xlabel('i*delta','FontSize',[10]);
ylabel('x_i','FontSize',[10]);
title(num2str(signal),'FontSize',[10])
set(s1,'FontSize',[10]);
```

```
s2=subplot(2,1,2);
axis('off')
text(0.3,0.8,['Mean signal: ',num2str(xmean)] ,'FontSize',[10]);
text(0.3,0.6,['Standard deviation: ',num2str(xstd)] ,'FontSize',[10]);
text(0.3,0.4,['Length of data: ',num2str(lengthx)] ,'FontSize',[10]);
text(0.3,0.2,['Deltat: ',num2str(deltat)] ,'FontSize',[10]);
set(s2,'FontSize',[10])
%=========================================
```

The program uses the data file called spec2.dat. You can read in other data file with 500 samples.

Autocorrelation check:

>> r0=sum(x_i.^2)/256; r(0)= 38.49

We are dividing by 256 since these are the number of points that we will have after we add zeros to the data in order to avoid circular correlation problems.

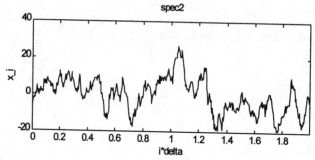

spec2

Mean signal: -2.147

Standard deviation: 8.888

Length of data: 500

Deltat: 0.004

```
%=========================================
str=str2mat(...
    'We now compute the autocorrelation function via the FFT.',...
    '',...
    'The result is the complete autocorrelation of the original data. Check to
see if the signal obtained',...
```

```
    'earlier for the zero lagged signal of the autocorrelation function
corresponds.');
disp(str);
x_izeros=zeros(size(1:1:1024));
x_izeros(1:M)=x_i;
f=fft(x_izeros,length(x_izeros)).*length(x_izeros);
h_j=f.*conj(f);
z=ifft(h_j)./(length(x_izeros)^2);
r0=real(z(1))/length(x_izeros);
z_j=real(z)/length(x_izeros);
clg;s1=subplot(1,1,1);
plot(1:1:1024,z_j,'r');axisn(1:1:1024);
xlabel('j','FontSize',[10]);
ylabel('z_j','FontSize',[10]);
title('Autocorrelation function','FontSize',[10]);
set(s1,'FontSize',[10])
%=======================================
```

We now compute the autocorrelation function via the FFT.

The result is the complete autocorrelation of the original data. Check to see if the signal obtained earlier for the zero lagged signal of the autocorrelation function corresponds.

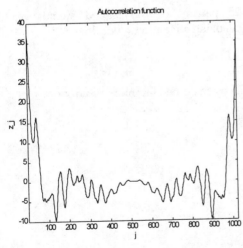

```
%=======================================
str=str2mat(...
    'We are only going to accept 25 percent of the lagged signals of the
autocorrelation function. This is done',...
    'using a Hamming window.',...
    ' ',...
```

```
'>> hamming(0.5*length(x_izeros))',...
' ',...
'We multiply the estimate of the ACF by the time Hamming window.');
disp(str);
w=hamming(0.5*length(x_izeros));
w_zeros=zeros(size(1:1:1024));
w_zeros(1:256)=w(257:512);
w_zeros(769:1024)=w(1:256);
y_kkk=w_zeros;
zz_kkk=z_j.*y_kkk;
clg;s1=subplot(2,1,1);
plot(1:1:1024,y_kkk,'r');axisn(1:1:1024);
xlabel('kkk','FontSize',[10]);
ylabel('y_kkk','FontSize',[10]);
title('Hamming window','FontSize',[10]);
set(s1,'FontSize',[10])
s2=subplot(2,1,2);
plot(1:1:1024,zz_kkk,'r', 1:1:1024,z_j,':k');axisn(1:1:1024);
xlabel('kkk','FontSize',[10]);
ylabel('z_j ...    zz_kkk _','FontSize',[10]);
title('Rxx: unwindowed(black) - windowed(red)','FontSize',[10]);
set(s2,'FontSize',[10])
%==========================================
```

We are only going to accept 25 percent of the lagged signals of the autocorrelation function. This is done using a Hammnig window.

```
>> hamming(0.5*length(x_izeros))
```

We multiply the estimate of the ACF by the time Hamming window.

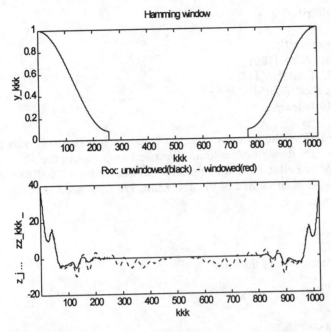

```
%=============================================
str=str2mat(...
     'Now compute the FFT of the windowed autocorrelation function and
thus an estimate of the spectral',...
     'density of the data. We need to compute the absolute signal and
multiply by deltat. Note that earlier',...
     'we made deltat = .004. We are only going to take the positive half of the
spectrum. We now need to',...
     'compute the frequency spacing.',...
     '',...
     '>> deltaf=1/(N*deltat)');
disp(str);
g=fft(zz_kkk,length(zz_kkk)).*length(zz_kkk);
deltaf=1/(1024*deltat);
ff_s=abs(g)*deltat/length(zz_kkk);
Ps=10*log10(ff_s);
check=sum(ff_s(1:512))*2*deltaf;
str=str2mat(...
     '',...
     'Compute now the last check of the units.',...
     '',...
     '>> Check=sum(ff_s(1:512))*2*deltaf',...
     '',...
     ['    Check: ',num2str(check),'        It corresponds!']);
disp(str);
```

```
f=deltaf*(0:1:length(g)/2-1);
clg;s1=subplot(1,1,1);
plot(f,Ps(1:512),'r');axisn(f);
xlabel('s*deltaf','FontSize',[10]);
ylabel('P_s        dB.','FontSize',[10]);
title('Spectrum via Rxx','FontSize',[10])
set(s1,'FontSize',[10]);clear;
```

Now compute the FFT of the windowed autocorrelation function and thus an
estimate of the spectral density of the data. We need to compute the absolute
signal and multiply by deltat. Note that earlier we made deltat = 0.004. We are
only going to take the positive half of the spectrum. We now need to compute
the frequency spacing.

>> deltaf=1/(N*deltat)

Compute now the last check of the units.

>> Check=sum(ff_s(1:512))*2*deltaf

Check: 39.69   It corresponds!

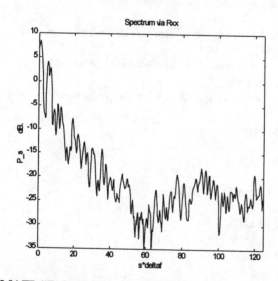

MATLAB Computer Example 5.6
**ESTIMATE OF THE SPECTRAL DENSITY OF A RANDOM PROCESS: USE
OF THE PERIODOGRAM**

In this program we will compute the estimate of the spectral density of a
random process using the periodogram method. We will use a known time-
limited sample function from a wide-sense stationary random process.

```
%=============================================
% Input parameters
                A=1.0;
                f0=5;
                deltat=0.01;
                theta=0.008;
                npsec = 32;
                overl = 0.5;
                knew = 16;
                M=128;
%=============================================
str=str2mat(...
    'In this program we will compute the estimate of the spectral density of a
random process using the',...
    'periodogram method. We will use a known time-limited sample
function from a wide-sense stationary',...
    'random process.',...
    ' ',...
    '>> x(i)=A*cos(2*pi*f0*i*detlat+ theta)+2*rand-1;');
disp(str);
i=0:1:127;
x_i=A*cos(2*pi*f0*i*deltat+ theta)+2*rand(1,128)-1;
xmean=mean(x_i);
xstd=std(x_i);
lengthx=length(x_i);
x_i=x_i-mean(x_i);
clg;s1=subplot(2,1,1);
plot(0:1:length(x_i)-1,x_i,'r');axisn(0:1:length(x_i)-1);
xlabel('i','FontSize',[10]);
ylabel('x_i','FontSize',[10]);
title('Noisy cosinusoid','FontSize',[10]);
set(s1,'FontSize',[10])
s2=subplot(2,1,2);
axis('off')
text(0.3,0.8,['Mean value: ',num2str(xmean)] ,'FontSize',[10]);
text(0.3,0.6,['Standard deviation: ',num2str(xstd)] ,'FontSize',[10]);
text(0.3,0.4,['Length of data: ',num2str(lengthx)] ,'FontSize',[10]);
set(s2,'FontSize',[10])
%=============================================
```

In this program we will compute the estimate of the spectral density of a random process using the periodogram method. We will use a known time limited sample function from a wide-sense stationary random process.

```
>> x(i)=A*cos(2*pi*f0*i*detlat+ theta)+2*rand-1;
```

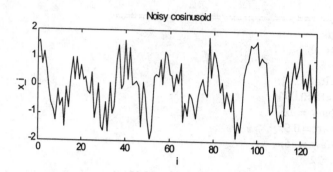

Mean value: 0.01764

Standard deviation: 0.8981

Length of data: 128

```
%============================================
str=str2mat(...
    'Noisy cosinusoid and generated Hamming window of length npsec per
section.',...
    ' ',...
    'Periodogram parameters:',...
    ' ',...
    '              npsec = 32',...
    '              overl = 0.5',...
    '              knew = 16');
disp(str);
kcompl = floor(M/npsec)*npsec;
isect = floor((kcompl-(npsec-knew))/knew);
w_i=hamming(npsec);
clg;s1=subplot(2,1,1);
plot(0:1:length(x_i)-1,x_i,'r');axisn(0:1:length(x_i)-1);
xlabel('i','FontSize',[10]);
ylabel('x_i','FontSize',[10]);
title('Noisy cosinusoid','FontSize',[10])
set(s1,'FontSize',[10]);
s2=subplot(212);
plot((0:1:31),w_i,'r');axisn(0:1:31);
xlabel('i','FontSize',[10]);
ylabel('w_i','FontSize',[10]);
```

```
title('Hamming window','FontSize',[10])
set(s2,'FontSize',[10])
%===========================================
```

Noisy cosinusoid and generated Hamming window of length npsec per section.

Periodogram parameters:
        npsec = 32
        overl = 0.5
        knew = 16

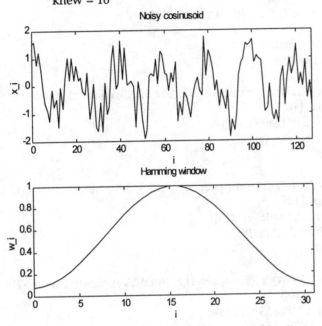

```
%===========================================
str=str2mat(...
    'There are seven overlapping sections. Beginning and endpoint number
of each section (note the 50 percent',...
    'overlap) : 0-31, 16-47, 32-63, 48-79, 64-95, 80-111,96-127.',...
    'Compute the normalizing factor due to the window we are using :',...
    '',...
    '>> u = sum(w.^2);');

disp(str);
str=str2mat(...
    '',...
    'Begin processing the different sections.',...
```

```
' ',...
'The solid line is the data prior to windowing, and the dashed line is
after multiplication by the',...
    'Hamming window.');
disp(str);
u=sum(w_i.^2);
z_ii=zeros(isect,npsec);f=zeros(isect,npsec);g_jj=zeros(1,npsec/2);
for i=0:1:isect-1;
z_ii(i+1,:)=x_i(i*npsec*overl+1:npsec+i*npsec*overl).*w_i';
f(i+1,:)=fft(z_ii(i+1,:),npsec);
g_jj=g_jj+f(i+1,1:npsec/2).*conj(f(i+1,1:npsec/2));
end
g_jj=(g_jj*deltat)/(u*isect);
deltaf=(1/(npsec*deltat));
clg;s1=subplot(221);
plot((0:1:npsec-1),x_i(1:32),'r',(0:1:npsec-1),z_ii(1,:),':k');axisn(0:1:npsec-1);
xlabel(' ii ','FontSize',[10]);
ylabel(' y_ii _    z_ii ....','FontSize',[10]);
title('Section # 1','FontSize',[10]);
set(s1,'FontSize',[10]);
s2=subplot(222);
plot((0:1:npsec-1),x_i(17:48),'r',(0:1:npsec-1),z_ii(2,:),':k');axisn(0:1:npsec-1);
xlabel(' ii ','FontSize',[10]);
ylabel(' y_ii _    z_ii ....','FontSize',[10]);
title('Section # 2','FontSize',[10]);
set(s2,'FontSize',[10]);
s3=subplot(223);
plot((0:1:npsec-1),x_i(33:64),'r',(0:1:npsec-1),z_ii(3,:),':k');axisn(0:1:npsec-1);
xlabel(' ii ','FontSize',[10]);
ylabel(' y_ii _    z_ii ....','FontSize',[10]);
title('Section # 3','FontSize',[10]);
set(s3,'FontSize',[10]);
s4=subplot(224);
plot((0:1:npsec-1),x_i(49:80),'r',(0:1:npsec-1),z_ii(4,:),':k');axisn(0:1:npsec-1);
xlabel(' ii ','FontSize',[10]);
ylabel(' y_ii_    z_ii ....','FontSize',[10]);
title('Section # 4','FontSize',[10]);
set(s4,'FontSize',[10])
%=========================================
```

There are seven overlapping sections. Beginning and end point number of each section (note the 50 percent overlap) : 0-31, 16-47, 32-63, 48-79, 64-95, 80-111,96-127.

Compute the normalizing factor due to the window we are using :

```
>> u = sum(w.^2);
```

Begin processing the different sections.

The solid line is the data prior to windowing and the dashed line is after multiplication by the Hamming window.

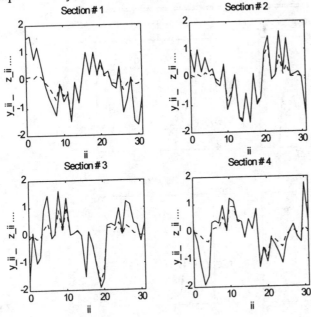

```
%============================================
clg;s1=subplot(221);
plot((0:1:npsec-1),x_i(65:96),'r',(0:1:npsec-1),z_ii(5,:),':k'),title('Section #
5','FontSize',[10])
xlabel(' ii ','FontSize',[10]),ylabel(' y_ii _   z_ii
....','FontSize',[10]),axisn(0:1:npsec-1);
set(s1,'FontSize',[10]);
s2=subplot(222);
plot((0:1:npsec-1),x_i(81:112),'r',(0:1:npsec-1),z_ii(6,:),':k');axisn(0:1:npsec-1);
xlabel(' ii ','FontSize',[10]);
ylabel(' y_ii _   z_ii ....','FontSize',[10]);
title('Section # 6','FontSize',[10]);
set(s2,'FontSize',[10])
s3=subplot(223);
plot((0:1:npsec-1),x_i(97:128),'r',(0:1:npsec-1),z_ii(7,:),':k');axisn(0:1:npsec-1);
xlabel(' ii ','FontSize',[10]);
ylabel(' y_ii _   z_ii ....','FontSize',[10]);
title('Section # 7','FontSize',[10]);
set(s3,'FontSize',[10])
s4=subplot(224);
```

```
plot((((0:1:(npsec/2)-1)*deltaf),g_jj,'k');axisn(0:1:npsec-1);
xlabel(' jj*deltaf ','FontSize',[10]);
ylabel('g_jj ','FontSize',[10]);
title('Periodogram','FontSize',[10]);
set(s4,'FontSize',[10]);clear;
%========================================
```

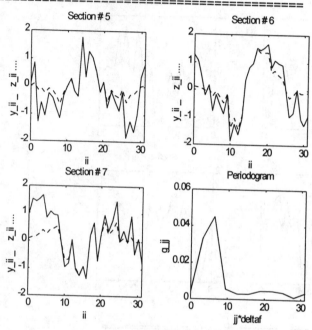

MATLAB Computer Example 5.7
ESTIMATE OF THE SPECTRAL DENSITY OF REAL DATA: USE OF THE PERIODOGRAM (1st case)

```
%============================================
% Input parameters
                signal='tonecz.dat';
                npsec=32;
                overl=0.5;
                knew=16;
%============================================
str=str2mat(...
    'In this program we will compute the estimate of the spectral density of
a real process called tonecz.dat',...
    'using the periodogram method. The length of the data is 128 samples.');
disp(str);
eval(['load ',signal]);
limit=length(signal)-4;
```

```
signal=signal(1:limit);
x=eval(signal);
xmean=mean(x);
xstd=std(x);
x=x-mean(x);x=x';
M=length(x);deltat=0.004;i=0:1:length(x)-1;
t=0:deltat:(M-1)*deltat;
if (M~=128)
str=str2mat(...
'            The length of the data must be 128!');
disp(str);
clear signal;
end
clg;s1=subplot(2,1,1);
plot(t,x,'r');axisn(t)
xlabel('i*deltat','FontSize',[10]);
ylabel('x_i ','FontSize',[10]);
title(num2str(signal),'FontSize',[10]);
set(s1,'FontSize',[10])
s2=subplot(2,1,2);
axis('off')
text(0.3,0.8,['Mean signal: ',num2str(xmean)] ,'FontSize',[10]);
text(0.3,0.6,['Standard deviation: ',num2str(xstd)] ,'FontSize',[10]);
text(0.3,0.4,['Length of data: ',num2str(M)] ,'FontSize',[10]);
text(0.3,0.2,['Deltat: ',num2str(deltat)] ,'FontSize',[10]);
set(s2,'FontSize',[10])
%=========================================
```

In this program we will compute the estimate of the spectral density of a real process called tonecz.dat using the periodogram method. The length of the data is 128 samples.

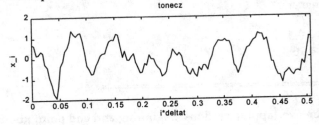

Mean signal: 0.159

Standard deviation: 0.6972

Length of data: 128

Deltat: 0.004

```
%================================================
str=str2mat(...
    'Original signal and generated Hamming window of length npsec per
section.',...
    ' ',...
    'Periodogram parameters:',...
    ' ',...
    '            npsec = 32',...
    '            overl  = 0.5',...
    '            knew  = 16');
disp(str);
kcompl = floor(M/npsec)*npsec;
isect = floor((kcompl-(npsec-knew))/knew);
w_i=hamming(npsec);
clg;s1=subplot(211);
plot(i,x,'r');axisn(i);
axisn(i),xlabel('i','FontSize',[10]);
ylabel('x_i','FontSize',[10]);
title(num2str(signal),'FontSize',[10]);
set(s1,'FontSize',[10])
s2=subplot(212);
plot((0:1:31),w_i,'r');axisn(0:1:31);
xlabel('i','FontSize',[10]);
ylabel('w_i','FontSize',[10]);
title('Hamming window','FontSize',[10]);
set(s2,'FontSize',[10])
%================================================
```

Original signal and generated Hamming window of length npsec per section.

Periodogram parameters:
            npsec = 32
            overl = 0.5
            knew = 16

```
%================================================
str=str2mat(...
    'There are seven overlapping sections. Beginning and end point number
of each section (note the 50 %',...
    'overlap) : 0-31, 16-47, 32-63, 48-79, 64-95, 80-111, 96-127.',...
    'Compute the normalizing factor due to the window we are using :',...
    ' ',...
    '>> u = sum(w.^2);');
disp(str);
u=sum(w_i.^2);
s=zeros(isect,npsec);f=zeros(isect,npsec);g=zeros(1,npsec/2);
```

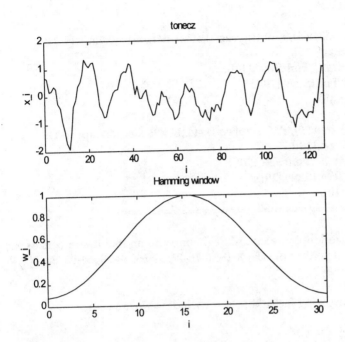

```
for i=0:1:isect-1;
s(i+1,:)=x(i*npsec*overl+1:npsec+i*npsec*overl).*w_i';
f(i+1,:)=fft(s(i+1,:),npsec);
g=g+f(i+1,1:npsec/2).*conj(f(i+1,1:npsec/2));
end
g=(g*deltat)/(u*isect);
deltaf=(1/(npsec*deltat));
str=str2mat(...
    ' ',...
    'Begin processing the different sections.',...
    ' ',...
    'The solid line is the data prior to windowing and the dashed line is after
multiplication by the',...
    'Hamming window.');
disp(str);
clg;s1=subplot(221);
plot((0:1:npsec-1),x(1:32),'r',(0:1:npsec-1),s(1,:),':k');axisn(0:1:npsec-1);
ylabel(' y_ii _ z_ii....','FontSize',[10]);
title('Section # 1','FontSize',[10]);
set(s1,'FontSize',[10])
s2=subplot(222);
plot((0:1:npsec-1),x(17:48),'r',(0:1:npsec-1),s(2,:),':k');axisn(0:1:npsec-1);
ylabel(' y_ii _ z_ii ....','FontSize',[10]);
title('Section # 2','FontSize',[10]);
```

```
set(s2,'FontSize',[10])
s3=subplot(223);
plot((0:1:npsec-1),x(33:64),'r',(0:1:npsec-1),s(3,:),':k');axisn(0:1:npsec-1);
xlabel(' ii ','FontSize',[10]);
ylabel(' y_ii _  z_ii ....','FontSize',[10]);
title('Section # 3','FontSize',[10]);
set(s3,'FontSize',[10])
s4=subplot(224);
plot((0:1:npsec-1),x(49:80),'r',(0:1:npsec-1),s(4,:),':k');axisn(0:1:npsec-1);
xlabel(' ii ','FontSize',[10]);
ylabel(' y_ii _  z_ii ....','FontSize',[10]);
title('Section # 4','FontSize',[10]);
set(s4,'FontSize',[10])
%=========================================
```

There are seven overlapping sections. Beginning and endpoint numbers of each section (note the 50 percent overlap) : 0-31, 16-47, 32-63, 48-79, 64-95, 80-111, 96-127.

Compute the normalizing factor due to the window we are using :

```
>> u = sum(w.^2);
```

Begin processing the different sections.

The solid line is the data prior to windowing and the dashed line is after multiplication by theHamming window.

```
%=========================================
clg;s1=subplot(221);
plot((0:1:npsec-1),x(65:96),'r',(0:1:npsec-1),s(5,:),':k');axisn(0:1:npsec-1);
ylabel(' y_ii _  z_ii ....','FontSize',[10]);
title('Section # 5','FontSize',[10]);
set(s1,'FontSize',[10])
s2=subplot(222);
plot((0:1:npsec-1),x(81:112),'r',(0:1:npsec-1),s(6,:),':k');axisn(0:1:npsec-1);
ylabel(' y_ii _  z_ii ....','FontSize',[10]);
title('Section # 6','FontSize',[10]);
set(s2,'FontSize',[10])
s3=subplot(223);
plot((0:1:npsec-1),x(97:128),'r',(0:1:npsec-1),s(7,:),':k');axisn(0:1:npsec-1);
xlabel(' ii ','FontSize',[10]);
ylabel(' y_ii _  z_ii ....','FontSize',[10]);
```

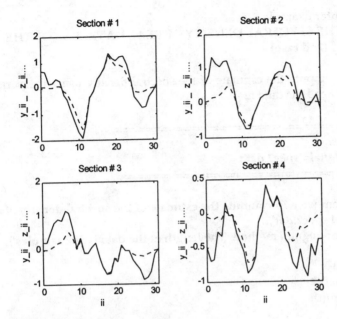

```
title('Section # 7','FontSize',[10]);
set(s3,'FontSize',[10])
s4=subplot(224);
plot(((0:1:(npsec/2)-1)*deltaf),g,'k');axisn(0:1:npsec-1);
xlabel(' jj*deltaf ','FontSize',[10]);
ylabel('g_jj ','FontSize',[10]);
title('Periodogram','FontSize',[10]);
set(s4,'FontSize',[10]);clear
%==========================================
```

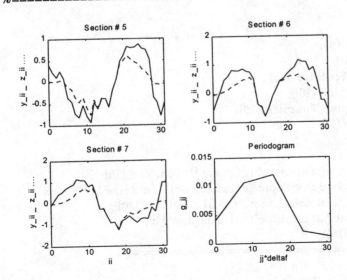

MATLAB Computer Example 5.8
ESTIMATE OF THE SPECTRAL DENSITY OF REAL DATA: USE OF THE
PERIODOGRAM (2nd case]

We are going to compute the estimate of the spectral density using the periodogram method. Read the data first.

```
%===========================================
% Input parameter
                signal='spec2.dat';
%===========================================
str=str2mat(...
    'In this program we will compute the estimate of the spectral density of a
real process called spec2.dat',...
    'using the periodogram method. The length of the data is 500 samples');
disp(str);
eval(['load ',signal]);
limit=length(signal)-4;
signal=signal(1:limit);
x=eval(signal);
xmean=mean(x);
xstd=std(x);
x=x-mean(x);x=x';
M=length(x);deltat=0.004;i=0:1:length(x)-1;npsec=64;overl=0.5;
knew=32;t=0:deltat:(M-1)*deltat;
if (M~=500)
str=str2mat(...
    '          The length of the data must be 500');
clear signal;
disp(str);
end
clg;s1=subplot(2,1,1);
plot(t,x,'r');axisn(t);
xlabel('i*deltat','FontSize',[10]);
ylabel('x_i','FontSize',[10]);
title(num2str(signal),'FontSize',[10])
set(s1,'FontSize',[10])
s2=subplot(2,1,2);
axis('off')
text(0.3,0.8,['Mean signal: ',num2str(xmean)] ,'FontSize',[10]);
text(0.3,0.6,['Standard deviation: ',num2str(xstd)] ,'FontSize',[10]);
text(0.3,0.4,['Length of data: ',num2str(M)] ,'FontSize',[10]);
text(0.3,0.2,['Deltat: ',num2str(deltat)] ,'FontSize',[10]);
set(s2,'FontSize',[10])
%===========================================
```

In this program we will compute the estimate of the spectral density of a real process called spec2.dat using the periodogram method. The length of the data is 500 samples.

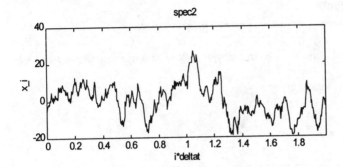

Mean signal: -2.147

Standard deviation: 8.888

Length of data: 500

Deltat: 0.004

```
%=============================================
str=str2mat(...
    'Original signal and generated Hamming window of length npsec per
section.',...
    ' ',...
    'Periodogram parameters:',...
    ' ',...
    '                npsec = 64',...
    '                overl = 0.5',...
    '                knew  = 32');
disp(str);
kcompl = floor(M/npsec)*npsec;
isect = floor((kcompl-(npsec-knew))/knew);
w_i=hamming(npsec);
clg;s1=subplot(211);
plot(i,x,'r');axisn(i);
xlabel('i','FontSize',[10]);
ylabel('x_i','FontSize',[10]);
title(num2str(signal),'FontSize',[10])
set(s1,'FontSize',[10])
s2=subplot(212);
```

```
plot((0:1:63),w_i,'r'); axisn(0:1:63);
xlabel('i','FontSize',[10]);
ylabel('w_i ','FontSize',[10]);
title('Hamming window','FontSize',[10])
set(s2,'FontSize',[10])
%==========================================
```

Original signal and generated Hamming window of length npsec per section.

Periodogram parameters:
             npsec = 64
             overl = 0.5
             knew = 32

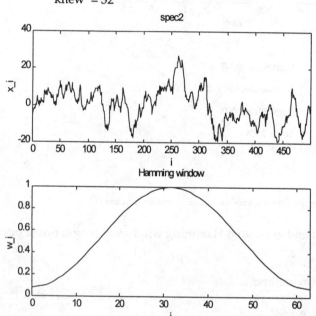

```
%===============================================
str=str2mat(...
     'There are seven overlapping sections. Beginning and endpoint numbers
of each section (note the 50 %',...
       'overlap) : 1:64, 65:128, 126:192, 193:256, 257:3203, 21:384, 385:448.',...
       'Compute the normalizing factor due to the window we are using :',...
       '',...
       '>> u = sum(w.^2);');
disp(str);
str=str2mat(...
       '',...
       'Begin processing the different sections.',...
```

```
' ',...
'The solid line is the data prior to windowing, and the dashed line is
after multiplication by the',...
'Hamming window.');
disp(str);
u=sum(w_i.^2);
s=zeros(isect,npsec);f=zeros(isect,npsec);g=zeros(1,npsec/2);
for i=0:1:isect-1;
s(i+1,:)=x(i*npsec*overl+1:npsec+i*npsec*overl).*w_i';
f(i+1,:)=fft(s(i+1,:),npsec);
g=g+f(i+1,1:npsec/2).*conj(f(i+1,1:npsec/2));
end
g=(g*deltat)/(u*isect);
deltaf=(1/(npsec*deltat));
clg;s1=subplot(221);
plot((0:1:npsec-1),x(1:64),':r',(0:1:npsec-1),s(1,:),'k'); axisn(0:1:npsec-1);
ylabel(' y_ii _ z_ii ....','FontSize',[10]);
title('Section # 1','FontSize',[10]);
set(s1,'FontSize',[10])
s2=subplot(222);
plot((0:1:npsec-1),x(65:128),':r',(0:1:npsec-1),s(3,:),'k'); axisn(0:1:npsec-1);
ylabel(' y_ii _ z_ii ....','FontSize',[10]);
title('Section # 3','FontSize',[10]);
set(s2,'FontSize',[10])
s3=subplot(223);
plot((0:1:npsec-1),x(129:192),':r',(0:1:npsec-1),s(5,:),'k'); axisn(0:1:npsec-1);
xlabel(' ii ','FontSize',[10]);
ylabel(' y_ii _ z_ii ....','FontSize',[10]);
title('Section # 5','FontSize',[10]);
set(s3,'FontSize',[10])
s4=subplot(224);
plot((0:1:npsec-1),x(193:256),':r',(0:1:npsec-1),s(7,:),'k'); axisn(0:1:npsec-1);
xlabel(' ii ','FontSize',[10]);
ylabel(' y_ii _ z_ii ....','FontSize',[10]);
title('Section # 7','FontSize',[10]);
set(s4,'FontSize',[10])
%=========================================
```

There are seven overlapping sections. Beginning and end point numbers of
each section (note the 50 percent overlap) : 1:64, 65:128, 126:192, 193:256,
257:3203, 21:384, 385:448.

Compute the normalizing factor due to the window we are using :

```
>> u = sum(w.^2);
```

Begin processing the different sections.
The solid line is the data prior to windowing and the dashed line is after multiplication by the
Hamming window.

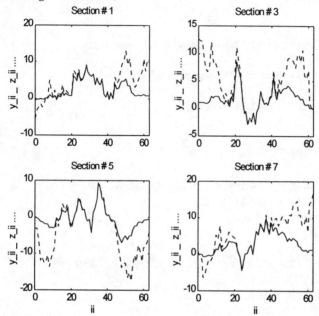

```
%=========================================
clg;s1=subplot(221);
plot((0:1:npsec-1),x(257:320),':r',(0:1:npsec-1),s(9,:),'k');axisn(0:1:npsec-1);
ylabel(' y_ii _ z_ii ....','FontSize',[10]);
title('Section # 9','FontSize',[10]);
set(s1,'FontSize',[10])
s2=subplot(222);
plot((0:1:npsec-1),x(321:384),':r',(0:1:npsec-1),s(11,:),'k');axisn(0:1:npsec-1);
ylabel(' y_ii _ z_ii ....','FontSize',[10]);
title('Section # 11','FontSize',[10]);
set(s2,'FontSize',[10])
s3=subplot(223);
plot((0:1:npsec-1),x(385:448),':r',(0:1:npsec-1),s(13,:),'k');axisn(0:1:npsec-1);
xlabel(' ii ','FontSize',[10]),
ylabel(' y_ii _ z_ii ....','FontSize',[10]);
title('Section # 13','FontSize',[10]);
p=10*log10(g);
set(s3,'FontSize',[10])
s4=subplot(224);
plot(((0:1:(npsec/2)-1)*deltaf),p,'k'); axisn(0:1:npsec-1);
xlabel(' jj*deltaf ','FontSize',[10]);
```

ylabel('P_jj   dB','FontSize',[10]);
title('Periodogram','FontSize',[10]);
set(s4,'FontSize',[10]);clear
%==

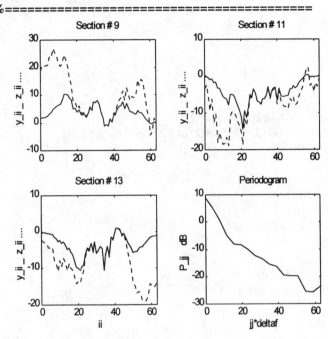

MATLAB Computer Example 5.9
AUTOREGRESSIVE ESTIMATION OF A THEORETICAL MODEL

```
%==============================================
% Input parameters
                order=6;
                A=0;
                M=101;
%==============================================
 str=str2mat(...
    'Estimation of the order of a theoretical model and computation of
 the  AR spectrum estimate.',...
    'Define the following time-series model.',...
    ' ',...
    '>> z(i)=0.75*z(i-1)-0.5*z(i-2) + A*rand');
disp(str);
noise=rand(M,1);
nmean=mean(noise);
noise=noise-nmean;
i=1:1:M;
```

```
z(1)=1+A*noise(1);z(2)=0.75+A*noise(2);
for i=3:1:M,
z(i)=0.75*z(i-1)-0.5*z(i-2)+A*noise(i);
end
FPE=zeros(size(1:1:6));
for i=1:1:6
th=ar(z,i+1,'yw');
FPE(i)=th(2,1);
end
FPE=FPE./FPE(1);
th=ar(z,order,'yw');a0=-th(3,1);a1=-th(3,2);
s_ij=2./(1+ a0^2 + a1^2 - 2*a0*(1 - a1)*cos(2*pi*(0:0.1:0.5))-
2*a1*cos(4*pi*(0:0.1:0.5))));
clg;s1=subplot(221);
plot(0:1:M-1,z,'r');axisn(0:1:M-1);
xlabel('i','FontSize',[10]);
ylabel('z_i','FontSize',[10]);
title('Theoretical data','FontSize',[10])
set(s1,'FontSize',[10])
s2=subplot(222);
plot((1:1:6),FPE,'r'); axisn(1:1:6);
xlabel('order','FontSize',[10]);
ylabel('FPE','FontSize',[10]);
set(s2,'FontSize',[10])
s3=subplot(223);
axis('off');
sub3_t1=text(0,0.8,' AR( 2 ) coefficients :','FontSize',[10]);
sub3_t2=text(0,0.6,['a(1) = ',num2str(-th(3,1))],'FontSize',[10]);
sub3_t3=text(0,0.4,['a(2) = ',num2str(-th(3,2))],'FontSize',[10]);
set(s3,'FontSize',[10]);
s4=subplot(224);
plot((0:0.1:0.5),s_ij);axisn(0:0.1:0.5);
xlabel('ij','FontSize',[10]);
ylabel('s_ij','FontSize',[10]);
title('Spectral density','FontSize',[10])
set(s4,'FontSize',[10]);clear
%============================================
```

Estimation of the order of a theoretical model and computation of the AR spectrum estimate. Define the following time-series model.

```
>> z(i)=0.75*z(i-1)-0.5*z(i-2) + A*rand
```

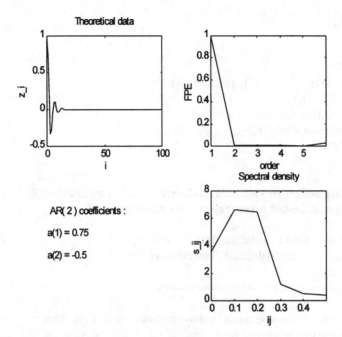

AR(2) coefficients :

a(1) = 0.75

a(2) = -0.5

MATLAB Computer Example 5.10
ESTIMATION OF THE CROSS-SPECTRAL DENSITY

```
%==============================================
% Input parameters
                deltat=0.01;
                theta=0.008;
                f0=5;
                f1=21;
                f2=21;
                A=1;
%==============================================
str=str2mat(...
    'In this program the cross-spectrum between two theoretical functions is
calculated. Define the',...    'following functions:',...
    ' ',...
    '>> x(i)=A*cos(2*pi*f1*i*detlat) + cos(2*pi*f0*i*detlat) + 2*rand;',...
    ' ',...
    '>> y(i)=cos(2*pi*f2*i*deltat);');
disp(str);
M=128;
i=0:1:M-1;
x_i=A*cos(2*pi*f1*i*deltat)+ cos(2*pi*f0*i*deltat)+2*rand(1,M);
y_i=cos(2*pi*f2*i*deltat);
xmean=mean(x_i);
```

```
ymean=mean(y_i);
x_i=x_i-xmean;
y_i=y_i-ymean;
clg;s1=subplot(211);
plot(0:1:M-1,x_i,'r', 0:1:M-1,y_i,':k'); axisn(0:1:M-1);
xlabel('i','FontSize',[10]);
ylabel('x _ y ....','FontSize',[10])
title('Cosinusoids','FontSize',[10]);
set(s1,'FontSize',[10])
s2=subplot(212);
axis('off');
text(0.3,0.8,['Cosinusoid mean value : ',num2str(ymean)] ,'FontSize',[10]);
text(0.3,0.6,['Noisy cosinusoid mean value : ',num2str(xmean)]
,'FontSize',[10]);
text(0.3,0.4,['Length of data : ',num2str(M)] ,'FontSize',[10]);
text(0.3,0.2,['Deltat : ',num2str(deltat)] ,'FontSize',[10]);
set(s2,'FontSize',[10])
%==============================================
```

In this program the cross-spectrum between two theoretical functions is calculated. Define the the following functions

>> x(i)=A*cos(2*pi*f1*i*detlat) + cos(2*pi*f0*i*detlat) + 2*rand;

>> y(i)=cos(2*pi*f2*i*deltat);

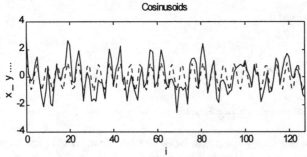

Cosinusoid mean value : -0.002389

Noisy cosinusoid mean value : 0.9695

Length of data : 128

Deltat : 0.01

```
%===============================================
str=str2mat(...
    'We now compute the ACF of each signal, extended with zeros, via
FFT.',...
    ' ',...
    'The result is the complete cross-correlation function. In order to estimate
the spectral density using the',...
    'cross-correlation function, we need to multiply the estimate of the cross-
correlation function by a time',...
    'domain window. We are going to accept only 25 percent of the lagged
values of the CCF.');
disp(str);
x=zeros(size(0:1:255));
y=zeros(size(0:1:255));
x(1:M)=x_i;
y(1:M)=y_i;
xx=fft(x,length(x)).*length(x);
yy=fft(y,length(y)).*length(y);
zz_j=xx.*conj(yy);
z=ifft(zz_j)./(length(x)^2);
z=real(z)/length(x);
w_k=hamming(0.5*length(x));
w_zeros=zeros(size(1:1:256));
w_zeros(1:64)=w_k(65:128);
w_zeros(193:256)=w_k(1:64);
y_kkk=w_zeros;
zz_kkk=z.*y_kkk;
t=(0:1:255);
clg;s1=subplot(211);
plot(0:1:255,y_kkk); axisn(0:1:255);
xlabel('kkk','FontSize',[10]);
ylabel('y_kkk','FontSize',[10]);
title('Hamming window','FontSize',[10]);
set(s1,'FontSize',[10])
s2=subplot(212);
plot(0:1:255,zz_kkk,':r',0:1:255,z,'k'); axisn(0:1:255);
xlabel('kkk','FontSize',[10]);
ylabel('z _   zz_kkk .....','FontSize',[10]);
set(s2,'FontSize',[10])
%===============================================
```

We now compute the ACF of each signal, extended with zeros, via FFT. The result is the complete cross-correlation function. In order to estimate the spectral density using the cross-correlation function, we need to multiply the estimate of the cross-correlation function by a time domain window. We are going to accept only 25 percent of the lagged values of the CCF.

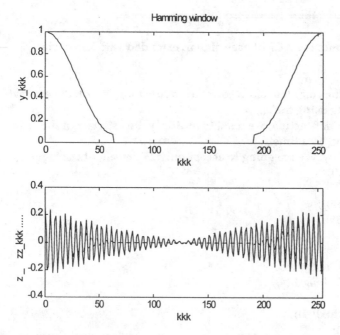

```
%===============================================
str=str2mat(...
    'We now compute the FFT of the windowed cross-correlation function
and thus an estimate of the',...
    'cross-spectrum. We need to compute the frequency spacing
deltaf=1/(N*deltat)');
disp(str);
g=fft(zz_kkk,length(zz_kkk)).*length(zz_kkk);
ff_s=abs(g)*deltat/256;
deltaf=1/(256*deltat);
f=deltaf*(0:1:length(g)/2-1);
clg;s1=subplot(2,2,1);
plot(1:1:128,y_i,'r');axisn(1:1:128);
title('Cosinusoid','FontSize',[10]),
set(s1,'FontSize',[10])
s2=subplot(2,2,2);
plot(1:1:128,x_i,'r');axisn(1:1:128);
title('Noisy cosinusoid','FontSize',[10]);
set(s2,'FontSize',[10])
s3=subplot(2,2,3);
plot(1:1:128,z(1:128),'r'); axisn(1:1:128);
xlabel('kkk','FontSize',[10]);
ylabel('z_kkk','FontSize',[10]);
title('Cross-correlation','FontSize',[10]);
set(s3,'FontSize',[10])
```

```
s4=subplot(2,2,4);
plot(f,ff_s(1:128),'k');axisn(f);
xlabel('s*deltaf','FontSize',[10]);
ylabel('ff_s','FontSize',[10]);
title('Cross-spectrum','FontSize',[10]);
set(s4,'FontSize',[10]);clear;
%===========================================
```

We now compute the FFT of the windowed cross-correlation function and thus an estimate of the cross-spectrum. We need to compute the frequency spacing deltaf=1/(N*deltat).

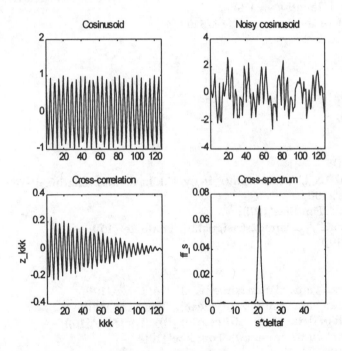

MATLAB Computer Example 5.11
CROSS-SPECTRAL DENSITY ESTIMATION

```
%===========================================
% Input parameters
                signal1='click3.dat';
                signal2='click4.dat';
%===========================================
str=str2mat(...
    'In this program the coherence function between two functions of time
recorded in a laboratory will be',...
    'obtained.');
```

```
disp(str);
eval(['load ',signal1]);
limit=length(signal1)-4;
signal1=signal1(1:limit);
x_i=eval(signal1);
eval(['load ',signal2]);
limit=length(signal2)-4;
signal2=signal2(1:limit);
y_i=eval(signal2);
lengthx=length(x_i);
lengthy=length(y_i);
if (lengthx ~=375  |  lengthy ~= 375)
disp('   The length of the data must be 375 !!!')
clear signal1;
clear signal2;
end
deltat=0.004;
xmean=mean(x_i);
ymean=mean(y_i);
x_i=x_i-xmean;
y_i=y_i-ymean;
clg;s1=subplot(211);
plot(0:1:length(x_i)-1,x_i,'r', 0:1:length(x_i)-1,y_i,':k');axisn(0:1:length(x_i)-1);
xlabel('i','FontSize',[10]);
ylabel('x_i _  y_i ....','FontSize',[10]);
title([num2str(signal1),' & ',num2str(signal2)], 'FontSize',[10]);
set(s1, 'FontSize',[10])
s2=subplot(212);
axis('off');
text(0.3,0.8,['x_i mean value : ',num2str(xmean)], 'FontSize',[10]);
text(0.3,0.6,['y_i mean value : ',num2str(ymean)], 'FontSize',[10]);
text(0.3,0.4,['Length of data : ',num2str(length(x_i))], 'FontSize',[10]);
text(0.3,0.2,['Deltat : ',num2str(deltat)], 'FontSize',[10]);
set(s2, 'FontSize',[10])
%==========================================
```

In this program the coherence function between two functions of time recorded in a laboratory will be obtained.

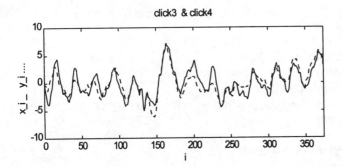

x_i mean value : -0.2951

y_i mean value : -1.219

Length of data : 375

Deltat : 0.004

```
%==========================================
str=str2mat(...
    'We now compute the autocorrelation and cross-correlation function, via
FFT of the signals extended',...
    'with zeros');
disp(str);
x=zeros(size(0:1:1023));
y =zeros(size(0:1:1023));
x(1:length(x_i))=x_i;
y(1:length(x_i))=y_i;
xx=fft(x,length(x)).*length(x);
yy=fft(y,length(y)).*length(y);
z_jj=xx.*conj(yy);
g_jj=xx.*conj(xx);
h_jj=yy.*conj(yy);
r_x=ifft(g_jj)./(length(x));
r_y=ifft(h_jj)./(length(y));
r_xy=ifft(z_jj)./(length(x));
r_xy=real(r_xy)/length(x);
r_x=real(r_x)/length(x);
r_y=real(r_y)/length(y);
clg;s1=subplot(221);
plot(0:1:length(x_i)-1,x_i,'r', 0:1:length(x_i)-1,y_i,':k');axisn(0:1:length(x_i)-1);
xlabel('i', 'FontSize',[10]);
ylabel('x_i _   y_i ....', 'FontSize',[10]),
```

```
title([num2str(signal1),' & ',num2str(signal2)] , 'FontSize',[10]);
set(s1, 'FontSize',[10])
s2=subplot(222);
plot(0:1:1023,r_x,'r');axisn(0:1:1023);
xlabel('jj', 'FontSize',[10]);
ylabel('r_x', 'FontSize',[10]);
title(['Correlation function ',num2str(signal1)] , 'FontSize',[10]);
set(s2, 'FontSize',[10])
s3=subplot(223);
plot(0:1:1023,r_y,'r');axisn(0:1:1023);
xlabel('jj', 'FontSize',[10]);
ylabel('r_y', 'FontSize',[10]);
title(['Correlation function ',num2str(signal2)] , 'FontSize',[10]);
set(s3, 'FontSize',[10])
s4=subplot(224);
plot(0:1:1023,r_xy,'r');axisn(0:1:1023);
xlabel('jj', 'FontSize',[10]);
ylabel('r_xy', 'FontSize',[10]);
title('Cross-correlation function', 'FontSize',[10]);
set(s4, 'FontSize',[10])
%==========================================
```

We now compute the autocorrelation and cross-correlation function, via FFT of the signals extended with zeros.

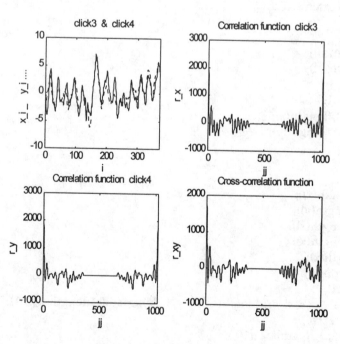

```
%=============================================
str=str2mat(...
   'In order to estimate the spectral density using the cross-correlation
function and the autocorrelation',...
   'functions, we need to multiply the correlation functions by a time
domain window. We are only going',...
   'to accept 25 percent of the lagged values of the ACF"s and CCF. We
compute the FFT of the windowed',...    'correlation functions and the
coherence function, with the estimator:',...
   ' ',...
   '                        cohxy=ff(s)./(ffx(s).*ffy(s));');
disp(str);
w_k=hamming(0.5*length(x));
w_zeros=zeros(size(1:1:1024));
w_zeros(1:256)=w_k(257:512);
w_zeros(769:1024)=w_k(1:256);
y_kkk=w_zeros;
zz_kkk=r_xy.*y_kkk;
rxx_kkk=r_x.*y_kkk;
ryy_kkk=r_y.*y_kkk;
g=fft(zz_kkk,length(zz_kkk)).*length(zz_kkk);
ff_s=(abs(g));ff_s=ff_s(1:512)/ length(zz_kkk);
gx=fft(rxx_kkk,length(rxx_kkk)).* length(rxx_kkk);
fx_s=abs(gx);fx_s=fx_s(1:512)/ length(rxx_kkk);
gy=fft(ryy_kkk,length(ryy_kkk)).* length(ryy_kkk);
fy_s=abs(gy);fy_s=fy_s(1:512)/ length(rxx_kkk);
deltaf=1/(1024*deltat);
cohxy=(ff_s.^2)./(fx_s.*fy_s);
f=deltaf*(0:1:length(g)/2-1);
clg;s1=subplot(2,2,1);
plot(f(1:256),ff_s(1:256),'r');axisn(f(1:256));
ylabel('| g_ii |', 'FontSize',[10]);
set(s1, 'FontSize',[10])
s2=subplot(2,2,2);
plot(f(1:256),fx_s(1:256),'r');axisn(f(1:256));
ylabel('| gx_ii |', 'FontSize',[10]);
set(s2, 'FontSize',[10])
s3=subplot(2,2,3);
plot(f(1:256),fy_s(1:256),'r');axisn(f(1:256));
xlabel('ii*deltaf', 'FontSize',[10]);
ylabel('| gy_ii |', 'FontSize',[10]);
set(s3, 'FontSize',[10])
s4=subplot(2,2,4);
plot(f(1:1:42),cohxy(1:42),'k');axisn(f(1:1:42));
xlabel('s*deltaf', 'FontSize',[10]);
ylabel('Coh_s', 'FontSize',[10]);
```

set(s4, 'FontSize',[10]);clear;
%==

In order to estimate the spectral density using the crosscorrelation function and
the autocorrelation functions, we need to multiply the correlations functions by
a time domain window. We are going to accept only 25 percent of the lagged
values of the ACF's and CCF. We compute the FFT of the windowed correlation
functions and the coherence function, with the estimator.

cohxy=ff(s)./(ffx(s).*ffy(s));

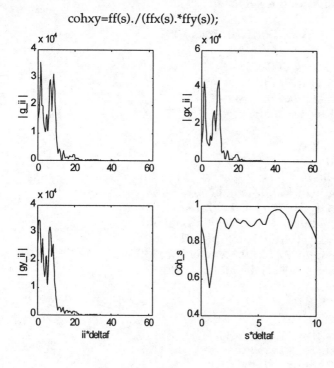

Linear Systems

6

6.0 INTRODUCTION

There are many applications of interest to engineers that may be theoretically modeled by the use of linear system theory. From linear system theory, we know that a system is linear if the principle of superposition holds. A commonly encountered example of a linear system is a filter. In this chapter we describe techniques to evaluate the response of linear systems to random processes.

Consider a system with given impulse response $h(t)$. From linear system theory we know that the output of the system $y(t)$ may be expressed as the input $x(t)$ convolved with the impulse response of the system $h(t)$. This relationship is given by the following equation:

$$y(t) = x(t) * h(t) = \int_{-\infty}^{\infty} x(\tau)h(t-\tau)\, d\tau \qquad (6.1)$$

Figure 6.1 illustrates this concept where $h(t)$ is the system impulse response, $x(t)$ is the input, and $y(t)$ is the output.

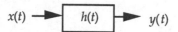

Figure 6.1 Linear system with impulse response $h(t)$, input $x(t)$ and output $y(t)$.

Assume now that the input is multiplied by a constant a; then the relationship between the output, the input, and the system impulse response may be found by using Equation (6.1).

$$y(t) = a[x(t) * h(t)] = \int_{-\infty}^{\infty} ax(\tau)h(t-\tau) \, d\tau = a\int_{-\infty}^{\infty} x(\tau)h(t-\tau) \, d\tau = ay(t) \quad (6.2)$$

The output $y(t)$ of the linear system is multiplied by the constant parameter a. Figure 6.2 illustrates this concept.

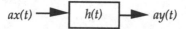

Figure 6.2 Linear system with impulse response $h(t)$, input $x(t)$, and output $y(t)$. In this case the input is multiplied by the constant parameter a.

Suppose now that the input is the sum of two inputs x_1 and x_2. Each of these inputs has been multiplied by constant parameters a and b. The output of the linear system is composed of the sum of the outputs to each of the respective inputs y_1 and y_2, multiplied by the respective parameters a and b. For a linear system, where the principle of superposition holds, the output equals the sum of the two inputs multiplied by the constants a and b, convolved with the impulse response of the system. Figure 6.3 illustrates this concept, and mathematically this relationship may be expressed by Equation (6.3):

$$y(t) = \int_{-\infty}^{\infty} x(\tau)h(t-\tau) \, d\tau = \int_{-\infty}^{\infty} [ax_1(\tau) + bx_2(\tau)]h(t-\tau) \, d\tau \quad (6.3)$$

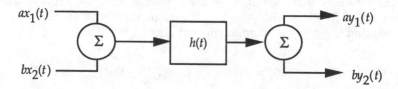

Figure 6.3 Linear system with impulse response $h(t)$. The input is now composed of the sum of two inputs ax_1 and bx_2. Since this is a linear system, the output is composed of ay_1 and by_2.

which simplifies to

$$y(t) = \int_{-\infty}^{\infty} ax_1(\tau)h(t-\tau)\, d\tau + \int_{-\infty}^{\infty} bx_2(\tau)h(t-\tau)\, d\tau \tag{6.4}$$

This yields

$$y_1(t) + y_2(t) = a[x_1(t) * h(t)] + b[x_2(t) * h(t)] \tag{6.5}$$

or

$$\begin{cases} y_1(t) = a[x_1(t) * h(t)] \\ y_2(t) = b[x_2(t) * h(t)] \end{cases} \tag{6.6}$$

6.1 PROPERTIES

Two properties of interest in linear systems that we will use throughout most of this chapter are time invariance and causality.

Time invariance is defined as that property of linear systems such that if the input is delayed by an amount τ, then the output will also be delayed by the same amount τ. Figure 6.4 illustrates this concept.

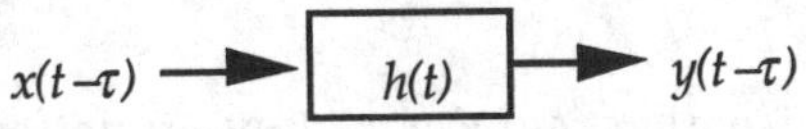

$$x(t-\tau) \longrightarrow \boxed{h(t)} \longrightarrow y(t-\tau)$$

Figure 6.4 Illustration of the property of time invariance in a linear system.

Causality is defined as the property of linear systems such that the response at time t depends only on past values of the input. This implies that $h(t) = 0$ for all $t < 0$. For a causal system Equation (6.1) changes as follows:

$$y(t) = x(t) * h(t) = \int_{0}^{\infty} x(\tau)h(t-\tau)\, d\tau \tag{6.7}$$

6.2 RANDOM INPUTS

When the input to a linear system is a random process, the output of the linear system will also be a random process. Consider now a stationary random process $X(t)$ which is the input to a linear system with impulse response $h(t)$. The output of the system is defined by the convolution integral.

$$Y(t) = X(t) * h(t) = \int_{-\infty}^{\infty} X(\tau)h(t-\tau) \, d\tau \tag{6.8}$$

Consider now the expected or mean value of the output:

$$E[Y(t)] = E[X(t) * h(t)] = E\left[\int_{-\infty}^{\infty} X(\tau)h(t-\tau) \, d\tau\right] \tag{6.9}$$

The integral and the expected value may be interchanged, yielding

$$E[Y(t)] = \int_{-\infty}^{\infty} E[X(\tau)]h(t-\tau) \, d\tau \tag{6.10}$$

which produces the following result:

$$E[Y(t)] = \mu_X \int_{-\infty}^{\infty} h(t-\tau) \, d\tau \tag{6.11}$$

The last element of this equation is called the *dc gain* of the linear system:

$$\text{dc gain} = \int_{-\infty}^{\infty} h(t-\tau) \, d\tau \tag{6.12}$$

Consider now the *output autocorrelation function*. We begin this derivation by considering the relationship between the input and the output:

$$Y(t) = X(t) * h(t) = \int_{-\infty}^{\infty} h(\alpha)X(t-\alpha) \, d\alpha \tag{6.13}$$

Multiply both sides by $Y(t + \rho)$ and take the expectations:

$$E[Y(t)Y(t+\rho)] = E\left[\int\int_{-\infty}^{\infty} h(\alpha)h(\beta)X(t-\alpha)X(t-\beta-\rho) \, d\alpha \, d\beta\right] \tag{6.14}$$

which yields

$$R_Y(\tau) = \int\int_{-\infty}^{\infty} h(\alpha)h(\beta)R_X(\tau+\alpha-\beta) \, d\alpha \, d\beta \tag{6.15}$$

Note that since the system is causal, we can rewrite this equation as

$$R_Y(\tau) = \iint\limits_{-\infty}^{\infty} h(\alpha)h(\beta)R_X(\tau + \alpha - \beta)\,d\alpha\,d\beta \tag{6.16}$$

Likewise, we compute the cross-correlation function between the input and the output. We proceed in a similar manner to Equation (6.14):

$$E[X(t)Y(t+\tau)] = E\left[\int_{-\infty}^{\infty} h(\alpha)X(t)X(t - \alpha + \tau)\,d\alpha\right] \tag{6.17}$$

which yields

$$R_{XY}(\tau) = \int_{-\infty}^{\infty} h(\alpha)R_X(\tau - \alpha)\,d\alpha \tag{6.18}$$

where we recognize this equation as a convolution operation. This equation states that the cross-correlation between the input and the output equals the convolution of the input autocorrelation function and the impulse response of the system.

These relationships are not easy to interpret in the time domain. It is always important to remember an important property of the relationships we just derived. The autocorrelation function of the output, given by Equation (6.16), is an *even* function of time, whereas the cross-correlation function, given by Equation (6.18), may not have any symmetry at all.

It will be easier to interpret these relationships in the frequency domain. The power spectral density of the output of the linear system $y(t)$ can be obtained as the Fourier transform of the autocorrelation function by

$$S_Y(f) = \int_{-\infty}^{\infty} R_Y(\tau)\exp(-j2\pi f\tau)\,d\tau \tag{6.19}$$

Substituting Equation (6.15) in the above equation, we obtain

$$S_Y(f) = \iiint\limits_{-\infty}^{\infty} h(\tau_1)h(\tau_2)R_X(\tau + \tau_2 - \tau_1)\exp(-j2\pi f\tau)\,d\tau_1\,d\tau_2\,d\tau \tag{6.20}$$

Substituting the argument of R_X, namely $\tau + \tau_2 - \tau_1$ for t, we can write

$$\tau = t + \tau_1 - \tau_2 \tag{6.21}$$

The expression for $S_Y(f)$ can be rewritten with the above substitutions as

$$S_Y(f) = \int\int\int_{-\infty}^{\infty} h(\tau_1)e^{-j2\pi f\tau_1}h(\tau_2)e^{j2\pi f\tau_2}R_X(t)e^{-j2\pi ft}\,d\tau_1\,d\tau_2\,dt \qquad (6.22)$$

The Fourier transforms of $h(t)$ can be denoted by $H(f)$, or the frequency response of the linear system. By using this notation $S_Y(f)$ can be simplified to

$$S_Y(f) = H(f)H(f)^* S_X(f) \qquad (6.23)$$

where $H(f)^*$ indicates the complex conjugate of $H(f)$. Therefore the relation between the input and output power spectral densities of a linear system can be expressed as

$$S_Y(f) = |H(f)|^2 S_X(f) \qquad (6.24)$$

Likewise, the cross-spectrum between input and output processes can be expressed in terms of $S_X(f)$ by

$$S_{XY}(f) = H(f)S_X(f) \qquad (6.25)$$

By definition, $S_Y(f)$ is a real-valued positive function. On the other hand, $S_{XY}(f)$ is a complex-valued function of frequency.

Example 6.1. Consider the linear system illustrated in Figure 6.5.

In the first case we assume white noise is applied as input to the system. By definition, such a random process has a mean value equal to zero and an autocorrelation function of the form

$$R_X(\tau) = A\delta(\tau) \qquad (6.26)$$

and where $\delta(\tau)$ is a delta function at the origin.

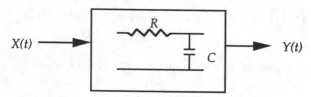

Figure 6.5 Linear system for Example 6.1.

The impulse response of the system (causal and time-invariant) is given by the expression

$$h(t) = \frac{1}{RC} \exp\left(\frac{-t}{RC}\right) \tag{6.27}$$

Consider now the mean value of the output. From Equation (6.11)

$$\mu_Y = \mu_X \int_{-\infty}^{\infty} h(t-\tau)\,d\tau = \mu_X \int_0^{\infty} h(t-\tau)\,d\tau = 0 \tag{6.28}$$

since by definition $\mu_X = 0$.

Consider now the autocorrelation function of the output. Using Equation (6.16) gives

$$R_{YY}(\tau) = \int\!\!\int_{-\infty}^{\infty} h(\alpha)h(\beta)R_{XX}(\tau + \alpha - \beta)\,d\alpha\,d\beta \tag{6.29}$$

which yields

$$R_{YY}(\tau) = A \int\!\!\int_{-\infty}^{\infty} h(\alpha)h(\beta)\delta(\tau + \alpha - \beta)\,d\alpha\,d\beta \tag{6.30}$$

$$R_{YY}(\tau) = A \int_0^{\infty} h(\alpha)\int_0^{\infty} h(\beta)\delta(\tau + \alpha - \beta)\,d\alpha\,d\beta \tag{6.31}$$

$$R_{YY}(\tau) = A \int_0^{\infty} h(\alpha)h(\alpha + \tau)\,d\alpha \tag{6.32}$$

For the present case

$$h(\alpha) = \frac{1}{RC} \exp\left(\frac{-t}{RC}\right) \tag{6.33}$$

which yields (let $A = 1$)

$$R_{YY}(\tau) = \frac{1}{(RC)^2} \int_0^{\infty} \exp\left(\frac{-\alpha}{RC}\right)\exp\left[\frac{-(\alpha+\tau)}{RC}\right]d\alpha \tag{6.34}$$

$$R_{YY}(\tau) = \frac{1}{(RC)^2} \exp\left[\frac{-\tau}{RC}\right] \int_0^\infty \exp\left[\frac{-2\alpha}{RC}\right] d\alpha \tag{6.35}$$

We need only consider values of $\tau \geq 0$, since the autocorrelation function is an even function of time.

$$R_{YY}(\tau) = \frac{1}{(RC)^2} \exp\left[\frac{-\tau}{RC}\right] \left.\frac{\exp\left[\frac{-2\alpha}{RC}\right]}{-2/RC}\right|_{\alpha=0}^{\alpha=\infty} \quad \tau \geq 0 \tag{6.36}$$

which yields

$$R_{YY}(\tau) = \frac{1}{2RC} \exp\left[\frac{-\tau}{RC}\right] \quad \tau \geq 0 \tag{6.37}$$

For $\tau \leq 0$, we simply use the fact that $R_{YY}(\tau)$ is an even function.

$$R_{YY}(\tau) = \frac{1}{2RC} \exp\left[\frac{\tau}{RC}\right] \quad \tau \leq 0 \tag{6.38}$$

and as a combined value,

$$R_{YY}(\tau) = \frac{1}{2RC} \exp\left[\frac{-|\tau|}{RC}\right] \quad \text{for all values of } \tau \tag{6.39}$$

A plot of this result is shown in Figure 6.6 for $RC = 1$.

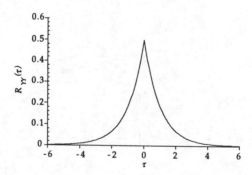

Figure 6.6 Plot of the autocorrelation function for $RC = 1$

The spectral density of the output is now calculated as the Fourier transform of the autocorrelation function. It is easy to show that:

$$F\{\exp(-\alpha|\tau|)\} = \frac{2\alpha}{\alpha^2 + f^2} \tag{6.40}$$

For our case,

$$F\left\{\frac{1}{RC}\exp\left(\frac{-|\tau|}{RC}\right)\right\} = \frac{1}{1 + f^2(RC)^2} \tag{6.41}$$

A plot of this result is shown in Figure 6.7 for $RC = 1$.

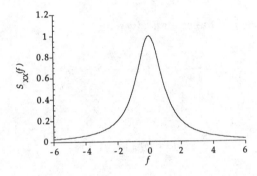

Figure 6.7 Plot of the spectral density function for $RC = 1$

In the second case, we assume that the input is composed of a periodic waveform added to white noise. The input is defined by

$$X(t) = A\cos(\omega_0 t + \theta) + N(t) \tag{6.42}$$

where A = constant
$\quad \omega_0$ = constant
$\quad N(t)$ = white noise
$\quad \theta$ = random variable uniformly distributed between 0 and 2π

We illustrate this second case by means of Figure 6.8. The autocorrelation function of the input is given by:

$$R_{XX}(\tau) = \frac{A^2}{2} \cos \omega_0 \tau + \delta(\tau) \tag{6.43}$$

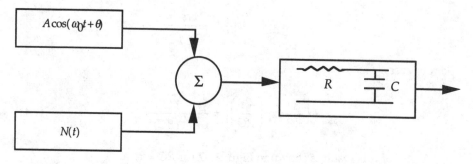

Figure 6.8 An input to a linear system consists of a periodic waveform and additive white noise.

and the spectral density of the input is given by

$$S_{XX}(f) = \frac{A^2}{4} \delta(f - f_0) + \frac{A^2}{4} \delta(f + f_0) + 1 \tag{6.44}$$

The spectral density of the output is given by

$$S_{YY}(f) = |H(f)|^2 S_{XX}(f) \tag{6.45}$$

where

$$H(f) = \frac{1}{1 + jfRC} \tag{6.46}$$

Therefore,

$$|H(f)|^2 = \frac{1}{1 + f^2(RC)^2} \tag{6.47}$$

and

$$S_{YY}(f) = \frac{1}{1 + f^2(RC)^2}\left[\frac{A^2}{4} \delta(f - f_0) + \frac{A^2}{4} \delta(f + f_0) + 1 \right] \tag{6.48}$$

6.3 ESTIMATE OF THE RESPONSE OF LINEAR SYSTEMS USING FREQUENCY DOMAIN TECHNIQUES

The output of a linear system is given by the convolution of the input with the impulse response of the system. Therefore the power spectral density of the output $y(t)$ is related to the power spectral density of the input $x(t)$ by relation (6.20), which is

$$S_Y(f) = S_X(f)|H(f)|^2 \qquad (6.49)$$

where $H(f)$ is the Fourier transform of the impulse response, generally referred to as the *frequency response* of the system. Often we have to evaluate the output of a linear system, for a given input random process. In addition, in most applications we deal with samples of a random process over a limited time interval. The following example demonstrates the procedure to obtain the output of a linear system using frequency domain techniques. Consider Computer Example 5.4. As mentioned in the example, we have a total of 128 data observations, sampled every 0.004 s. To the data for this particular example we add 60 Hz noise. A plot of the data is shown in Figure 6.9.

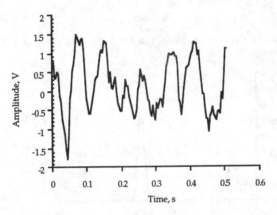

Figure 6.9 A plot of the data from Computer Example 5.4.

A plot of the same data after addition of 60 Hz noise is shown in Figure 6.10.

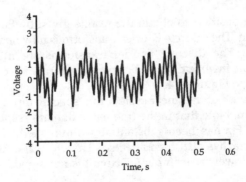

Figure 6.10 A plot of the data from Example 5.4 after adding 60 Hz noise.

The estimate of the spectral density (positive half only) prior to adding 60 Hz is shown in Figure 6.11. Following the addition of 60 Hz noise, the estimate of the spectral density was calculated and is shown in Figure 6.12. Note the peak at 60 Hz. Now let us consider the response of the *RC* circuit shown in Figure 6.8 to this input signal. For an *RC* constant of .032 the transfer function of the linear system depicted in Figure 6.8 is shown in Figure 6.13.

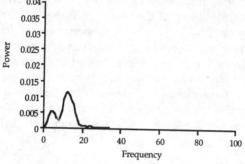

Figure 6.11 A plot of the spectral density of the data from Computer Example 5.4 prior to adding of 60 Hz noise.

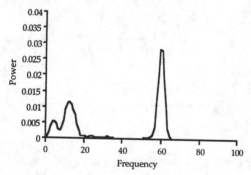

Figure 6.12 A plot of the spectral density of the data from Computer Example 5.4 after adding of 60 Hz noise.

We are now in a position to obtain the results given by Equation (6.1) in the frequency domain. The discrete Fourier transform is now obtained for both sets of discrete data. The discrete Fourier transforms are multiplied by each other, and then the inverse discrete Fourier transforms are computed, yielding the result shown in Figure 6.14.

Thus the reponse of the linear system to a specific input is obtained in the frequency domain. Note that in the frequency domain response, even though the power at 60 Hz has been substantially diminished, the power spectral density of the data has also been affected. Therefore the *RC* circuit acts as a low-pass filter. However, a simple *RC* circuit is not a good low-pass filter.

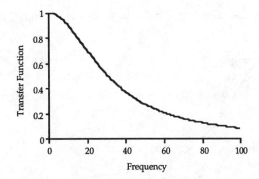

Figure 6.13 A plot of the transfer function of the linear system depicted in Figure 6.8 for RC = .032.

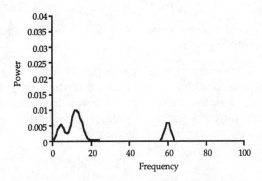

Figure 6.14 A plot of the spectral density of the data from Computer Example 5.4 after passing (convolving) the data through the linear filter.

A relatively simple *maximally flat transfer function* filter or *Butterworth* filter may be specified in terms of its transfer function as

$$H(\omega) = \frac{\omega_c}{\omega_c + j\omega} \tag{6.50}$$

where ω_c is the cutoff frequency or half-power frequency and

$$|H(\omega)|^2 = \frac{1}{1 + (\omega/\omega_c)^2} \tag{6.51}$$

The well-known family of Butterworth filters may be created by changing the value of n in the following equation:

$$|H(\omega)|^2 = \frac{1}{1 + (\omega/\omega_c)^{2n}} \tag{6.52}$$

A family of transfer functions for $\omega_c = 5$ and values of n ranging from 1 to 5 is shown in Figure 6.15.

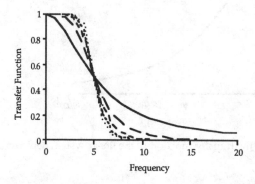

Figure 6.15 A plot of the transfer function of the Butterworth filter for cutoff frequency = 5.0 and values of n ranging from 1 to 5.

We now repeat the same exercise, except that the RC filter is replaced by a Butterworth filter of order 3 with $f_c = 30$. Figures 6.16 through 6.18 show, in order, the estimate of the spectral density of the input, the transfer function of the Butterworth filter (order = 3) with cutoff frequency equal to 30.0 Hz, and the estimate of the spectral density of the output.

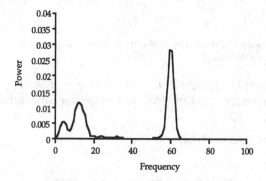

Figure 6.16 A plot of the spectral density of the data from Computer Example 5.4 after addition of 60 Hz noise.

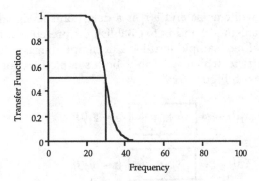

Figure 6.17 Plot of the transfer function of the filter for cutoff frequency = 30.0 and order = 6.

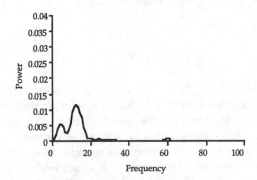

Figure 6.18 A plot of the spectral density of the data from Computer Example 5.4 after the data have been filtered with the Butterworth filter.

We can see from the results of Figure 6.18 that the 60-Hz noise is virtually eliminated and the original signal is not perturbed.

6.4 MATCHED FILTERS

We explore now the concept of a matched filter since it is a simple yet powerful example of linear system concepts. Suppose we have two sample functions of a random process. The first sample function is composed of pure random white noise. The second sample function is composed of pure random white noise with a deterministic signal of known characteristics added to it. These two sample functions may be represented as follows:

$$\text{Sample function 1} \quad x_1(t) = n(t)$$

$$\text{Sample function 2} \quad x_2(t) = n(t) + v(t)$$

(6.53)

where $n(t)$ is random white noise and $v(t)$ is a deterministic signal of known characteristics. The problem at hand is to establish the presence or absence of the waveform $v(t)$ in either sample function 1 or sample function 2. Consider now a linear system $h(t)$ to which we apply both sample functions. We can illustrate this process with Figure 6.19.

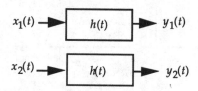

Figure 6.19 Two sample functions are passed through the same linear filter $h(t)$ where $y_1(t)$ and $y_2(t)$ are the respective outputs of the linear operations.

Choosing $h(t)$ such that

$$h(t) = kv(t_0 - t) \tag{6.54}$$

provides a very effective filter. This type of filter is called the *matched filter*. The matched filter is an optimum linear system such that the system response is matched to the input signal. The matched filter of a signal $v(t)$ of duration t_0 is characterized by the system whose impulse response is $v(t_0 - t)$. The matched filter output at $t = t_0$ maximizes the signal-to-noise power ratio. This filter will help one determine the presence or absence of the signal $v(t)$. The relative effectiveness of this filter depends primarily on the strength (or variance) of the noise. We will show examples of this filter for increasing amounts of noise, but the complete study of this type of filter is beyond the scope of this textbook.

Example 6.2 Consider the case where

$$x_1(t) = \text{white noise}$$
$$x_2(t) = \text{white noise} + \text{time-limited triangular waveform} \tag{6.55}$$

Figures 6.20, 6.21 and 6.22 are illustrations of $x_1(t)$, $v(t)$ and $x_2(t)$ respectively.

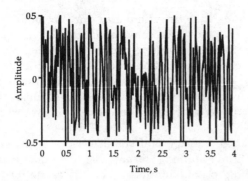

Figure 6.20 Illustration of $x_1(t)$.

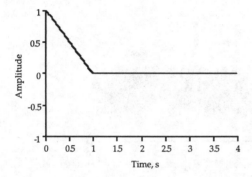

Figure 6.21 Illustration of time-limited triangular waveform $v(t)$.

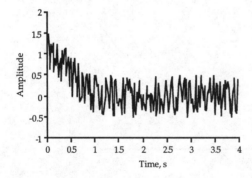

Figure 6.22 Illustration of $x_2(t)$.

For this particular case the amplitude (equal to 1) of the additive noise is such that it is easy to visually identify the triangular waveform within $x_2(t)$. The problem at hand is to decide, using a linear system $h(t)$, whether the waveform is present or absent. It may seem at first that this is a trivial problem, since it is obvious that the triangular waveform in question is present in $x_2(t)$ and not in $x_1(t)$. However, consider Figure 6.23.

Our task now perhaps is becoming more daunting. Although we can still visually inspect the waveform shown and (for the present level of noise) determine the location of $v(t)$, that is, the known triangular waveform, the added amount of time (now we have 20 s as opposed to 10) makes the identification task more difficult. We will return to the case illustrated by Figure 6.22.

Since we know a-priori the shape and characteristics of the signal, i.e., deterministic, triangular, time-limited, we are able to immediately define the shape of $h(t)$. According to Equation (6.54), we can define the shape for $h(t)$. Figure 6.24 represents the shape of $h(t)$.

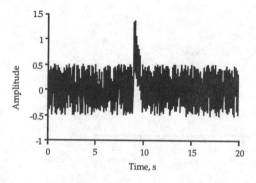

Figure 6.23 For this particular case, the waveform $v(t)$ has been added at an arbitrary place within a 20-s waveform.

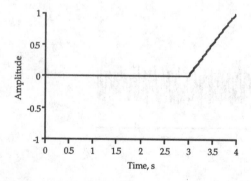

Figure 6.24 Illustration of $h(t)$ to be used for this case.

We are now able to proceed with the frequency domain technique to compute the complete convolution between $x_2(t)$ and $h(t)$. The discrete Fourier transforms of each waveform and their product are obtained. The inverse discrete Fourier transform of this product will yield the complete convolution of the two functions of time. Figure 6.25 shows the result of the convolution process using the frequency domain technique. Both functions of time need to be augmented with zeros in order to avoid the circular problems outlined earlier. We can see that at $t = 4$ (or t_0) the output of the convolution indicates a peak value, typical of a matched filter output.

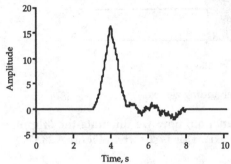

Figure 6.25 Results obtained by convolving $x_2(t)$ with $h(t)$.

The results can also be interpreted as the results of a convolution process. It is easier to begin with a simple example like the one used here. By *simple* it is meant that we know that the signal $v(t)$ is located precisely at $t = 0$. The location of the peak shown in Figure 6.25 is related to the actual position of the signal $v(t)$ within the noise sample. This simple exercise will help us understand other, more difficult cases to be examined next.

From linear system theory we know that in order to convolve one time signal with another, we need to "flip" one of them around the y axis. To better comprehend this process, we illustrate $h(t)$ and $x_2(t)$ after zeros have been added, prior to computing their discrete Fourier transforms. These functions are shown in Figures 6.26 and 6.27.

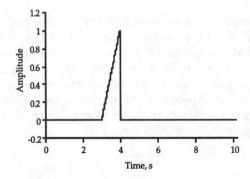

Figure 6.26 Function $h(t)$ after adding zeros.

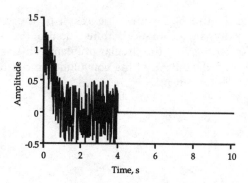

Figure 6.27 Function $x_2(t)$ after adding zeroes.

Continuing with the explanation, now we illustrate the "flipping" of one of these functions about the y axis. This is shown in Figure 6.28.

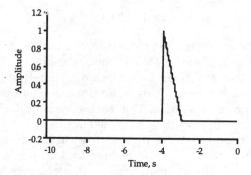

Figure 6.28 Function $h(t)$ after it has been "flipped," resulting in $h(-t)$.

The convolution process is one of cross-multiplying and shifting. The first cross-multiplication occurs when the functions first overlap. This occurs when $x_2(t)$ and $h(-t)$ are both at $t = 0$. After this result, $h(-t)$ is shifted by Δt. When the high point in $h(-t)$ at $t = -3.98$ s matches $x_2(t)$ at $t = 0$, the maximum value results. This is illustrated in Figure 6.29 and corresponds to the maximum shown in Figure 6.25. Therefore, the output of the convolution not only tells us that the signal is present, it also provides information (for the amount of noise shown) as to the location of the signal.

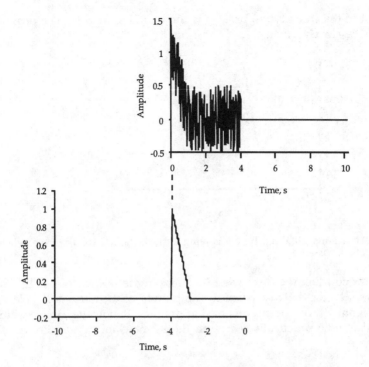

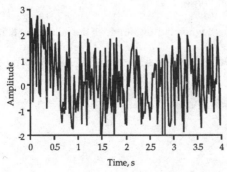

Figure 6.29 High point in $h(-t)$ at t = -3.98 s matches $x_2(t)$ at $t = 0$

Example 6.3 This example is really a continuation of Example 6.2, as we now consider the cases where the amplitude of the noise is 4 and 10. Figure 6.30 shows $x_2(t)$ for the first case when the amplitude of the noise is 4. We know that the signal $v(t)$ has been immersed at point $t = 0$; however, the position of the signal is becoming less visible with the increase in the amplitude of the noise.

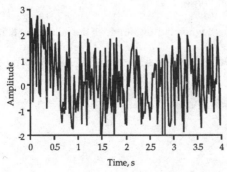

Figure 6.30 This figure shows $x_2(t)$ when the amplitude of the noise equals 4.

The result of the convolution process is shown in Figure 6.31. The position of the signal is clearly visible.

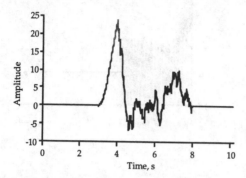

Figure 6.31 Results obtained by convolving $x_2(t)$ with $h(t)$ for the case when the amplitude of the noise is 4.

We now consider the cases where the amplitude of the noise is 10. Figure 6.32 shows $x_2(t)$ for the case when the amplitude of the noise is 10. We know that the signal $v(t)$ has been immersed at point $t = 0$; however, the position of the signal is now practically lost with the increase in the amplitude of the noise.

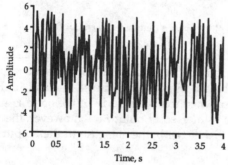

Figure 6.32 This figure shows x₂(t) when the amplitude of the noise equals 10.

The result of the convolution process is shown in Figure 6.33. The position of the signal is still visible.

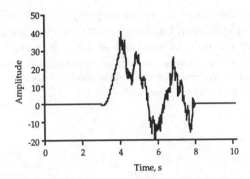

Figure 6.33 Results obtained by convolving $x_2(t)$ with $h(t)$ for the case when the amplitude of the noise is 10.

Suppose that we select the wrong waveform to convolve with. The transfer function of the filter, namely $h(t)$, remains the same, but the input signal unknown to us is $x_1(t)$ as opposed to $x_2(t)$. This is indeed part of the processing with a matched filter that we need to consider. It is obvious that at times we are going to try to find the presence or absence of a waveform where there is none present.

Example 6.4 Consider now the case when the input function to the filter is pure noise. We proceed as before, except that now we use $x_1(t)$ as the input to the matched filter. In this case we have selected the amplitude of the noise to be equal to 1. Similar results could be obtained for cases where the amplitude of the noise is larger yet. Figure 6.34 shows the input function to the filter, or $x_1(t)$. Comparing this figure to Figure 6.22, we see that it is impossible to tell whether a signal is present. One of the options could be that the amplitude of the noise is so large that it has obscured the signal.

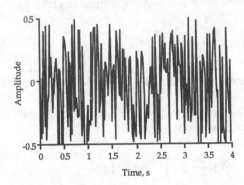

Figure 6.34 Input to the matched filter composed of pure noise.

The result of the convolution process is shown in Figure 6.35. To make the comparison clearer, the convolution has been plotted to the same scale as Figure 6.25. Comparing this figure to Figure 6.25 and to Figures 6.31 and 6.33, we would have to conclude that there is no signal present in this segment of data. Thus we can see that a matched filter can be used for detecting a signal $v(t)$ when it is corrupted by additive noise.

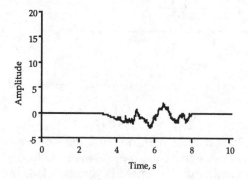

Figure 6.35 Results obtained by convolving $x_1(t)$ with $h(t)$ for the case when the amplitude of the noise is 1.

6.5 CHAPTER SUMMARY

A linear system may be represented by this figure:

$$x(t) \longrightarrow \boxed{h(t)} \longrightarrow y(t)$$

and the relationship between the input and the output is given by

$$y(t) = x(t) * h(t) = \int_{-\infty}^{\infty} x(\tau)h(t-\tau)\,d\tau$$

Two of the most fundamental properties of a linear system are that it be *time-invariant* and *causal*.

When the input to a linear system is a random process, the output is also random. The mean value of the output is given by

$$E[Y(t)] = \mu_X \int_{-\infty}^{\infty} h(t-\tau)\,d\tau$$

where the last element of this equation, or the integral, is called the *dc gain* of the linear system.

The output autocorrelation function is given by

$$R_{YY}(\tau) = \int\limits_{-\infty}^{\infty}\!\!\int h(\alpha)h(\beta)R_{XX}(\tau+\alpha-\beta)\,d\alpha\,d\beta$$

The cross-correlation between the input and the output is given by the following equation:

$$R_{XY}(\tau) = \int_{-\infty}^{\infty} h(\alpha)R_{XX}(\tau-\alpha)\,d\alpha$$

It may be easier to interpret these last two equations from a frequency point of view. The spectral density of the output of the system is given by

$$F\{R_{YY}(\tau)\} = S_{YY}(f) = |H(f)|^2 S_{XX}(f)$$

Likewise, the cross-spectral density between the input and the output is given by

$$F\{R_{XY}(\tau)\} = S_{XY}(f) = H(f)S_{XX}(f)$$

One of the simplest yet most powerful linear filters is known as a *matched filter*. A matched filter helps determine the presence or absence of a signal when the test sample we are filtering has additive noise. The relative effectiveness of this filter depends primarily on the variance of the noise. Choosing $h(t)$ such that

$$h(t) = kv(t_0 - t)$$

provides a very effective filter.

6.6 PROBLEMS

1. An ergodic random process having an auto-correlation function of the form

$$R_X(\tau) = 10\delta(\tau) + 25$$

is applied to the input of a linear system having an impulse response of

the form

$$h(t) = \begin{cases} 4(1-t) & 0 \le t \le 1 \\ 0 & \text{elsewhere} \end{cases}$$

Find the mean value of the process at the output of the system.

2. A random process with autocorrelation function

$$R_X(\tau) = 36\delta(\tau) + 64$$

is applied to a system whose impulse response is

$$h(t) = \left(10e^{-2t} - 5e^{-t}\right)u(t)$$

Find the mean value of the output.

3. The input to the circuit shown below is a white noise process with spectral density A. What is the output autocorrelation function?

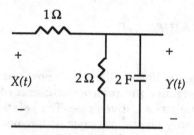

4. White noise is applied to the input of a linear system whose impulse response is shown below. Determine and sketch the autocorrelation function of the output.

5. Determine the cross-correlation function $R_{XY}(\tau)$ for a system having white noise with spectral density 1 V^2/Hz and impulse response given by

$$h(t) = u(t) - u(t-1)$$

6. Band-limited white noise having a spectral density of

$$S_X(\omega) = \begin{cases} 10 & |\omega| \le 10\pi \\ 0 & |\omega| > 10\pi \end{cases}$$

is passed through a system having a transfer function of

$$|H(\omega)|^2 = \begin{cases} 1 - \dfrac{|\omega|}{20\pi} & |\omega| \le 20\pi \\ 0 & |\omega| > 20\pi \end{cases}$$

What is the variance of the system output?

7. Consider the system given below. The input to the system can be considered to be white noise having a spectral density of 8 V^2/Hz. If the system has an impulse response given by

$$h(t) = \begin{cases} e^{-2t^2} & 0 < t < \infty \\ 0 & \text{elsewhere} \end{cases}$$

$$X(t) \longrightarrow \boxed{h(t)} \longrightarrow Y(t)$$

determine the output spectral density of the system, using frequency domain techniques.

8. An ergodic random process has a spectral density of the form

$$S_X(\omega) = 98\pi\delta(\omega) + 37$$

It is applied to a system whose transfer function is

$$H(s) = \frac{4}{s+28}$$

Find the magnitude of the mean value of the system output.

9. In the system shown below, the input signals are

$$X_1(t) = 4 + 10\cos(62t + \theta)$$

and

$$X_2(t) = 6 + 10\cos(6.2t + \phi)$$

where the two phase angles are *both* random variables that are uniformly distributed between 0 and 2π, and for which

$$f(\theta, \phi) = f(\theta)f(\phi)$$

Find the output of the crosscorrelator $R_{Y_1Y_2}(\tau)$.

(Note: A *numerical* value is desired, *not* a function of R and C.)

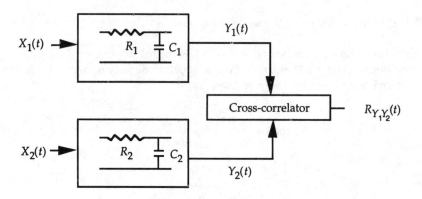

10. In the network shown below, $X(t)$ is the input and $Y(t)$ is the output.

(a) If $X(t)$ is from a stationary random process with a mean value of 10, what is the mean value of $Y(t)$?

(b) If the input is white noise with a spectral density of 5 V^2/Hz, what is the variance of the output?

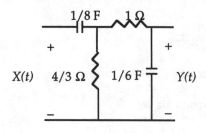

11. Find the autocorrelation function of the output of the circuit in Problem 10 if the input is white noise with a spectral density of 5 V^2/Hz.

12. Determine the cross-correlation function $R_{XY}(\tau)$ for a system having white noise (of spectral density 8 V^2/Hz) as input and impulse response given by

$$h(t) = \begin{cases} 0.5(1-8t^3) & 0 < t < 0.5 \\ 0 & \text{elsewhere} \end{cases}$$

13. Consider the system given below. The input to this system can be considered to be white noise having spectral density 8 V^2/Hz.

$$X(t) \longrightarrow \boxed{h(t)} \longrightarrow Y(t)$$

If the system has an impulse response given by

$$h(t) = \begin{cases} e^{-2t^2} & 0 < t < \infty \\ 0 & \text{elsewhere} \end{cases}$$

find the autocorrelation function of the output. Use strictly time domain techniques.

14. Solve Problem 13, using strictly frequency domain techniques.

15. White noise is applied to the input of a linear system whose impulse response is shown below. Determine and sketch the autocorrelation function of the output.

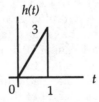

16. In the system shown below, X(t) is from a white noise process having a spectral density of 5 V^2/Hz. Find the cross-correlation function $R_{XY}(\tau)$ for all values of τ.

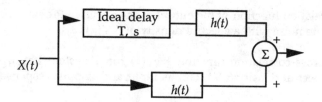

17. Assume that you have a time-limited triangular waveform as shown.

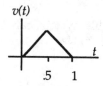

This waveform is added to white noise which is 5 s long. The waveform is immersed (added) at $t = 1$ s of the noise waveforms. Using MATHCAD or MATLAB, show the output of a matched filter (i.e., the convolution process) designed to find whether the triangular waveform is present or not.

18. Using the same filter that was designed for Problem 17, now obtain the output convolution for the case when the input is pure noise. Compare to the result obtain for Problem 17.

19. Repeat Problem 17, but now increase the amount of the input noise. Determine the point where it is impossible to tell whether the signal is present or not.

20. A single complex of an electrocardiographic signal is defined by its major peaks as the P, QRS, T complex. Using the Internet or a book of Physiology, obtain a representation of such a signal. These should be numbers representing the electrocardiographic complex. To define a heart rate detector, assume that you have a total of 5 of these complexes representing an average heart rate of 70 beats per minute. The heart rate (beats per minute) is defined as the time difference between adjacent R waves. Define a linear filter (matched filter) that will detect the QRS complex and will produce as an output the average heart rate of the 5 complexes.

6.7 COMPUTER EXAMPLES

MATHCAD Computer Example 6.1
LINEAR SYSTEMS AND CONVOLUTION

This example begins in a manner very similar to Computer Example 5.4. Read the data file.

$i := 0..127$

$M := 128$

$deltat = 0.004$

$y_i := \cos(2 \cdot \pi \cdot 60 \cdot i \cdot deltat)$

We now add 60-Hz noise to the data.

$x_i := READ(\text{tonecz}) + y_i$

Remove the dc components of the data.

$$a := \frac{\displaystyle\sum_i x_i}{M}$$

$x_i := x_i -$

$a = 0.16$

Plot the data:

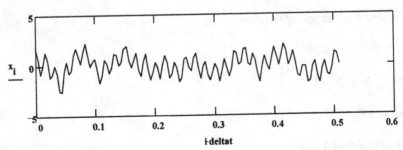

As we can appreciate, the data are heavily contaminated with 60-Hz noise.

Let's proceed as before and compute its spectral density to verify the presence of 60 Hz. We need to extend the array with zeros. This is necessary since we are going to compute the complete ACF of the data.

$j := 0, 1 .. 255$

$x_{M+i} := 0$

$length(x) = 256$

Compute now the FFT of the data:

$f := CFFT(x) \cdot length(x)$

$h_j := (|f_j|)$

Compute the inverse FFT.

$z := \dfrac{ICFFT(h)}{length(x)}$

The result is the complete autocorrelation function of the original data.

$\dfrac{z_0}{256} = 0.4$

$z_j := \dfrac{z_j}{256}$

$N \equiv 256$

Generate a Hamming window:

$k := 0 .. (N) \cdot (0.25)$

$kk := N \cdot 0.25 - 1$

$kk = 63$

$jj := 1 .. (N) \cdot (0.25)$

$$w_k := 0.54 + \left(0.46 \cdot \cos\left(\pi \cdot \frac{k}{kk}\right)\right)$$

$kkk := 0..255$

$y_{kkk} := 0$

$y_k := w$

$y_{256-jj} := w_j$

We now multiply the estimate of the autocorrelation function by the time window:

$zz_{kkk} := 0$

$zz_{kkk} := z_{kkk} \cdot y_{kk}$

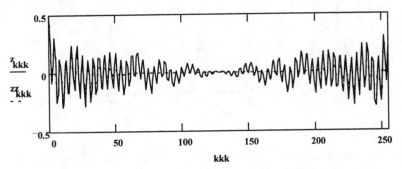

Consider the above plot of the autocorrelation function. There is a frequency present throughout the autocorrelation function. Can you tell which frequency this is?

We now compute the FFT and thus an estimate of the spectral density of the data.

$g := CFFT(zz) \cdot length(x)$

We now need to compute the absolute value of g and multiply by Δt. We are only going to take the positive half of the spectrum since the negative half is identical.

$$MM := \frac{length(g)}{2} + 1$$

$$MM = 129$$

$$s := 0 .. MM - 1$$

$$ff_s := \left| g_s \right| \cdot delta$$

$$length(g) = 256$$

$$deltaf := \frac{1}{N \cdot delta}$$

The estimate of the spectral density is plotted in a regular linear scale.

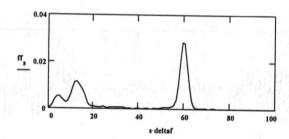

The 60-Hz noise added is plainly visible. You may want to consider why the 60-Hz noise is not a delta function at 60 Hz.

We now describe the transfer function of a single-stage *RC* circuit as a low-pass filter.

$$a_s := \frac{1}{1 + s^2 \cdot 0.001}$$

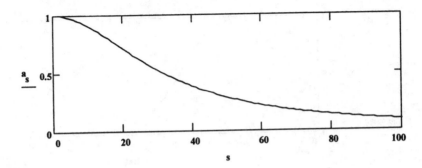

$$g_s := ff_s \cdot a_s$$

We perform the convolution in the frequency domain by multiplying the two frequency functions.

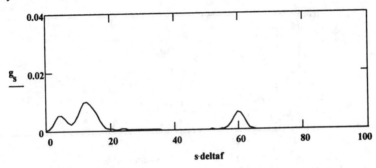

As we can see, this is not a good filter. Although the 60-Hz interference has diminished, the power in the part of the spectrum which corresponds to the signal has also diminished. We now use a Butterworth filter of order 4 as described in the text.

$$a_s := \frac{1}{1 + \left(\dfrac{s}{40}\right)^8}$$

And a plot of the transfer function shows how much sharper this filter is than the single-stage *RC* filter.

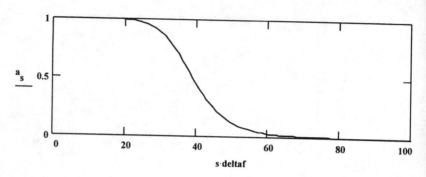

The output of the filter is now computed:

$$g_s := ff_s \cdot a_s$$

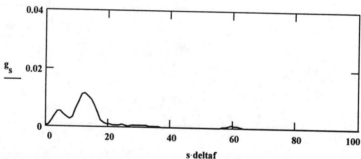

and a much better result is obtained.

MATHCAD Computer Example 6.2
MATCHED FILTERS

Let us generate the first 200 values of random noise. For this particular case, we will let the amplitude of the noise be equal to A, which will equal 1.

$$A := 1$$

$$B := \frac{A}{2}$$

$$i := 0 .. 199$$

$$x_i := (A \cdot rnd(1.0)) - B$$

By subtracting $A/2$, the mean value of the set of numbers will approximate 0.
Plot the data:

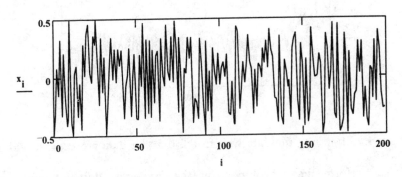

We now generate the "signal" we want to test. This is a triangular waveform similar to the one in the text.

$y_i := 0.$

$j := 50, 51 .. 100$

$y_{j-50} := 2.0 - 0.02 \cdot$

Plot the data:

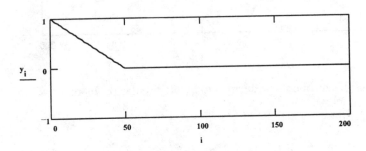

Now we add the noise and the signal. Plot the result:

$z_i := x_i + y_i$

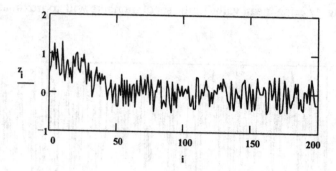

By simply observing the data we can tell that the data are present within the noise.

To perform the convolution (in the frequency domain), we need to convolve the signal plus noise with the reverse of the signal.

$x_i := 0$

$x_i := y_{199-i}$

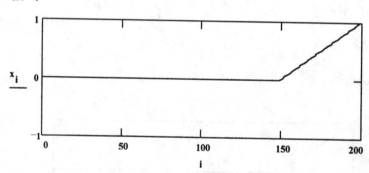

We extend both sets of data with zeros to avoid circular problems.

$M := 200$

$j := 1 .. 311$

$x_{M+j} := 0$

$\text{length}(x) = 512$

Note that this is a power of 2.

$z_{M+j} := 0$

$k := 0 .. 511$

$length(z) = 512$

Compute now the FFT of each set of data.

$f := CFFT(x) \cdot length(x)$

$g := CFFT(z) \cdot length(z)$

$N := 512$

Compute the product.

$h_k := f_k \cdot g$

$kk := 0 .. 511$

$$a := \frac{ICFFT(h)}{N} + 1$$

$length(a) = 512$

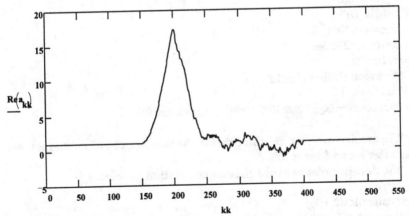

Why do we plot the real value of a? Is this necessary?

As we can see, the presence of the signal is clearly visible. Following the explanation in the text, explain why there is a peak at point 200.

MATLAB Computer Example 6.1
LINEAR SYSTEMS AND CONVOLUTION

```
%=============================================
% Input parameter
                signal='tonecz.dat';
                i=0:1:127;
                M=128;
                deltat=0.004;
                t=0:deltat:(M-1)*deltat;
                y_i=cos(2*pi*60*i*deltat);
%=============================================
eval(['load ',signal]);
limit=length(signal)-4;
signal=signal(1:limit);
x_i=eval(signal);
lengthx=length(x_i);
if (lengthx ~=128)
str=str2mat(...
    '           The length of the data must be 128 !!!');
clear signal;
disp(figNumber,str);
end
x_i=x_i+y_i';
xmean=mean(x_i);
x_i=x_i-xmean;
xstd=std(x_i);
lengthx=length(x_i);
r0=sum(x_i.^2)/256;
str=str2mat(...
    'Autocorrelation check:',...
    '',...
    ['>> r0=sum(x_i.^2)/256;','    r(0)= ',num2str(r0)],...
    '',...
    'We are dividing by 256 since this is the number of points that we will
have after we add zeros to',...
    'the data in order to avoid circular correlation problems.');
disp(str);
clg;s1=subplot(2,1,1);
plot(t,x_i,'r');axisn(t);
xlabel('i*deltat','FontSize',[8]);
ylabel('x_i','FontSize',[8]);
title(num2str(signal),'FontSize',[8]);
set(s1,'FontSize',[8]);
s2=subplot(2,1,2);
axis('off')
```

```
text(0.3,0.8,['Mean signal: ',num2str(xmean)] ,'FontSize',[8]);
text(0.3,0.6,['Standard deviation: ',num2str(xstd)] ,'FontSize',[8]);
text(0.3,0.4,['Length of data: ',num2str(lengthx)] ,'FontSize',[8]);
text(0.3,0.2,['Deltat: ',num2str(deltat)] ,'FontSize',[8]);
set(s2,'FontSize',[8])
```

Autocorrelation check:

>> r0=sum(x_i.^2)/256; r(0)= 0.4976

We are dividing by 256 since this is the number of points that we will have after we add zeros to the data in order to avoid circular correlation problems.

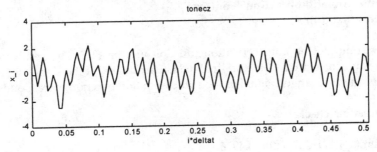

Mean signal: 0.1596

Standard deviation: 1.001

Length of data: 128

Deltat: 0.004

```
str=str2mat(...
    'We now compute the autocorrelation function via the FFT.',...
    ' ',...
    'The result is the complete autocorrelation of the original data. Check to
see if the signal obtained',...
    'earlier for the zero lagged signal of the autocorrelation function
corresponds.');
disp(str);
x_izeros=zeros(size(0:1:255));
x_izeros(1:M)=x_i;
f=fft(x_izeros,length(x_izeros)).*length(x_izeros);
h_j=f.*conj(f);
z=ifft(h_j)./(length(x_izeros)^2);
```

```
r0=real(z(1))/length(x_izeros);
z_j=real(z)/length(x_izeros);
str=str2mat(...
    ' ',...
    'Autocorrelation check:',...
    ' ',...
    ['>> r0=sum(x_i.^2)/256;','   r(0)= ',num2str(r0)]);
disp(str);
clg;s1=subplot(1,1,1);
plot(0:1:255,z_j,'r');axisn(0:1:255);
xlabel('j','FontSize',[8]);
ylabel('z_j','FontSize',[8]);
title('Autocorrelation function','FontSize',[8])
set(s1,'FontSize',[8])
```

We now compute the autocorrelation function via the FFT. The result is the complete autocorrelation of the original data. Check to see if the signal obtained earlier for the zero lagged signal of the autocorrelation function corresponds.

Autocorrelation check:

>> r0=sum(x_i.^2)/256; r(0)= 0.4976

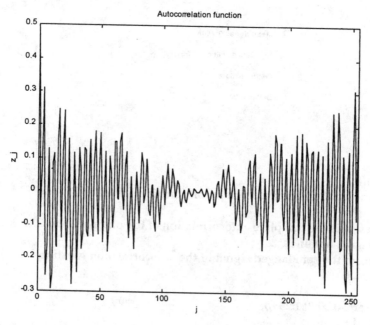

```
str=str2mat(...
```

6.7 MATLAB Computer Examples

```
    'We are only going to accept 25% of the lagged signals of the
autocorrelation function. This is done',...
    'using a Hamming window.',...
    ' ',...
    '>> hamming(0.5*length(x_izeros))',...
    ' ',...
    'We multiply the estimate of the ACF by the time Hamming window.');
disp(str);
vec=1:1:256;
w=hamming(0.5*length(x_izeros));
w_zeros=zeros(size(vec));
w_zeros(1:64)=w(65:128);
w_zeros(193:256)=w(1:64);
y_kkk=w_zeros;
zz_kkk=z_j.*y_kkk;
clg;s1=subplot(2,1,1);
plot(0:1:255,y_kkk);axisn(0:1:255);
xlabel('kkk','FontSize',[8]);
ylabel('y _kkk','FontSize',[8]);
title('Hamming window','FontSize',[8]);
set(s1,'FontSize',[8]);
s2=subplot(2,1,2);
plot(0:1:255,zz_kkk,'r', 0:1:255,z_j,':k');axisn(0:1:255);
xlabel('kkk','FontSize',[8]);
ylabel('z_j ...    zz_kkk _','FontSize',[8]);
title('Rxx: unwindowed(black)  -  windowed(red)','FontSize',[8]);
set(s2,'FontSize',[8])
```

We are only going to accept 25 percent of the lagged signals of the auto-correlation function. This is done using a Hammnig window.

```
>> hamming(0.5*length(x_izeros))
```

We multiply the estimate of the ACF by the time Hamming window.
```
str=str2mat(...
    'Now compute the FFT of the windowed autocorrelation function and
thus an estimate of the spectral',...
    'density of the data. We need to compute the absolute signal and
multiply by deltat. Note that earlier we',...
    'made deltat = 0.004. We are only going to take the positive half of the
spectrum. We now compute the',...
    'frequency spacing.',...
    ' ',...
    '>> deltaf=1/(N*deltat)');
disp(str);
g=fft(zz_kkk,length(zz_kkk)).*length(zz_kkk);
```

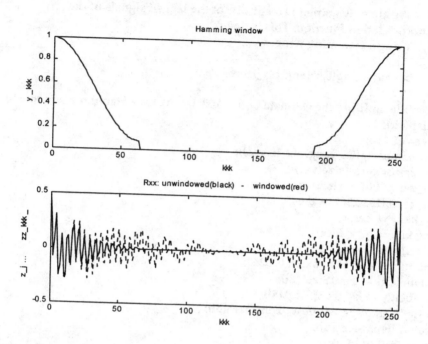

```
ff_s=abs(g)*deltat/256;
deltaf=1/(256*deltat);
check=sum(ff_s(1:128))*2*deltaf;
f=deltaf*(0:1:length(g)/2-1);
clg;s1=subplot(1,1,1);
plot(f(1:1:100),ff_s(100),'r'); axisn(f(1:1:100));
xlabel('s*deltaf','FontSize',[8]);
ylabel('ff_s','FontSize',[8]);
set(s1,'FontSize',[8])
str=str2mat(...
    '',...
    'Compute now the last check of the units.',...
    '',...
    '>> Check=sum(ff_s(1:128))*2*deltaf/N',...
    '',...
    ['    Check: ',num2str(check),'        It corresponds !!!']);
disp(str);
```

Now compute the FFT of the windowed autocorrelation function and thus an estimate of the spectral density of the data. We need to compute the absolute signal and multiply by deltat. Note that earlier we made deltat = 0.004. We are only going to take the positive half of the spectrum. We now compute the frequency spacing.

>> deltaf=1/(N*deltat)

Compute the last check of the units.

>> Check=sum(ff_s(1:128))*2*deltaf/N

Check: 0.4979   It corresponds !!!

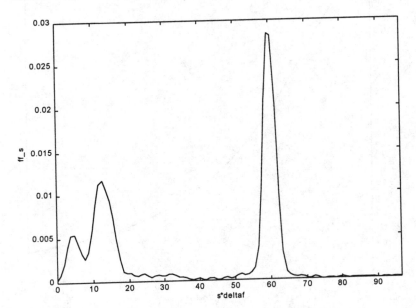

str=str2mat(...
    'We now describe the transfer function of a single stage RC circuit as a
lowpass filter',...
    ' ',...
    '>>a_s=1 / (1 + 0.001*s^2)',...
    ' ');
disp(str);
a_s=1*ones(size(f))./(1*ones(size(f))+0.001*f.^2);
clg;s1=subplot(1,1,1);
plot(f(1:1:100),a_s(1:100),'r'); axisn(f(1:1:100));
xlabel('s*deltaf','FontSize',[8]);
ylabel('a_s','FontSize',[8]);
set(s1,'FontSize',[8])

We now describe the transfer function of a single-stage *RC* circuit as a low-pass filter

```
>>a_s=1 / ( 1 + 0.001*s^2)
```

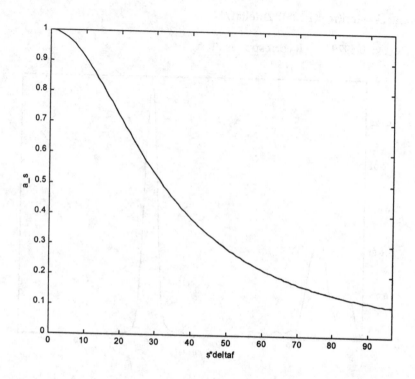

```
str=str2mat(...
    'We perform the convolution in the frequency domain by multiplying
the two frequency functions',...
    ' ',...
    '>>g_s=ff_s*a_s',...
    ' ');
disp(str);
g_s=ff_s(1:128).*a_s;
clg;s1=subplot(1,1,1);
plot(f(1:1:100),g_s(1:100),'r'); axisn(f(1:1:100));v=axis;v(3)=0;v(4)=0.04;axis(v)
xlabel('s*deltaf','FontSize',[8]);
ylabel('g_s','FontSize',[8]);
set(s1,'FontSize',[8])
```

We perform the convolution in the frequency domain by multiplying the two frequency functions.

```
>>g_s=ff_s*a_s
```

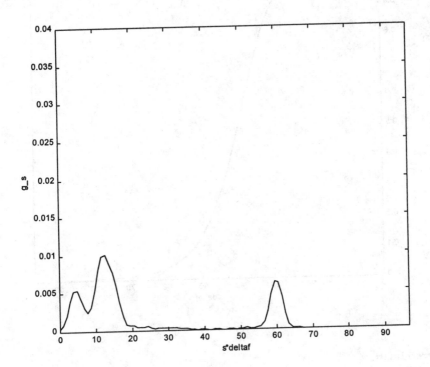

```
str=str2mat(...
      'We now use a Butterworth filter of order 4:',...
      ' ',...
      '>>a_s=1 / ( 1 + (s/40)^8)',...
      ' ');
disp(str);
a_s=1*ones(size(f))./(1*ones(size(f))+(f/40).^8);
clg;s1=subplot(1,1,1);
plot(f(1:1:100),a_s(1:100),'r'); axisn(f(1:1:100));
xlabel('s*deltaf','FontSize',[8]);
ylabel('a_s','FontSize',[8]);
set(s1,'FontSize',[8])
```

We now use a Butterworth filter of order 4:

```
>>a_s=1 / ( 1 + (s/40)^8)
```

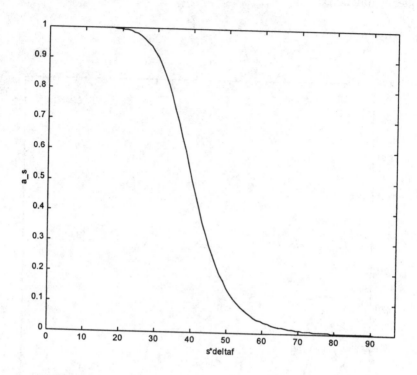

```
str=str2mat(...
    'The output of the filter is now computed:',...
    ' ',...
    '>>g_s=ff_s*a_s',...
    ' ');
disp(str);
g_s=ff_s(1:128).*a_s;
clg;s1=subplot(1,1,1);
plot(f(1:1:100),g_s(1:100),'r'); axisn(f(1:1:100));v=axis;v(3)=0;v(4)=0.04;axis(v)
xlabel('s*deltaf','FontSize',[8]);
ylabel('g_s','FontSize',[8]);
set(s1,'FontSize',[8])
```

The output of the filter is now computed:

```
>>g_s=ff_s*a_s
```

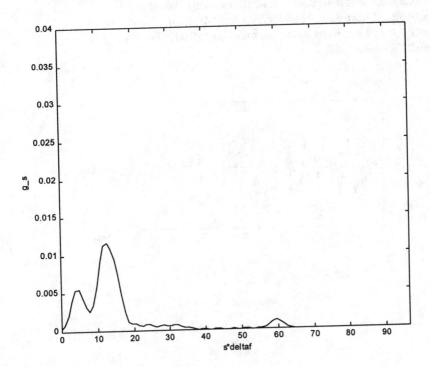

MATLAB Computer Example 6.2
MATCHED FILTERS

```
%=========================================
% Input parameter
                A=1;
                i=0:1:199;
                M=200;
%=========================================

x_i=A*randn(M,1);
x_i=x_i-mean(x_i); xmean=mean(x_i);
xstd=std(x_i);
lengthx=M;
clg;s1=subplot(2,1,1);
plot(i,x_i,'r');axisn(i);
xlabel('i','FontSize',[8]);
ylabel('x_i','FontSize',[8]);
set(s1,'FontSize',[8]);
s2=subplot(2,1,2);
axis('off')
```

```
text(0.3,0.8,['Mean signal: ',num2str(xmean)] ,'FontSize',[8]);
text(0.3,0.6,['Standard deviation: ',num2str(xstd)] ,'FontSize',[8]);
text(0.3,0.4,['Length of data: ',num2str(lengthx)] ,'FontSize',[8]);
set(s2,'FontSize',[8])
```

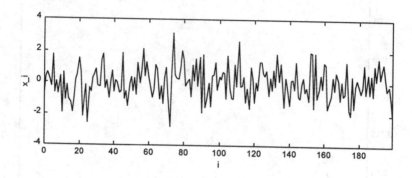

Mean signal: 3.331e-017

Standard deviation: 1.038

Length of data: 200

```
str=str2mat(...
    'We now generate the "signal" we want to test.');
disp(str);
j=50:1:100;
temp=1:149;temp=zeros(size(temp));
y_i=2-0.02*j;y_i=[y_i temp];
clg;s1=subplot(1,1,1);
plot(i,y_i,'r');axisn(i);v=axis;v(3)=-1;v(4)=1;axis(v);
xlabel('i','FontSize',[8]);
ylabel('y_i','FontSize',[8]);
set(s1,'FontSize',[8]);
```

We now generate the "signal" we want to test.

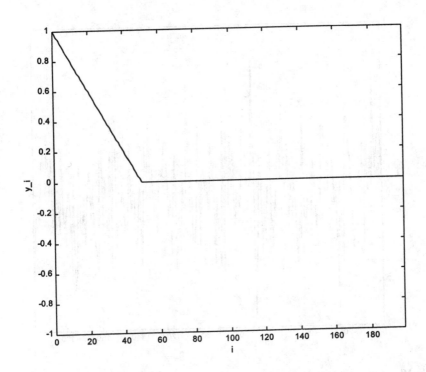

```
str=str2mat(...
     'Now we add the noise and the signal.');
disp(str);
z_i=x_i+y_i';
clg;s1=subplot(1,1,1);
plot(i,z_i,'r');axisn(i);
xlabel('i','FontSize',[8]);
ylabel('z_i','FontSize',[8]);
set(s1,'FontSize',[8]);
```
Now we add the noise and the signal.
```
str=str2mat(...
     'In order to perform the convolution (in the frequency domain), we need
to convolve the signal plus noise.',...
     'with the reverse of the signal');
disp(str);
x_i=fliplr(y_i);
clg;s1=subplot(1,1,1);
plot(i,x_i,'r');axisn(i); v=axis;v(3)=-1;v(4)=1;axis(v);
xlabel('i','FontSize',[8]);
```

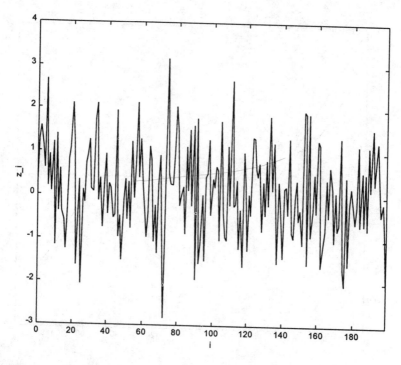

```
ylabel('x_i','FontSize',[8]);
set(s1,'FontSize',[8]);
```

In order to perform the convolution (in the frequency domain), we need to convolve the signal plus noise with the reverse of the signal

```
str=str2mat(...
    'We need to extend both sets of data with zero to avoid circular problems
and compute the FFT of each',...
    'set of data');
disp(str);
temp=1:312;temp=zeros(size(temp));
x=[x_i temp];
z=[z_i' temp];
f=fft(x,length(x)).*length(x);
g=fft(z,length(z)).*length(z);
h=f.*g;
z=ifft(h)./(length(h)^2);
clg;s1=subplot(1,1,1);
plot((1:512),real(z),'r');axisn(1:512);
xlabel('kk','FontSize',[8]);
ylabel('Re(a_kk)','FontSize',[8]);
set(s1,'FontSize',[8]);
```

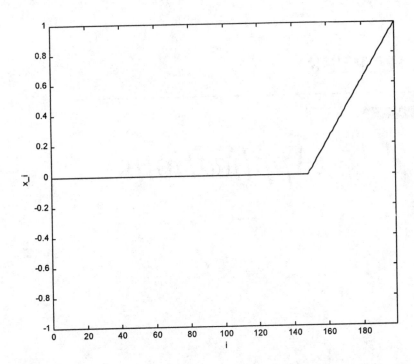

We need to extend both sets of data with zero to avoid circular problems and compute the FFT of each set of data

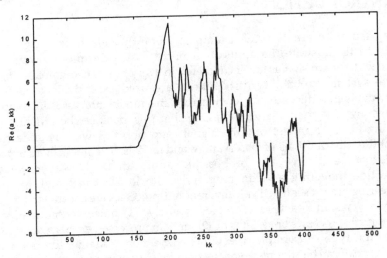

CHAPTER

7 *Applications*

7.0 INTRODUCTION

This chapter presents some examples from the electrical engineering discipline to illustrate the application of the basic principles introduced in this book. Random signals are encountered in every radar, communication, and control system. Probability distributions are used extensively to describe observations in semiconductor engineering and model processes in computer engineering. The basic material described in this book can be utilized as a building block to model random signals and noise as well as understand probability distributions of observations and models. The intent here is not to provide an extensive list of possible applications. However, several applications covering a wide range of topics in electrical and computer engineering are presented. Typically, random processes are encountered in two possible ways in practical examples, namely, (1) noise contamination of deterministic processes or systems or (2) random processes, events, and signals. In the first type of example, the effort is spent on modeling the noise such that either it can be suppressed or the deterministic processes can be detected in the presence of noise. In the case of random processes, events, and signals, the type of application may deal with characterizing the process itself. The following sections cover examples in the areas of radar systems, communication systems, computer systems, semiconductor engineering, and biomedical engineering.

7.1 RADAR SYSTEMS

Weather radar systems are very good examples of the application of random processes and signals. Radars are used to detect the presence, type, and range of objects from the receivers. This is done by transmitting an electromagnetic wave and then observing the signal scattered back to the radar by the objects (or targets) encountered by the transmitted electromagnetic wave. The distance to the target is obtained from the time delay of the back-scattered signal, and the type or size of target is obtained from the strength of the back-scattered signal. In a weather radar system the targets are composed of a large number of precipitation particles in the atmosphere such as raindrops, hailstones, and ice crystals. These particles have a random size distribution, are randomly located in space, and move around with a random distribution of velocities. Figure 7.1 shows a simplified block diagram of a weather radar. The stable local oscillator generates very stable sinusoidal signals, which are modulated and amplified by a high-power amplifier to generate high-power microwaves for transmission from the antenna. The transmit/receive switch connects the transmitter during a short transmit period and connects the receiver to the antenna during the rest of the time.

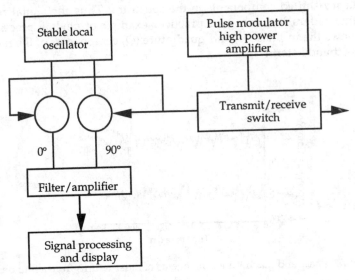

Figure 7.1 Simplified block diagram of a weather radar.

The signal returned from the target scatterers is passed through a pair of mixers to obtain the amplitude and phase of the received signal (coherent detection). The signal is subsequently processed by a signal processor, and the results are shown on a radar display. The signal back-scattered from a single precipitation particle, such as a raindrop, received by the radar is dependent on the size of the particle and the distance from the particle to the radar. The phase of the returned signal is controlled by the distance, and it changes by 360° for distances of each wavelength. Typical operating wavelengths of microwave radars are on the order of a few centimeters. The total signal received by the radar from weather targets is the sum of individual random phasors (or sinusoids) from a large number of particles. Therefore the in-phase (I) and quadrature (Q) components of the radar received signals are gaussian distributed (invoking the central limit theorem). The in-phase and quadrature components are also independent. The voltage signal of the radar returns at a time instant t, or $V_t = (I^2 + Q^2)^{1/2}$, is Rayleigh distributed, and the signal power $P_t = I^2 + Q^2$ has an exponential distribution.

With the passage of time the precipitation particles (scatterers) move and rearrange themselves in space, which causes the radar received signals to fluctuate in time. The actual received voltage at time instant $t + \Delta t$ will depend on the position of the scatterers at time $t + \Delta t$. For sufficiently small Δt the radar signals at t and $t + \Delta t$ will be correlated, and the correlation structure depends on the spectrum of the radar signal. The spectral nature of radar signals is determined completely by the velocity of the scatterers related by the Doppler effect. Faster-moving particles introduce higher frequency-shifted components in the spectrum. Thus the signal return from weather radars is a very good practical example of random processes. Figure 7.2 shows the in-phase (I) and quadrature (Q) components of the radar returns from a thunderstorm

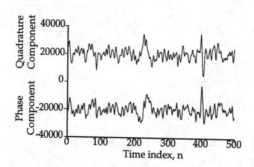

Figure 7.2 Phase and quadrature components of radar returns from a thunderstorm.

Figure 7.3 shows the time record of the signal power. Figure 7.4 shows the histogram of the *I* (Figure 7.4a), *Q* (Figure 7.4b), and *P* (Figute 7.4c) signals.

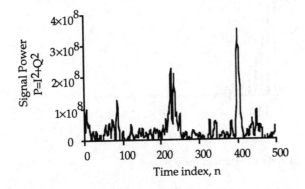

Figure 7.3 Time record of the signal power.

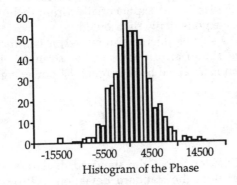

Figure 7.4a Normalized histogram of the phase component.

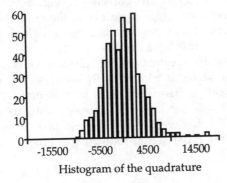

Figure 7.4b Normalized histogram of the quadrature component.

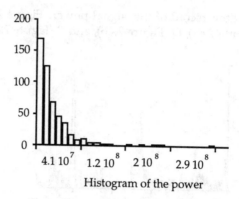

Histogram of the power

Figure 7.4c Normalized histogram of the power.

We can see from the histograms that I and Q are approximately gaussian distributed with zero mean and the same variance. The power signal constructed as $I^2 + Q^2$ is distributed exponentially (from theory). The histogram of P shows that it is exponentially distributed.

The radar power signals fluctuate; however, it is important to estimate the mean power to obtain an estimate of the strength of the target. The estimate of the mean is obtained by averaging the power signals:

$$\hat{P} = \sum_{i=1}^{n} P_i \tag{7.1}$$

We have seen in Chapter 5 that the standard deviation in the estimate of the mean is smaller than the standard deviation of P_i by a factor $1/(N_I)^{0.5}$ where N_I is the number of independent samples. In this case N_I will be less than n because the samples are correlated. Figure 7.5 shows the auto-correlation function of the voltage signals from the radar. We can see from the autocorrelation structure that the signals have a high degree of correlation, which decreases as the time lag increases.

Figure 7.6 shows the spectrum of the coherent voltage. The spectrum of the coherent voltage signal is plotted with velocities (in meters per second) on the X axis instead of frequency. This transformation is done using the Doppler effect ($f = 2v/\lambda$) to relate the frequency shifts to velocities. Thus we can see from Figure 7.6 that we can make an estimate of the mean velocity of the scatterers from the radar signal returns. This process essentially makes a Doppler radar.

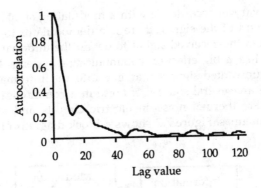

Figure 7.5 Autocorrelation of the coherent voltage signal.

Observing the spectrum shown in Figure 7.6 more carefully, we see that the minimum spectral level of about 55 dB (compared to the peak of 90 dB) corresponds to the noise floor of the radar system, which is white noise. The signal spectrum sits on the white noise pedestal. The area under the spectrum above the noise level indicates the signal power, and the area under the noise level indicates the noise power. Thus we can see that a weather radar system is a good application of most of the material discussed in this book.

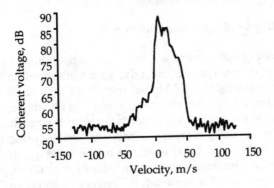

Figure 7.6 Spectrum of the coherent voltage.

7.2 COMMUNICATION SYSTEMS

In a communication system, signals containing information (baseband) are transmitted through a channel. These signals are received on the other end of the channel by a receiver. The signals at baseband frequencies are modulated so that the resulting signal is suitable for transmission over the channel. The modulation process enables a shift in the range of frequencies of the information-bearing signal. In the modulation process, the characteristics of a carrier signal vary in accordance with a modulating signal (or baseband). In the receiving end of the signal, there is a demodulator to retrieve the baseband signal from the received signal to obtain the information content in the signal. Noise has a big effect on communication systems. Noise essentially describes all unwanted signals that get into the information transmission process. There are several sources of noise in a communication system. Some such sources are thermal noise in electric circuits and atmospheric noise received by antennas. Figure 7.7 shows a block diagram of a communication system.

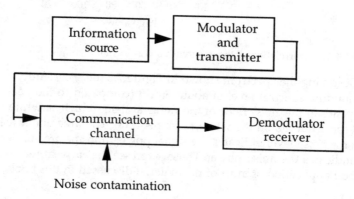

Figure 7.7 Block diagram of a communication system.

In the following subsections we will study the effect of noise in two widely used modulation systems, namely, amplitude modulation (AM) and frequency modulation (FM). In AM the amplitude of a sinusoidal wave is varied in accordance with the baseband signal. In FM the frequency of the sinusoidal carrier wave is varied in accordance with the baseband signal. The most commonly used model for noise is *additive white gaussian* noise. This implies that the noise signal $n(t)$ is added to the message signal $s(t)$ at the receiver. In addition, the power spectral density of the noise signal $n(t)$ is at a constant value $N_0/2$, and the noise signal density is gaussian. The signal at the input of the demodulator in a communication system can be expressed as

$$x(t) = s(t) + n(t) \tag{7.2}$$

where $s(t)$ is the modulated waveform and $n(t)$ is the noise signal. Typically all communication and radar receivers have a low-pass filter at the front end. The bandwidth B of this low-pass filter is determined based on the baseband signal bandwidth and the modulation technique.

The signal-to-noise ratio (SNR) is a useful measure of noise contamination in a signal. Typically the SNR is calculated at the input and output of a receiver, to assess the ability of the receiver to recover the information content in the signal in the presence of noise. A receiver *figure of merit* can be defined as the ratio of input to output SNRs.

Noise in Amplitude Modulation Systems

The most common amplitude modulation system is one that transmits the amplitude-modulated carrier without removal of either sideband frequencies or the carrier itself. This AM signal is received by using an envelope detector. The full AM signal can be represented by

$$s(t) = A_c \left[1 + kx(t) \right] \cos \omega_c t \qquad (7.3)$$

where $A_c \cos \omega_c t$ is the carrier wave, $x(t)$ is the baseband modulating signal, and k is the modulating constant, which determines the amount of modulation. The performance of the AM receiver is evaluated by computing the input and output SNRs. The signal power at the input of the AM receiver is composed of the power on the carrier component and the modulating signal. The average carrier power is given by

$$.5A_c^2$$

The average power in the modulation signal is given by

$$\frac{A_c^2 k^2 P_x}{2}$$

where P_x is the average power of $x(t)$. Therefore the total power of the modulating signal at the input of the receiver is

$$\frac{A_c^2 \left(1 + k^2 P_x \right)}{2}$$

The average power of noise is obtained from the noise density $N_0/2$ and the bandwidth B of the modulating signal as $N_0 B$.

The input SNR is given by

$$\mathrm{SNR}_i = \frac{A_c^2 \left(1 + k^2 P_x \right)}{2 N_0 B} \qquad (7.4)$$

The envelope detector used in simple AM receivers is a nonlinear device. We use an intuitive approach to obtain the output SNR. The envelope of the AM signal follows the peak of the sinusoid, oscillating with an angular frequency ω_c. The information content of the signal is essentially $A_c k x(t)$. Therefore the output signal power is

$$.5 A_c^2 k^2 P_x$$

When the noise contamination is not significant, only the component of the noise that is in phase with the baseband signal will be detected in an envelope detector, which is only half the noise power. Therefore, the output SNR can be obtained as

$$SNR_o = \frac{A_c^2 k^2 P_x}{2 N_0 B} \tag{7.5}$$

The ratio of output SNR to input SNR for AM systems can be written from Equations (7.4) and (7.5) as

$$\frac{SNR_o}{SNR_i} = \frac{k^2 P_x}{1 + k^2 P_x} \tag{7.6}$$

Thus we can see that the figure of merit of AM is always less than 1. However, slightly more complex AM systems that avoid transmission of the carrier can achieve a figure of merit of unity.

Noise in Frequency Modulation Systems

FM systems offer better noise performance than do AM systems. Figure 7.8 shows the block diagram of an FM system. In an FM system the frequency of the carrier wave is changed in accordance with the baseband signal.

The FM signal can be represented as

$$s(t) = A_c \cos\left[\omega_c t + k_f \int x(t)\, dt \right] \tag{7.7}$$

where $A_c \cos \omega_c t$ is the carrier signal, $x(t)$ is the baseband signal, and k_f is the modulation constant. We can see from Equation (7.7) that an FM signal waveform is not analytically simple and it is not straightforward to obtain the bandwidth of the FM signal.

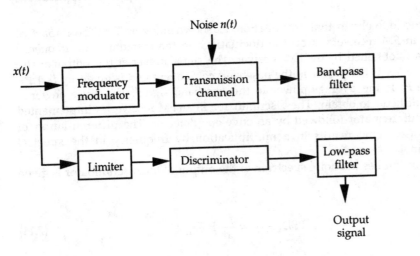

Figure 7.8 Block diagram of an FM system.

In the following we consider the simpler case of a monotone baseband signal i.e., the baseband signal is of the form $x(t) = A \cos \omega_m t$, to define the properties of an FM communication system. With a monotone baseband signal the FM signal can be written as

$$s(t) = A_c \cos\left(\omega_c t + \beta \sin 2\pi f_m t\right) \tag{7.8}$$

where $\beta = k_f A_m / f_m$ is called the *modulation index*. The quantity $k_f A_m = \Delta f$ is called the *frequency deviation*. The bandwidth of the FM signal can vary over a wide range depending on the value of β. When $\beta \ll 1$, the system is called *narrowband* FM, and when $\beta \gg 1$, the system is called *wideband* FM. The case of narrowband FM is very similar to that of AM. The bandwidth of wideband FM can be approximated by the formula called *Carlson's rule* where the bandwidth B is given by

$$B_f = 2\Delta f\left(1 + \frac{1}{\beta}\right) \tag{7.9}$$

The input signal-to-noise ratio is given by

$$SNR_i = \frac{A_c^2}{2BN_0} \tag{7.10}$$

where B is the bandwidth of the message signal. Since the information in an FM signal is in the instantaneous frequency of the signal, FM receivers are constructed to obtain that information, as shown in Figure 7.8. The impact of noise in FM receivers is on the fluctuations in the instantaneous frequency estimates obtained by the FM receiver. The signal power at the output of the FM receiver is the power in the baseband component and is given by $(k_f)^2 P_x$, where P_x is the average power in the baseband signal. The noise power is more difficult to obtain. The discriminator in an FM receiver is implemented by a differentiator followed by an envelope detector. The differentiation of the noise signal results in a multiplication by frequency in the spectral domain.

This implies the noise spectrum at the output of the discriminator is given by

$$\text{PSD}_{\text{OD}}(f) = \frac{f^2}{A_c^2} \text{PSD}_{\text{ID}}(f) \tag{7.11}$$

where PSD_{OD} represents the power spectral density at the output of the discriminator and PSD_{ID} represents the power spectral density at the input of the discriminator. The term A_c in Equation (7.11) is obtained as part of the differentiation process. In addition, we need to know that only the quadrature component of the noise contribution is significant at the input of the discriminator. This noise power spectral density at the output needs to be integrated only over the message bandwidth, because the noise outside the message bandwidth will be removed by the low-pass filter at the end of the FM receiver system. This is important because the bandwidth of the FM signal can be much larger than the bandwidth of the baseband signal. Integrating the noise power density given by Equation (7.11) over the message bandwidth of B, we get for the output noise power

$$\frac{2N_0 B^3}{3A_c^2}$$

Finally, the SNR_o of the FM receiver is obtained as

$$\text{SNR}_o = \frac{3A_c^2 k_f^2 P_x}{2N_0 B^3} \tag{7.12}$$

and subsequently the figure of merit of the FM system is obtained as

$$\frac{\text{SNR}_o}{\text{SNR}_i} = \frac{3k_f^2 P_x}{B^2} \tag{7.13}$$

This can be simplified in the case of single tone modulation to

$$\frac{SNR_o}{SNR_i} = \frac{3\beta^2}{2} \tag{7.14}$$

We can see that FM systems have very good performance in the presence of noise when we increase the bandwidth. In other words, in FM systems we can obtain increased noise performance by increasing the bandwidth.

7.3 SOLID-STATE ELECTRONICS

Probability distribution and density functions are used extensively in solid-state electronics such as energy-level distribution of electrons and intensity distribution of the laser. In this example we describe the application that deals with the distribution of electrons.

The electrons in a semiconducting crystal can occupy various energy levels. One problem of interest is to know the electron distribution between the various levels at any temperature T. The fundamental equation of statistical mechanics gives the probability that any particular level will be occupied by an electron

$$P_e(E) = \frac{1}{\exp\left[(E - E_F)/kT + 1\right]} \tag{7.15}$$

where k is Boltzmann's constant and T is the absolute temperature. Here E_F is a constant obtained such that the mean number of electrons is equal to the total number. This function is called the *Fermi-Dirac distribution*. Figure 7.9 shows the distribution.

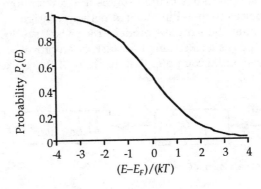

Figure 7.9 Fermi-Dirac distribution.

From Equation (7.15) we can see that if $E - E_F >> kT$, then the probability of occupation of that level is small. Similarly, if $E - E_F$ is large with respect to kT and negative, the level is almost certainly occupied. If $E = E_F$, then $P_e(E)$ is .5. If there is an allowed level E_F, then that level is equally likely to be occupied or empty. The levels where $E > E_F$ are more likely to be empty, and the levels when $E < E_F$ are more likely to be occupied. The energy level corresponding to E_F is called the *Fermi level*. The Fermi-Dirac distribution can be applied to determine the distribution of electrons in a semiconductor.

7.4 COMPUTER ENGINEERING

Computer networks connect many computer systems so that they can communicate and exchange data. With the ever-increasing number of computer networks and computer systems, it is important to model and study the performance of the data networks. There are fundamentally two different types of networks, namely, packet-switched networks and circuit-switched networks. In packet-switched networks, blocks of data are sent over a network to a destination over a path specified by the design of the system. In the circuit-switched networks, the connection between the transmitter and receiver is set up by the transmitter-receiver pair. In the following we demonstrate the application of probability theory in packet-switched networks.

The blocks of data that travel in a network (or packets) are processed at each network node to determine the next transmission path from the node. The transmission links have limited capacity, and many packets share the same hardware transmission link. Therefore it becomes necessary to keep the packets waiting at nodes as part of the network design and management strategy. Thus this problem naturally lends itself to the probabilistic modeling of waiting in a queue. The arrival of the packets, size of packets, and arbitrary source-to-destination routes are all random and have to be modeled appropriately to aid in efficient network design.

Let us consider the simplest model of a packet service node where we have a buffer (such as an incoming file stack in an office) where the incoming packets are stored and a computer that disposes of the packets after servicing them (Figure 7.10).

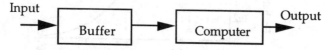

Figure 7.10 Packet service node model.

Such a model is very similar to a bank teller service with only one service window open. The success of the above service scheme depends on the arrival rate of customers and the service rate of the window clerk. The same structure holds for the network service. Let λ_1 be the packet arrival rate and λ_2 be the packet service rate. Obviously when $\lambda_1 > \lambda_2$ the contents of the buffer will increase without any limit. It may be necessary to block new additions to a buffer if it grows without limit. The Poisson random variable described in Chapter 2 is extensively used to model the arrival of packets to a buffer, or arrival of customers to a bank. The Poisson arrival process can be derived based on a few simple assumptions:

1. There exists a positive quantity λ such that the probability that exactly one arrival event will occur in a small time interval Δt is approximately equal to $\lambda \Delta t$.
2. The probability of more than one arrival in a small interval Δt is negligible.
3. The arrivals in nonoverlapping time intervals are independent. It can be shown that, based on the assumptions, the number of arrivals n over an interval t is Poisson distributed and given by

$$p_n(t) = \frac{(\lambda t)^n e^{-\lambda t}}{n!} \quad n = 0, 1, 2, \ldots \tag{7.16}$$

In addition to the number of packet arrivals in time t, the interarrival times are important. We need to note here that the number of packet arrivals is a discrete random variable whereas the time between arrivals is a continuous random variable. The distribution of the interarrival time T can be derived by considering the probability of no arrivals in time t.

Now $P(T > t)$ is the probability that no arrivals happen up to time t. This is given by Equation (7.16) as $e^{-\lambda t}$, where λ is the rate of arrival. The distribution function of random variable T is given by

$$F_T(t) = P(T \leq t) = 1 - P(T > t) = 1 - e^{-\lambda t} \quad t > 0 \tag{7.17}$$

Therefore we can see from the above equation that the distribution function of interarrival times is given by $1 - e^{-\lambda t}$, corresponding to an exponentially distributed random variable.

The simple network system can be described as the one with a single server, with the number of packets arriving as a Poisson distribution, and the inter-arrival times distributed exponentially. In addition, we need a service discipline. One of the most commonly encountered service disciplines is first-come, first-served. If the service capacity λ_2 of the server is smaller than the packet arrival rate λ_1, the buffer contents will grow without limit. A parameter $\mu = \lambda_1 / \lambda_2$ is defined called the *traffic intensity*. This parameter plays a critical role in studying the performance of the simple network model. This model is the same as the model of a queue in queuing theory. The above-mentioned simple model is called $M/M/1$ *queue*, where the symbol M denotes *Markov process*, which in this case is the Poisson process. The first M indicates the arrival process, the second M indicates the service process (exponential service times), and the number 1 corresponds to the number of servers. The other commonly encountered types of arrival processes and service processes are indicated by the letters G and D, where G (general) indicates the arrivals or service times that are arbitrary and D (deter-ministic) indicates they are con-stant.

One of the important quantities of interest in the analysis of network queues is the number of packets or customers in the queue. Since we are dealing with random processes, these are described in terms of probabilities. Let p_n be the probability that there are n packets in the buffer waiting to be processed. It can be shown for an $M/M/1$ model with infinite buffer size that the equilibrium value of p_n is given by

$$p_n = (1-\mu)\mu^n \qquad\qquad (7.18)$$

where $\mu = \lambda_1 / \lambda_2 < 1$. When $\mu = .5$, that is, arrival rate is half of the service rate, the probability of the buffer's being empty is .5, which is the same as the probability that the buffer is nonempty. If the length of the buffer is limited to size L, then the probability that there are n packets in the buffer is given by

$$p_0 = \frac{1-\mu}{1-\mu^{L+1}} \qquad\qquad (7.20)$$

In this case the probability that the buffer is empty p_0 and full p_L is given by

$$p_0 = \frac{(1-\mu)}{1-\mu^{L+1}} \qquad\qquad (7.20)$$

and

$$p_L = \frac{(1-\mu)\mu^L}{1-\mu^{L+1}} \qquad\qquad (7.21)$$

Another quantity of design interest is the expected number of packets waiting to be served. This can be evaluated from the definition of expectation as

$$E(n) = \frac{\mu}{1-\mu} \qquad (7.22)$$

All the parameters derived here for the simple system are the quantities of interest even for complicated network systems.

7.5 BIOMEDICAL SYSTEMS

Random signal analysis techniques are extensively used in biomedical systems. Electrocardiography is a well-known procedure to monitor the functioning of a person's heart. Cardiovascular variables such as the heart rate, arterial blood pressure, and shape of electrocardiographic waveforms all fluctuate on a beat-by-beat basis. These variations are the response to the regulatory mechanisms of the cardiovascular system. The analysis of these variations is an important tool in physiology, experimental psychology, neonatal care, and the diagnosis and prediction of neurologic and cardiovascular disorders. In this section we present the analysis of heart rate and respiratory amplitudes that are extensively used in biomedical systems. In order to understand the biomedical interpretation of the example presented here, we introduce two medical terms, namely, *sympathetic* and *parasympathetic*. A sympathetic response will increase the heart rate and the strength of the contraction of the heart, whereas a parasympathetic response decreases the heart rate.

The heart rate and the respiratory amplitude, which are two random processes, are typically used in the analysis of the heart rate variability. The respiratory amplitude is used because the respiration produces changes in the heart rate that are mediated by the parasympathetic function. Figure 7.11 presents an example of heart rate and respiratory amplitude signals for a person at rest (7.11a, 7.11b) and during exercise (7.11c, 7.11d).

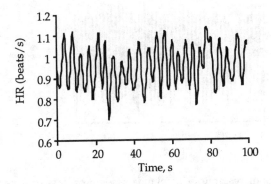

Figure 7.11a Example of heart rate (HR) random process of a person at rest.

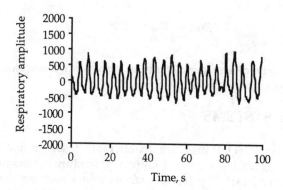

Figure 7.11b Example of respiratory amplitude random process of a person at rest.

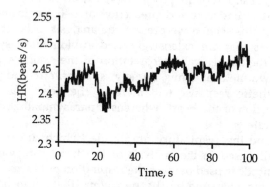

Figure 7.11c Example of heart rate (HR) random process of a person during exercise.

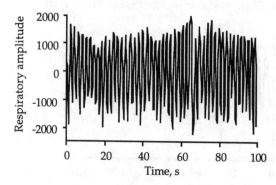

Figure 7.11d Example of respiratory amplitude random process of a person during exercise.

The effect of the respiratory amplitude over the heart rate variability can be easily observed in the rest condition. However, during exercise, the parasympathetic influence is reduced (a higher heart rate is needed to pump blood faster), and although the respiratory amplitude and frequency have increased, the effect of the respiration on the heart rate has decreased. The random processes shown in Figure 7.11 can be considered as wide-sense stationary during records of short duration (minutes) when the person is in stable condition. Sometimes, a linear trend has to be subtracted from the heart rate to eliminate the variations of the mean in the random process. In the following we present an example of the frequency domain analysis of the heart rate.

The frequency domain analysis (power spectral analysis) of heart rate variability has been shown to be a powerful tool because of its capability of discriminating the effect of several mechanisms involved in the control process. Figure 7.12 presents the power spectral analysis of heart rate for a person at rest and during exercise. Figure 7.12a shows the power spectral density of the heart rate variability of the person at rest. Two main bands can be observed: a high-frequency band (0.15 to 0.35 Hz), and a low-frequency band (<0.15 Hz).

The high-frequency fluctuations are purely parasympathetically mediated, so they are closely related to the respiration. The low-frequency fluctuations are jointly mediated by the sympathetic and parasympathetic systems. They have been related with the response to variation in the control of the blood pressure, temperature, and other physiological variables related to the vascular system. Figure 7.12b shows the spectrum of the respiratory amplitude. In Figure 7.12d, the power spectral density of the heart rate during exercise is presented. In this case, the amplitude of the high-frequency band is low compared with the amplitude of the other band. This is due to the reduction in the parasympathetic response of the system.

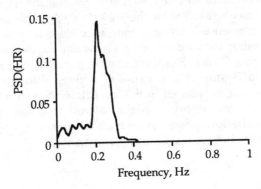

Figure 7.12a Power spectral density (PSD) of the heart rate (HR) at rest.

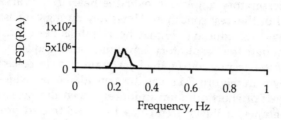

Figure 7.12b Power spectral density (PSD) of the respiratory amplitude (RA) at rest.

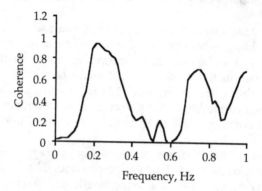

Figure 7.12c The coherence function between the heart rate and the respiratory amplitude at rest.

The main components of the spectra of the respiratory amplitude at rest (Figure 7.12b) and during exercise (Figure 7.12e) correspond to the respiratory frequency. During exercise, the respiratory frequency increases, so the main components in the PSD are in higher frequencies. It can be observed that the high-frequency bands of the heart rate PSD and of the respiratory signals are correlated for measurements taken with the subject at rest. This relation can be observed via the coherence function, where it is close to 1 in the high-frequency band, indicating a strong correlation between the two processes at those frequencies (Figure 7.12c). However, during exercise, the correlation decreases, as shown in Figure 7.12f, because it is given through the parasympathetic response that is reduced in this condition. Note that the total energy of the heart rate variability is higher at rest than during exercise. The total energy and the low-frequency-to-high-frequency energy ratio at rest are indexes of the cardiovascular control performance.

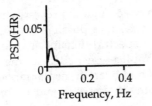

Figure 7.12d Power spectral density (PSD) of the heart rate (HR) and respiratory amplitude (RA) during exercise. The coherence function between those processes is also presented.

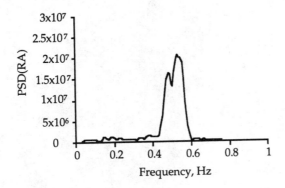

Figure 7.12e Power spectral density (PSD) of the heart rate (HR) and respiratory amplitude (RA) during exercise. The coherence function between those processes is also presented.

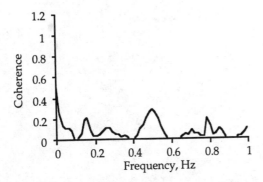

Figure 7.12f Power spectral density (PSD) of the heart rate (HR) and res-piratory amplitude (RA) during exercise. The coherence function between those processes is also presented.

Thus, we can see that analyzing the two random processes, namely, by the heart rate and the respiratory amplitude, it is possible to assess the cardiovascular control. In particular, power spectral density analysis has allowed the use of this technique in clinical applications, such as the evaluation of the recovery in myocardial infarction and heart transplantation, of maturity in newborns, the diagnoses of neuropathy in diabetes, and of congestive heart failure.

Appendix

Introduction to the Fast Fourier Transform

A.1 INTRODUCTION

The fast Fourier transform (FFT) is a fast computer algorithm used to compute the discrete Fourier transform, or DFT. Its usefulness lies in the fact that, with increasingly larger numbers of data points, the speed of computation of the FFT is greatly reduced compared to the time needed for the computation of the straightforward DFT. It is simple to compute the straightforward DFT, and later in this chapter the algorithm for the computation of the DFT will be explained. We will observe that a number of the computations required for the DFT are redundant; i.e., they are computed several times. This observation prompted the development of the fast Fourier transform. The fast algorithmic speed permitted the performance of complex operations in the frequency domain which were virtually impossible to perform in the time domain. By using well-known time-frequency relationships, the operation was first transformed to the frequency domain via the FFT, the operation was performed in the frequency domain, and the results were transformed back to the time domain via the inverse FFT. Some of these operations will be illustrated later in the appendix.

A.2 RELATIONSHIP BETWEEN THE CONTINUOUS FOURIER TRANSFORM (CFT) AND THE DISCRETE FOURIER TRANSFORM (DFT)

Given a nonperiodic signal $f(t)$, it can be represented as an integral of complex exponential components

$$f(t) = \frac{1}{2\pi} \int_{-\infty}^{\infty} F(\omega) \exp\left(j\omega t\right) d\omega \qquad (A.1)$$

where

$$F(\omega) = \int_{-\infty}^{\infty} f(t) \exp\left(-j\omega t\right) dt \qquad (A.2)$$

This is the well-known Fourier transform pair and is often expressed by the relationship

$$F[f(t)] = F(\omega) \qquad (A.3)$$

and

$$F^{-1}[F(\omega)] = f(t) \qquad (A.4)$$

where $F[\cdots]$ is the symbol used for the Fourier transform operation.

Assume now that a function of time $f(t)$ has been sampled in time. This sampling must be performed according to the sampling theorem which states that if a function of time has no energy beyond a specific frequency f_c, then $f(t)$ can be uniquely determined from a knowledge of its sampled values. This sampling must be performed at a rate not larger than

$$\Delta t = \frac{1}{2f_c} \qquad (A.5)$$

The discrete Fourier transform of $f(t)$ is given by the following equation:

$$F(n) = \sum_{k=0}^{N-1} f(k\,\Delta t) \exp\left(\frac{-j2\pi n k}{N}\right) \qquad (A.6)$$

However,

$$F(n) = \frac{1}{\Delta t} F\left(\omega = \frac{2\pi n}{N \Delta t}\right) \qquad (A.7)$$

This equation states that the discrete Fourier transform of the function $f(k\Delta t)$ is the same as the value of the continuous transform $\mathcal{F}(\omega)$ at the values

$$\omega = \frac{2\pi n}{N \Delta t} \qquad (A.8)$$

This is an important result as it yields the correct units for any computation of the discrete Fourier transform. And F can be represented with an argument n. However, when we deal with sampled signals, F can also have an argument equal to

$$\frac{2\pi n}{N \Delta t}$$

to indicate the relation to the frequencies of the signal being sampled. We have used both representations in this appendix depending on the context.

Example A.1 Let us now consider Equation (A.6). If asked to compute the DFT of a set of data points, we could use this equation and obtain the results as follows. Assume first that the function of time is purely real and that it has been sampled according to the sampling theorem. Also assume that there are $N = 50$ values of data. We may represent these values as follows:

$$f(0), f(\Delta t), f(2\,\Delta t), f(3\,\Delta t), \ldots, f(49\,\Delta t) \qquad (A.9)$$

We now use Euler's equation to compute Equation (A.6):

$$\exp j\alpha = \cos\alpha + j\sin\alpha \qquad (A.10)$$

which in our case yields

$$\exp\left(-\frac{j2\pi nk}{50}\right) = \cos\frac{2\pi nk}{50} - j\sin\frac{2\pi nk}{50} \qquad (A.11)$$

Now we expand Equation (A.6) as follows:

For $n = 0$,

$$F(0) = \Delta t \left[f(0) + f(\Delta t) + f(2\,\Delta t) + \cdots + f(49\,\Delta t) \right] \qquad (A.12)$$

For $n = 1$,

$$F\left(\frac{2\pi}{50\,\Delta t} \right) = \Delta t \left[f(0) + f(\Delta t)\left(\cos\frac{2\pi}{50} - j\sin\frac{2\pi}{50} \right) \right.$$

$$\left. + f(2\,\Delta t)\left(\cos\frac{4\pi}{50} - j\sin\frac{4\pi}{50} \right) + \cdots \right] \qquad (A.13)$$

For $n = 2$,

$$F\left(\frac{2\cdot 2\pi}{50\,\Delta t} \right) = \Delta t \left[f(0) + f(\Delta t)\left(\cos\frac{4\pi}{50} - j\sin\frac{4\pi}{50} \right) \right.$$

$$\left. + f(2\,\Delta t)\left(\cos\frac{8\pi}{50} - j\sin\frac{8\pi}{50} \right) + \cdots \right] \qquad (A.14)$$

etc.

In practice, this is a very easy algorithm to program. We only need to separate the real and the imaginary parts and treat them separately. In an actual example suppose that we are given the following eight real values:

$$
\begin{aligned}
x(\,0\,) &= 1. \\
x(1\Delta t) &= 0. \\
x(2\Delta t) &= 0. \\
x(3\Delta t) &= 1. \\
x(4\Delta t) &= 1. \\
x(5\Delta t) &= 2. \\
x(6\Delta t) &= 0. \\
x(7\Delta t) &= 0.
\end{aligned}
$$

Computing the above operations, we obtain the following results:

$$F(0) = (5.00, 0.00) \qquad F\left(\frac{2\pi}{8\,\Delta t}\right) = (-2.12, .707)$$

$$F\left(\frac{2 \cdot 2\pi}{8\,\Delta t}\right) = (2.00, -1.00) \qquad F\left(\frac{3 \cdot 2\pi}{8\,\Delta t}\right) = (2.12, .707)$$

$$F\left(\frac{4 \cdot 2\pi}{8\,\Delta t}\right) = (-1.00, .0) \qquad F\left(\frac{5 \cdot 2\pi}{8\,\Delta t}\right) = (2.12, -.707)$$

$$F\left(\frac{6 \cdot 2\pi}{8\,\Delta t}\right) = (2.00, 1.00) \qquad F\left(\frac{7 \cdot 2\pi}{8\,\Delta t}\right) = (-2.12, .707)$$

As anticipated, the first value is purely real and equal to the sum of the input values. Two multiplications are required to compute each value of frequency.

One of the most important consequences of the operations of time-limiting, sampling, and computation of the Fourier transform of a waveform is the apparent periodicity of the discrete time waveform. This result must be taken into account when we use the FFT to perform operations such as convolutions and correlation. In order to understand this result, it is necessary to realize that every operation in the time/frequency domain results in a corresponding operation in the frequency/time domain. For example, the sampling operation in the time domain at sampling rate Δt results in a sampling operation in the frequency domain at sampling rate $1/\Delta t$. This sampling operation in the time-domain may be modeled by multiplying the time domain waveform by a periodic impulse train of rate Δt. Consider a periodic impulse train and its corresponding Fourier transform

$$f(t) = \sum_{m=-\infty}^{\infty} \delta(t - m\,\Delta t) \qquad\qquad (A.15)$$

$$F(\omega) = \omega_0 \sum_{n=-\infty}^{\infty} \delta(\omega - n\omega_0) \qquad\qquad (A.16)$$

where

$$\omega_0 = \frac{2\pi}{\Delta t} \qquad\qquad (A.17)$$

Note that a periodic impulse train in the time domain transforms to a periodic impulse train in the frequency domain. Likewise, the operation of multiplication in the time domain, performed in order to sample the waveform, corresponds to a convolution operation in the frequency domain.

The end result of all these operations is that the time-limited waveform and the band-limited Fourier transform are considered to be periodic waveforms. Equation (A.6) shows the computation of the DFT for values $n = 0, 1, \ldots, N-1$. Additional values of the index n yield no new information as:

$$F\left(\frac{2\pi n}{N\,\Delta t}\right) = F\left(\frac{2\pi(n+N)}{N\,\Delta t}\right) \tag{A.18}$$

A.3 THE INVERSE DISCRETE FOURIER TRANSFORM (IDFT)

The inverse discrete Fourier transform is given by

$$f(k\,\Delta t) = \frac{1}{N}\sum_{n=0}^{N-1} F\left(\frac{2\pi n}{N\,\Delta t}\right)\exp\frac{j2\pi nk}{N} \quad k = 0, 1, \ldots, N-1 \tag{A.19}$$

Note that Equations (A.6) and (A.19) are nearly identical except for a sign on the exponential. This formula exhibits the same type of periodicity exhibited by Equation (A.6).

A.4 THE FFT ALGORITHM, COMPUTATIONAL SPEED, AND EFFECTS OF ZERO PADDING

A comprehensive discussion of the FFT algorithm is beyond the scope of this text. Suffice it to say that when the number of values of data is a power of 2, the algorithm uses basic redundancies in the computation to significantly reduce the number of multiplications and additions needed. This result is illustrated in Example A.2.

Example A.2 In order to understand better the FFT algorithm and the considerable time savings associated with its use, assume that we have four data points and that we wish to to compute their discrete Fourier transform. Assume, for the sake of simplicity, that the index of $F(\ )$ is simply a function of n. We now expand Equation (A.6):

For $n = 0$,

$$F(0) = f(0) + f(1) + f(2) + f(3) \qquad \text{(A.20)}$$

For $n = 1$,

$$F(1) = f(0) + f(1) \exp\left(\frac{-j2\pi}{4}\right) + f(2) \exp\left(\frac{-j4\pi}{4}\right) + f(3) \exp\left(\frac{-j6\pi}{4}\right) \qquad \text{(A.21)}$$

For $n = 2$,

$$F(2) = f(0) + f(1) \exp\left(\frac{-j4\pi}{4}\right) + f(2) \exp\left(\frac{-j8\pi}{4}\right) + f(3) \exp\left(\frac{-j12\pi}{4}\right) \qquad \text{(A.22)}$$

For $n = 3$,

$$F(3) = f(0) + f(1) \exp\left(\frac{-j6\pi}{4}\right) + f(2) \exp\left(\frac{-j12\pi}{4}\right) + f(3) \exp\left(\frac{-j18\pi}{4}\right) \qquad \text{(A.23)}$$

It is customary (Bergland, 1969; Elliott, 1982; Brigham, 1988) to substitute

$$W_N = \exp\left(\frac{-j2\pi}{N}\right)$$

Consequently, our set of equations becomes (for $N = 4$)

$$F(0) = f(0)W_4^0 + f(1)W_4^0 + f(2)W_4^0 + f(3)W_4^0 \qquad \text{(A.24)}$$

$$F(1) = f(0)W_4^0 + f(1)W_4^1 + f(2)W_4^2 + f(3)W_4^3 \qquad \text{(A.25)}$$

$$F(2) = f(0)W_4^0 + f(1)W_4^2 + f(2)W_4^4 + f(3)W_4^6 \qquad \text{(A.26)}$$

$$F(3) = f(0)W_4^0 + f(1)W_4^3 + f(2)W_4^6 + f(3)W_4^9 \qquad \text{(A.27)}$$

These equations may be expressed in matrix form as follows:

$$\begin{bmatrix} F(0) \\ F(1) \\ F(2) \\ F(3) \end{bmatrix} = \begin{bmatrix} W_4^0 & W_4^0 & W_4^0 & W_4^0 \\ W_4^0 & W_4^1 & W_4^2 & W_4^3 \\ W_4^0 & W_4^2 & W_4^4 & W_4^6 \\ W_4^0 & W_4^3 & W_4^6 & W_4^9 \end{bmatrix} \begin{bmatrix} f(0) \\ f(1) \\ f(2) \\ f(3) \end{bmatrix} \tag{A.28}$$

Note that:

$$W_4^0 = 1 \quad W_4^6 = W_4^2 \quad W_4^4 = W_4^0 = 1 \quad \text{and} \quad W_4^9 = W_4^1$$

and the equations now become

$$\begin{bmatrix} F(0) \\ F(1) \\ F(2) \\ F(3) \end{bmatrix} = \begin{bmatrix} 1 & 1 & 1 & 1 \\ 1 & W_4^1 & W_4^2 & W_4^3 \\ 1 & W_4^2 & 1 & W_4^2 \\ 1 & W_4^3 & W_4^2 & W_4^1 \end{bmatrix} \begin{bmatrix} f(0) \\ f(1) \\ f(2) \\ f(3) \end{bmatrix} \tag{A.29}$$

The new reduction in terms in (A.29) is due to the fact that (Brigham, 1988)

$$W_N^{nk} = W_N^{nk \bmod(N)} \tag{A.30}$$

where $nk \bmod(N)$ is the remainder upon division of nk by N. For example, $N = 4$ and let $n = 3$ and $k = 3$. Then $9 \bmod(4) = 1$. Therefore

$$W_4^9 = W_4^1 \tag{A.31}$$

It is also noted that

$$W_4^2 = \exp\left(\frac{-2 \cdot j2\pi}{4}\right) = \exp(-j\pi) = -1 \tag{A.32}$$

Consequently, the set of equations is further reduced to:

$$\begin{bmatrix} F(0) \\ F(1) \\ F(2) \\ F(3) \end{bmatrix} = \begin{bmatrix} 1 & 1 & 1 & 1 \\ 1 & W_4^1 & -1 & W_4^3 \\ 1 & -1 & 1 & -1 \\ 1 & W_4^3 & -1 & W_4^1 \end{bmatrix} \begin{bmatrix} f(0) \\ f(1) \\ f(2) \\ f(3) \end{bmatrix} \tag{A.33}$$

Note that this is specific to this case, since $N = 4$. The set of equations expressed by (A.33) may now be expressed as follows:

$$F(0) = f(0) + f(1) + f(2) + f(3)$$
$$F(1) = f(0) + W_4^1 f(1) - f(2) + W_4^3 f(3)$$
$$F(2) = f(0) - f(1) + f(2) - f(3) \qquad\qquad (A.34)$$
$$F(3) = f(0) + W_4^3 f(1) - f(2) + W_4^1 f(3)$$

Comparing the original four Equations (A.24), (A.25), (A.26), and (A.27) with the new set of Equations (A.34), we observe that the original set of equations required 16 complex multiplications and 12 complex additions. Computation via the reduced set represented by the last set of equations requires only 4 complex multiplications and 8 complex additions. The basic redundancies found by Equation (A.30) and the substitution given by Equation (A.32) require a power of 2 in the number of data values.

Computational Speed

The computational speed of the FFT is approximately directly proportional to the number of complex multiplications. The table below shows the approximate number of complex additions and multiplications required for a power-of-2 FFT and the number of complex additions and multiplications required for the corresponding DFT.

Number of Points	DFT		FFT	
	Complex Additions	Complex Multiplications	Complex Additions	Complex Multiplications
4	16	16	8	2
8	64	64	24	8
16	256	256	64	24
32	1024	1024	160	64
64	4096	4096	384	160
128	16384	16384	896	384

One assumption is that the speed of computation is mostly dependent on the machine cycles required to perform a complex multiplication. For $N = 128$ the ratio of the number of complex multiplications using the DFT to that of the FFT is greater than 43. So the savings in time may be quite substantial.

A logical question to follow the preceding development is, How limiting is the need to use a power-of-2 number of points? For example, suppose that the number of data points is $N = 1000$. The closest power-of-2 number is 1024, or 2^{10}, so what are our choices? Should we simply use the first 512 points of data ($512 = 2^9$)? We are indeed losing information if we discard 488 data values. A number of algorithms which do not require a power of 2 are also available. For example, the *relative prime factor algorithm* involves the use of the *Chinese remainder theorem* (Burrus, 1977, 1988). Others achieve efficiency by expressing the discrete Fourier transform in terms of polynomial multiplication (Winograd, 1978).

Zero Padding

Suppose that the data being analyzed are not a power of 2. In such cases we have to make one of several choices. Suppose that the number of points is $N = 1000$. The next power of 2 is 1024. The data could be restricted to the closest power of 2 without exceeding the actual number of data points, that is, $N = 512$. An obvious problem with this procedure is that part of the data would not be utilized in the analysis process. Another solution would be to pad the data with enough values to reach the next closest power of 2. In this case, such a value would be $N = 1024$. The effects of padding the data with zero values are explained with the following example.

Example A.3 We now analyze the effects of padding the data with zeros. The DFT of a discrete function of time is given by

$$F\left(\omega = \frac{2\pi n}{N \Delta t}\right) = \sum_{k=0}^{N-1} f(k \Delta t) \exp\left(-\frac{j2\pi nk}{N}\right) \quad n = 0, 1, \ldots, N-1 \quad \text{(A.35)}$$

We now define a new discrete function of time $g(k\Delta t)$ such that:

$$g(k \Delta t) = f(k \Delta t) \quad 0 \leq k \leq N-1$$
$$g(k \Delta t) = 0 \quad\quad N-1 \leq k \leq 2N-1 \quad\quad \text{(A.36)}$$

and the DFT of the new discrete function is as follows:

$$G\left(\omega = \frac{2\pi n}{N \Delta t}\right) = \sum_{k=0}^{N-1} g(k \Delta t) \exp\left(-\frac{j2\pi nk}{N}\right) \quad n = 0, 1, \ldots, N-1 \quad \text{(A.37)}$$

Using the definition given by Equation (A.36), we can rewrite the preceding equation as follows:

$$G\left(\omega = \frac{2\pi n}{2N\,\Delta t}\right) = \sum_{k=0}^{N-1} f(k\,\Delta t)\,\exp\left(-\frac{j2\pi nk}{2N}\right) \quad n = 0,\,1,\,\dots,\,2N-1 \quad \text{(A.38)}$$

Note that $g(k\Delta t) = 0$ for $k > N - 1$. In order to illustrate the effects of adding zeros to the data, let us assume that $N = 2$, and expand Equations (A.35) and (A.38).

Let $f(k\Delta t) = [f(0), f(1)]$

and let

$g(k\Delta t) = [f(0), f(1),0,0]$

We now expand Equation (A.35):

$$F(0) = f(0)\,\exp\left(\frac{-j\cdot 2\pi\cdot 0\cdot 0}{2}\right) + f(1)\,\exp\left(\frac{-j\cdot 2\pi\cdot 0\cdot 1}{2}\right) \qquad \text{(A.39)}$$

and

$$F(1) = f(0)\,\exp\left(\frac{-j\cdot 2\pi\cdot 1\cdot 0}{2}\right) + f(1)\,\exp\left(\frac{-j\cdot 2\pi\cdot 1\cdot 1}{2}\right) \qquad \text{(A.40)}$$

which yields

$$F(0) = f(0) + f(1) \qquad \text{(A.41)}$$

and

$$F(1) = f(0) + f(1)\,\exp\left(-j\pi\right) \qquad \text{(A.42)}$$

We now expand Equation (A.39):

$$G(0) = g(0)\,\exp\left(\frac{-j\cdot 2\pi\cdot 0\cdot 0}{4}\right) + g(1)\,\exp\left(\frac{-j\cdot 2\pi\cdot 0\cdot 1}{4}\right)$$
$$+ g(2)\,\exp\left(\frac{-j\cdot 2\pi\cdot 0\cdot 2}{4}\right) + g(3)\,\exp\left(\frac{-j\cdot 2\pi\cdot 0\cdot 3}{4}\right) \qquad \text{(A.43)}$$

$$G(1) = g(0) \exp\left(\frac{-j \cdot 2\pi \cdot 1 \cdot 0}{4}\right) + g(1) \exp\left(\frac{-j \cdot 2\pi \cdot 1 \cdot 1}{4}\right)$$
$$+ g(2) \exp\left(\frac{-j \cdot 2\pi \cdot 1 \cdot 2}{4}\right) + g(3) \exp\left(\frac{-j \cdot 2\pi \cdot 1 \cdot 3}{4}\right) \tag{A.44}$$

$$G(2) = g(0) \exp\left(\frac{-j \cdot 2\pi \cdot 2 \cdot 0}{4}\right) + g(1) \exp\left(\frac{-j \cdot 2\pi \cdot 2 \cdot 1}{4}\right)$$
$$+ g(2) \exp\left(\frac{-j \cdot 2\pi \cdot 2 \cdot 2}{4}\right) + g(3) \exp\left(\frac{-j \cdot 2\pi \cdot 2 \cdot 3}{4}\right) \tag{A.45}$$

$$G(3) = g(0) \exp\left(\frac{-j \cdot 2\pi \cdot 3 \cdot 0}{4}\right) + g(1) \exp\left(\frac{-j \cdot 2\pi \cdot 3 \cdot 1}{4}\right)$$
$$+ g(2) \exp\left(\frac{-j \cdot 2\pi \cdot 3 \cdot 2}{4}\right) + g(3) \exp\left(\frac{-j \cdot 2\pi \cdot 3 \cdot 3}{4}\right) \tag{A.46}$$

which yields

$$G(0) = g(0) + g(1) \tag{A.47}$$

$$G(1) = g(0) + g(1) \exp\left(\frac{-j\pi}{2}\right) \tag{A.48}$$

$$G(2) = g(0) + g(1) \exp\left(-j\pi\right) \tag{A.49}$$

$$G(3) = g(0) + g(1) \exp\left(\frac{-j \cdot \pi \cdot 3}{2}\right) \tag{A.50}$$

Note that $g(0) = f(0)$, $g(1) = f(1)$, $g(2) = 0$, and $g(3) = 0$. Therefore,

$$G(0) = f(0) + f(1) \tag{A.51}$$

$$G(1) = f(0) + f(1) \exp\left(\frac{-j\pi}{2}\right) \tag{A.52}$$

$$G(2) = f(0) + f(1) \exp\left(-j\pi\right) \tag{A.53}$$

$$G(3) = f(0) + f(1) \exp\left(\frac{-j \cdot \pi \cdot 3}{2}\right) \tag{A.54}$$

Comparing these equations with (A.41) and (A.42), we note the following:

$$G(0) = f(0) + f(1) = F(0) \tag{A.55}$$

$$G(1) = f(0) + f(1) \exp\left(\frac{-j\pi}{2}\right) \tag{A.56}$$

$$G(2) = f(0) + f(1) \exp\left(-j\pi\right) = F(1) \tag{A.57}$$

$$G(3) = f(0) + f(1) \exp\left(\frac{-j\pi 3}{2}\right) \tag{A.58}$$

The values of $G(0)$ and $G(2)$ are identical to the values of $F(0)$ and $F(1)$. The values of $G(1)$ and $G(3)$ are actually values *interpolated* from the two other values. The effect of adding zeros to the data in the time domain is *equivalent* to interpolating the frequency-domain data.

A.5 PRACTICAL ASPECTS OF THE USE OF THE FFT ALGORITHM

Efficient use of the FFT requires careful attention to detail. Because of its speed, the FFT tends to be used and abused routinely. The FFT is often used to analyze data collected from random processes even though its use in such cases may lead to serious errors of interpretation. Let us explain the mechanics of the use of the FFT. It is always a good idea, when "canned" subroutines are available, to test them using known data to yield known results.

1. In order to obtain the appropriate units, the results of the FFT must be multiplied by Δt.
2. As a rule, the number of data values must be a power of 2. Although a number of algorithms exist that use other techniques, the power-of-2 algorithms are the best known and most widely utilized.
3. Given N data values, there will be, upon return from the FFT, N complex values in the frequency domain. These N values represent both the positive and negative values of frequency (see item 6 below).
4. The spacing between frequencies is specified by Equation (A.6). This equation is repeated here for completeness:

$$F\left(\omega = \frac{2\pi n}{N\,\Delta t}\right) = \sum_{k=0}^{N-1} f(k\,\Delta t) \exp\left(\frac{-j2\pi nk}{N}\right) \quad n = 0, 1, \ldots, N-1 \tag{A.59}$$

The spacing between frequencies will be found at multiples of

$$\Delta f = \frac{n}{N\,\Delta t} \qquad (A.60)$$

Consequently, the spacing between frequencies is determined directly by the number of data values. The more data values for a constant Δt, the closer the spacing between discrete frequencies.

5. The highest discrete frequency present is dictated by the spacing between the time samples, that is, Δt. According to the sampling theorem, the highest frequency present will be

$$f_M = \frac{1}{2\,\Delta t} \qquad (A.61)$$

6. The values resulting from the computation of the FFT are distributed as follows:

    (a) The first $N/2 + 1$ samples in the frequency domain are the positive frequency components having frequencies 0, $1/(N\Delta t)$, $2/(N\Delta t)$, $3/(N\Delta t)$, $\dots$, $1/(2\Delta t)$ Hz.

    (b) The last $N/2$ samples are the negative frequency components having frequencies of $-1/(2\Delta t), \dots, -1/(N\Delta t)$.

The following example will illustrate some of these concepts:

Example A.4 A discrete waveform has the following characteristics:

$$x(k\,\Delta t) = \begin{cases} \cos(\omega_0 k\,\Delta t) & 0 \le k \le 127 \\ 0 & \text{elsewhere} \end{cases}$$

For the purposes of illustration we will let

$$\omega_0 = 16\pi \quad \text{and} \quad k\,\Delta t = \frac{k}{128}$$

This implies that $f = 8$ Hz, and therefore $\Delta t = .078$ s more than meets the sampling theorem criteria. We will now add random noise to the data. The new discrete signal will then have the following charac-teristics:

$$x(k\,\Delta t) = \cos(\omega_0 k\,\Delta t) + 2\,\mathrm{rnd}(2) - 2$$

The function rnd(2) is a MATHCAD function which generates random numbers distributed between 0 and 2. Subtracting the value of 2 causes the random number generated to be distributed between -2 and 2. A total of 128

A.5 Practical Aspects of the use of the FFT

values (1 s) were generated by this process, and they are shown in Figure A.1.

We compute the FFT of $x(k\Delta t)$ and plot the magnitude of the results, using the FFT function in MATHCAD. This function computes Equation (A.6) and multiplies the results by $1/n$. The results are shown in Figure A.2. Note that we plot from $f = 0$ to $f = 1/(2\Delta t)$ Hz. For this particular case, the following parameters apply:

$$\Delta t = \frac{1}{128} \quad \text{or} \quad .0078\,s$$

$$\Delta f = \frac{n}{N\,\Delta t} = \frac{n}{128 \cdot \dfrac{1}{128}} = n\,\text{Hz} \quad \text{and} \quad f_M = \frac{1}{2\,\Delta t} = \frac{1}{2 \cdot \dfrac{1}{128}} = 64\,\text{Hz}$$

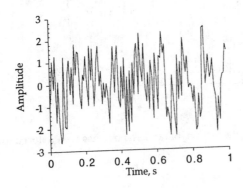

Figure A.1 Plot of $x(k\Delta t)$.

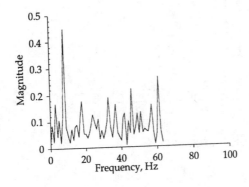

Figure A.2 Plot of the magnitude of the FFT.

Suppose that we decrease the amount of random noise. We redefine $x(k\Delta t)$ as follows:

$$x(k \, \Delta t) = \cos \, (\omega_0 k \, \Delta t) + \text{rnd}(2) - 1$$

Since the amount of random noise has been decreased we might expect to observe the cosinusoid in the plot. The plot for $x(k\Delta t)$ is shown in Figure A.3. The results of computing the FFT of $x(k\Delta t)$ are shown in Figure A.4.

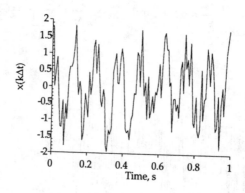

Figure A.3 Plot of $x(k\Delta t)$, the amount of random noise has been reduced for this example.

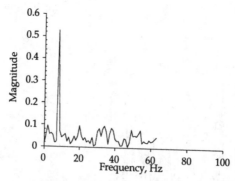

Figure A.4 Plot of magnitude of the FFT.

Bibliography

Akaike, H.: Fitting Autoregressions for Predictions, *Annals of the Institute of Statistical Mathematics*, Tokyo, pp. 243-247, 1969.

Bendat, J. S., and Piersol, A.G.: *Random Data, Analysis and Measurement Procedures*, John Wiley and Sons, New York, 1986.

Box, G. E. P., and Jenkins, G. M.: *Time Series Analysis: Forecasting and Control*, Holden-Day, Oakland, CA, 1976.

Burrus, C. S.: Index Mappings for Multidimensional Formulation of the DFT and Convolution, *IEEE Trans. on Acoustics, Speech, and Signal Processing*, vol. ASSP-25, pp. 239-242, 1977.

Burrus, C. S.: "Efficient Fourier Transform and Convolution Algorithms," in *Advanced Topics in Signal Processing*, J. S. Lim and A. V. Oppenheim, Eds., Prentice-Hall, Englewood Cliffs, NJ, 1988.

Hamming, R. W.: *Numerical Methods for Scientists and Engineeers*, McGraw-Hill Book Company, New York, 1962.

Papoulis, A.: *Probability, Random Variables, and Stochastic Processes*, McGraw-Hill Book Company, New York, 1968.

Press, W.H., Flannery, B.P., Teukolsky, S.A., and Vetterling, W.T.: *Numerical Recipes in C, The Art of Scientific Computing*, Cambridge University Press, Cambridge, MA, 1988.

Shanmugan, K. S., and Breipohl, A. M.: *Random Signals: Detection, Estimation and Data Analysis*, John Wiley and Sons, New York, 1988.

ADDITIONAL REFERENCES

Brigham, E. O.: *The Fast Fourier Transform*, Prentice-Hall, Englewood Cliffs, NJ, 1974.

Cooper, G. R., and McGillem, C. D.: *Probabilistic Methods of Signal and Systems Analysis*, Holt, Rinehart and Winston, New York, 1986.

Leon-Garcia, A.: *Probability and Random Processes for Electrical Engineers*, Addison-Wesley, Reading, MA, 1989.

Peebles, P. Z.: *Probability, Random Variables and Random Signal Principles*, McGraw-Hill Book Company, New York, 1980.

Stark, H., and Woods, J. W.: *Probability, Random Processes, and Estimation Theory for Engineers*, Addison-Wesley, Reading, MA, 1994.

Index

Addition rule, 10
Additive white gaussian noise, 498
Amplitude-modulated carrier, 499
Autocorrelation function, 212
Autoregressive, 294

Bernoulli trials, 21
Binary transmission channel, 24
Binomial distribution, 23
Butterworth filter, 451

Causality, 441
Central limit theorem, 138
Central moments, 53
Certain event, 2, 4
Chi-Square test, 150
Coherence, 315
Communication system, 497
Complement, 5
Conditional density, 121
Conditional probability, 12
Continuous random variable, 41
Continuous sample spaces, 12
Convergence in distribution, 139
Convolution theorem, 277
Correlation coefficient, 116, 118
Correlation, 112
Covariance, 115
Cross-correlation function, 225
Cross-spectral density, 307
Cumulative distribution , 42

Deterministic random process, 211

DFT, 513
Discrete random variable, 41
Discrete sample spaces, 11
Doppler effect, 494

Einstein-Wiener-Khinchinetheorem, 270
Electrocardiography, 507
Ergodicrandom processes, 211
Ergodicity, 224
Expected value, 50

Fast Fourier transform, 513
Fermi level, 504
Fermi-Dirac distribution, 503
FFT, 51
Final prediction error, 301
FM systems, 500
Frequency deviation, 501
Frequency response, 449

Gaussian random variables, 57

Hamming window, 282

Impossible event, 4
Independence of events, 18
Intersection, 5

Joint distribution function, 106
Joint density function, 108

Kolmogoroff-Smirnoff test, 155

Law of total probability, 14
Levinson recursion, 295
Logical product, 5
Logical sum, 4

M/M/1 model, 506
Matched filter, 453
Mean square value, 52
Mean value, 50
Median, 153
Model-Based Approach, 294
Modulation index, 501
Moment, 52
Mutually exclusive., 5

Narrowband FM, 501

Orthogonal, 112

Packet service node, 504
Periodogram Method, 284
Poisson approximation, 24
Power spectral density, 269
Probability density function, 47
Pseudorandom number generators, 147

Queue, 506

Radar systems, 493
Random process, 204
Random processes, 203
Random variable, 41
Range, 41
Rayleigh distributed, 494
Rms, 53

Sample function, 204
Signal-to-noise ratio, 498
Space of elementary events, 4
Standard deviation, 153
Stationary random process., 207
Statistically independent, 112
Sum of two random variables, 124

Target scatterers, 494
Theorem of inverse probability, 16
Time invariance, 441
Time-average autocorrelation, 224
Töeplitz matrix, 295
Traffic intensity, 505

Uncorrelated, 112
Uniform random variables, 56
Union, 4

Venn diagrams, 6

White noise, 274
Wide-sense process, 207

Wideband FM, 501

Yule-Walker equations, 295

Zero padding, 518